Telecommunications Optimization: Heuristic and Adaptive Techniques

Telecommunications Optimization: Heuristic and Adaptive Techniques

Edited by

David W. Corne
School of Computer Science, Cybernetics and Electronic Engineering, University of Reading, UK

Martin J. Oates
BT Research Laboratories, Martlesham Heath, UK

George D. Smith
School of Information Systems, University of East Anglia, UK

JOHN WILEY & SONS, LTD
Chichester · New York · Weinheim · Brisbane · Singapore · Toronto

Copyright © 2000 John Wiley & Sons, Ltd
Baffins Lane, Chichester,
West Sussex, PO19 1UD, England

National   01243 779777
International   (+44) 1243 779777

e-mail (for orders and customer service enquiries): cs-books@wiley.co.uk

Visit our Home Page on http://www.wiley.co.uk or http://www.wiley.com

All Rights Reserved. No part of this publication may be reproduced, stored in a retrieval system, or transmitted, in any form or by any means, electronic, mechanical, photocopying, recording, scanning or otherwise, except under the terms of the Copyright, Designs and Patents Act 1988 or under the terms of a licence issued by the Copyright Licensing Agency Ltd, 90 Tottenham Court Road, London W1P 0LP, UK, without the permission in writing of the Publisher and the copyright owner, with the exception of any material supplied specifically for the purpose of being entered and executed on a computer system, for the exclusive use by the purchaser of the publication.

Neither the authors nor John Wiley & Sons, Ltd accept any responsibility or liability for loss or damage occasioned to any person or property through using the material, instructions, methods or ideas contained herein, or acting or refraining from acting as a result of such use. The authors and Publisher expressly disclaim all implied warranties, including merchantability or fitness for any particular use.

Designations used by companies to distinguish their products are often claimed as trademarks. In all instances where John Wiley & Sons is aware of a claim, the product names appear in initial capital or all capital letters. Readers, however, should contact the appropriate companies for more complete information regarding trademarks and registration.

Other Wiley Editorial Offices

John Wiley & Sons, Inc., 605 Third Avenue,
New York, NY 10158-0012, USA

Wiley-VCH Verlag GmbH
Pappelallee 3, D-69469 Weinheim, Germany

Jacaranda Wiley Ltd, 33 Park Road, Milton,
Queensland 4064, Australia

John Wiley & Sons (Asia) Pte Ltd, 2 Clementi Loop #02-01,
Jin Xing Distripark, Singapore 129809

John Wiley & Sons (Canada) Ltd, 22 Worcester Road,
Rexdale, Ontario, M9W 1L1, Canada

Library of Congress Cataloging-in-Publication Data

Telecommunications optimization : heuristics and adaptive techniques / edited by David W. Corne, Martin J. Oates, George D. Smith.
    p. cm.
  Includes bibliographical references and index.
  ISBN 0-471-98855-3 (alk. paper)
    1. Telecommunications systems—Management. 2. Heuristic programming. I. Corne, David. II. Oates, Martin J. III. Smith, George D., Dr.

TK5102.5 T3963 2000
621 .382—dc21   00-032092

British Library Cataloguing in Publication Data

A catalogue record for this book is available from the British Library

ISBN 0-471-98855-3

Produced from PostScript files supplied by the authors.
Printed and bound in Great Britain by Antony Rowe, Chippenham, Wiltshire.
This book is printed on acid-free paper responsibly manufactured from sustainable forestry in which at least two trees are planted for each one used for paper production.

To Bessie Corne and Mervyn Joseph Corne

DWC

To Maureen Gillian Oates and Jeffrey Oates

MJO

To Gordon, Kate and Sophie

GDS

Contents

Contributors ix
Preface xiii
Acknowledgements xv

1 Heuristic and Adaptive Techniques in Telecommunications: An Introduction 1

Part One: Network Planning and Design 15

2 Evolutionary Methods for the Design of Reliable Networks 17
3 Efficient Network Design using Heuristic and Genetic Algorithms 35
4 Tabu Search and Evolutionary Scatter Search for 'Tree-Star' Network Problems, with Applications to Leased Line Network Design 57
5 Addressing Optimization Issues in Network Planning with Evolutionary Computation 79
6 Node-Pair Encoding Genetic Programming for Optical Mesh Network Topology Design 99
7 Optimizing The Access Network 115

Part Two: Routing and Protocols 135

8 Routing Control in Packet Switched Networks using Soft Computing Techniques .. 137
9 The Genetic Adaptive Routing Algorithm 151
10 Optimization of Restoration and Routing Strategies 167
11 GA-based Verification of Network Protocols Performance 185
12 Neural Networks for the Optimization of Runtime Adaptable Communication Protocols 199

Part Three: Software, Strategy and Traffic Managenent 221

13 Adaptive Demand-based Heuristics for Traffic Reduction in Distributed
Information Systems 223

14 Exploring Evolutionary Approaches to Distributed Database Management 235

15 The Automation of Software Validation using Evolutionary Computation 265

16 Evolutionary Game Theory Applied to Service Selection
and Network Ecologies 283

17 Intelligent Flow Control Under Game Theoretic Framework 307

18 Global Search Techniques for problems in Mobile Communications 331

19 An Effective Genetic Algorithm for Fixed Channel Assignment 357

References 373

Index 393

Contributors

M. Baldi
Politecnico di Torino, Dipartimento di Automatica e Informatica, Torino, Italy

George Bilchev
BT Research Laboratories, Martlesham Heath, Ipswich, Suffolk, UK

Jeffrey Blessing
Department of Electrical Engineering and Computer Science, University of Wisconsin-Milwaukee, USA
blessing@cs.uwm.edu

David Brittain
University of Bristol, Bristol, UK

Brian Carse
Faculty of Engineering, University of the West of England, Bristol, UK
Brian.Carse@uwe.ac.uk

Huimin Chen
Department of Electrical Engineering, The University of Connecticut, Storrs, Connecticut, USA

Steve Y. Chiu
GTE Laboratories, Inc., 40 Sylvan Road, Waltham, Massacheusetts, USA

David W. Corne,
School of Computer Science, Cybernetics and Electronic Engineering, University of Reading, Reading, UK
D.W.Corne@reading.ac.uk

F. Corno
Politecnico di Torino, Dipartimento di Automatica e Informatica, Torino, Italy

Justin C. W. Debuse
School of Information Systems, University of East Anglia, Norwich, UK
jcwd@sys.uea.ac.uk

Berna Dengiz
Department of Industrial Engineering, Gazi University, 06570 Maltepe,
Ankara, Turkey
berna@rorqual.cc.metu.edu.tr

Robert S. Fish
Teleca Ltd, Reading, Berkshire, UK
Rfish@teleca.co.uk

Terence C. Fogarty
Department of Computer Studies, Napier University, Edinburgh, UK
T.Fogarty@dcs.napier.ac.uk

Fred Glover
Graduate School of Business, University of Colorado at Boulder, Boulder,
Colorado, USA

Brian Jones
University of Glamorgan, Wales

Bhaskar Krishnamachari
Wireless Multimedia Laboratory, School of Electrical Engineering, Cornell
University, USA

Yanda Li
Department of Automation, Tsinghua University, Beijing, China

Roger J. Loader
School of Computer Science, Cybernetics and Electronic Engineering,
University of Reading, Reading, UK
Roger.Loader@reading.ac.uk

Masaharu Munetomo
Information and Data Analysis, Graduate School of Engineering, Hokkaido
University, Sapporo, Japan
munetomo@main.eng.hokudai.ac.jp

Alistair Munro
Department of Electrical and Electronic Engineering, University of Bristol,
Bristol, UK
Alistair.Munro@bristol.ac.uk

Contributors

Martin J. Oates
BT Research Laboratories, Martlesham Heath, Ipswich, Suffolk, UK
moates@btinternet.com

Sverrir Olafsson
BT Research Laboratories, Martlesham Heath, Ipswich, Suffolk, UK
sverrir.olafsson@bt.com

K.F. Poon
School of Computing and Engineering Technology, University of Sunderland, Sunderland, UK

M. Rebaudengo
Politecnico di Torino, Dipartimento di Automatica e Informatica,Torino, Italy

Mark D. Ryan
School of Information Systems, University of East Anglia, Norwich, UK

Mark C. Sinclair
Department of Electronic Systems Engineering, University of Essex, Wivenhoe Park, Colchester, Essex, UK
mcs@essex.ac.uk

Alice E. Smith
Department of Industrial and Systems Engineering, Auburn University, AL, USA
aesmith@eng.auburn.edu

George D. Smith
School of Information Systems, University of East Anglia, Norwich, UK
gds@sys.uea.ac.uk

M. Sonza Reorda
Politecnico di Torino, Dipartimento di Automatica e Informatica,Torino, Italy

Giovanni Squillero
Politecnico di Torino, Dipartimento di Automatica e Informatica,Torino, Italy

John Tindle
School of Computing and Engineering Technology, University of Sunderland, Sunderland, UK

Brian C. H. Turton
Cardiff School of Engineering, University of Wales, Cardiff, UK
Turton@cardiff.ac.uk

Ian M. Whittley
School of Information Systems, University of East Anglia, Norwich, UK

Stephen B. Wicker
Wireless Multimedia Laboratory, School of Electrical Engineering, Cornell University, USA

Jon Sims Williams
University of Bristol, Bristol, UK

Jiefeng Xu
Delta Technology Inc., 1001 International Boulevard, Atlanta, Georgia, USA

Preface

Each of us is interested in optimization, and telecommunications. Via several meetings, conferences, chats, and other opportunities, we have discovered these joint interests and decided to put together this book.

It certainly wasn't easy. What made it difficult is partly the reason for this book in the first place: things are moving fast! As researchers in both industry and academia, we are under constant pressure to keep up with new technologies and opportunities. Part of our day jobs involves developing good algorithms for a steady surge of new problems arising from industry. Another part of our day jobs involves finding realistic problems which might yield to some of the algorithms we have developed, which we particularly like, but which don't work on anything else. The remainder of our day jobs tends to involve lecturing (for two of us), project management (for all of us), and things like exam-setting and marking. Finally, for two of us, childcare takes up a substantial part of our night jobs. Naturally, the book was finished when we worked out how to stuff 46 hours into a single day, and keep that up for a few successive weeks.

Seriously, though, telecommunicatons is growing and changing very quickly, and all of this involves the increasing availability of new real-world problems for optimization researchers to address. Telecommunications engineers tend to know all about the problems involved, but are often not aware of developments in computer science and artificial intelligence which might go a long way towards solving those problems. We therefore argue, at the beginning of Chapter 1, that there needs to be more collaboration between the communities, so that both sides can be equipped for the challenges ahead. Indeed, this book is the result of one such collaboration.

We have oriented it slightly towards the telecommunications engineer who knows little about the computational techniques, rather than the other way around. This is mainly because it is much easier to describe the techniques than it is to describe the telecommunications issues; the latter is a much larger collection. However, individual chapters go into some detail, often at an introductory level, in describing particular kinds of problem. Also, we provide a brief introduction to the essence of the problems at the beginning of each part.

Let us know what you think. In particular, if you are a telecommunications engineer, and have been inspired by a chapter or two in this book to try out some techniques that you otherwise wouldn't have, please get in touch: your story may make it into the next edition!

DWC, MJO, GDS, July 2000

Acknowledgements

In the course of getting this book together, several people were very helpful and deserve our sincere thanks. We are all grateful for the patience and encouragement of Mark Hammond, Sarah Hinton, Sarah Lock and Pak-Hang Wang at Wiley, and for the careful job done by Jeremy Thompson, the proof reader. We are also grateful to Chris Hines for technical support, but we don't want him to know that. DWC would like to thank his wife Anna and son Stephen for providing a world outside academia, which he occasionally was able to visit during this project. MJO would like to thank his wife Glenys and sons Gareth and David for their ongoing support. GDS would like to take this opportunity to thank Nortel Networks (Harlow, UK) for sponsoring much of his telecommunications-related research, and also to say thank you to the many individuals at the Nortel Networks Harlow SNT Lab with whom he has had the pleasure of working for many years.

1

Heuristic and Adaptive Computation Techniques in Telecommunications: an Introduction

David Corne, Martin Oates and George Smith

1.1 Optimization Issues in Telecommunications

The complexity and size of modern telecommunications networks provide us with many challenges and opportunities. In this book, the challenges that we focus on are those which involve *optimization*. This simply refers to scenarios in which we are aiming to find something approaching the 'best' among many possible candidate solutions to a problem. For example, there are an intractably large number of ways to design the topology of a private data network for a large corporation. How can we find a particularly good design among all of these possibilities? Alternatively, we may be trying to find a good way to assign frequency channels to the many users of a mobile network. There are a host of complex constraints involved here, but it still remains that the number of possible candidate solutions which meet the main constraints is still too large for us to hope to examine each of them in turn. So, again, we need some way of finding good solutions among all of these possibilities.

These challenges present opportunities for collaboration between telecommunications engineers, researchers and developers in the computer science and artificial intelligence communities. In particular, there is a suite of emerging software technologies specifically aimed at optimization problems which are currently under-used in industry, but with great potential for profitable and effective solutions to many problems in telecommunications.

Much of this book focuses on these optimization techniques, and the work reported in the forthcoming chapters represents a good portion of what is currently going on in terms of applying these techniques to telecommunications-related problems. The techniques employed include so-called 'local search' methods such as simulated annealing (Aarts and Korst, 1989) and tabu search (Glover, 1989; 1989a), and 'population-based' search techniques such as genetic algorithms (Holland, 1975; Goldberg, 1989), evolution strategies (Schwefel, 1981; Bäck, 1996), evolutionary programming (Fogel, 1995) and genetic programming (Koza, 1992). Section 1.3 gives a brief and basic introduction to such techniques, aimed at one type of reader: the telecommunications engineer, manager or researcher who knows all too much about the issues, but does not yet know a way to address them. Later chapters discuss their use in relation to individual problems in telecommunications.

1.2   Dynamic Problems and Adaptation

A fundamental aspect of many optimization issues in telecommunications is the fact that they are *dynamic*. What may be the best solution now may *not* be the ideal solution in a few hours, or even a few minutes, from now. For example, the provider of a distributed database service (such as video-on-demand, web-caching services, and so forth) must try to ensure good quality of service to each client. Part of doing this involves redirecting clients' database accesses to different servers at different times (invisibly to the client) to effect appropriate load-balancing among the servers. A good, modern optimization technique can be used to distribute the load appropriately across the servers, however this solution becomes invalid as soon as there is a moderate change in the clients' database access patterns. Another example is general packet routing in a large essentially point-to-point network. Traditionally, routing tables at each node are used to look up the best 'next hop' for a packet based on its eventual destination. We can imagine an optimization technique applying to this problem, which looks at the overall traffic pattern and determines appropriate routing tables for each node, so that general congestion and delay may be minimized, i.e. in many cases the best 'next hop' might not be the next node on a shortest path, since this link may be being heavily used already. But, this is clearly a routine which needs to be run over and over again as the pattern of traffic changes.

Iterated runs of optimization techniques are one possible way to approach dynamic problems, but it is often a rather inappropriate way, especially when good solutions are needed very fast, since the environment changes very quickly. Instead, a different range of modern computational techniques are often appropriate for such problems. We can generally class these as 'adaptation' techniques, although those used later in this book are actually quite a varied bunch. In particular, later chapters will use neural computation, fuzzy logic and game theory to address adaptive optimization in dynamic environments, in some cases in conjunction with local or population-based search. Essentially, whereas an optimization technique provides a fast and effective way to find a good solution from a huge space of possibilities, an adaptive technique must provide a good solution almost *instantly*. The trick here is that the methods usually employ 'off-line' processing to learn about the problem they are addressing, so that, when a good, quick result is required, it is ready to deliver. For example, an adaptive approach to optimal packet routing in the face of changing traffic patterns would involve some continual but minimal processing which continually updates the routing tables at each node based on current information about delays and traffic levels.

In the remainder of this chapter, we will briefly introduce the optimization and adaptation techniques which we have mentioned. More is said about each of these in later chapters. Then, we will say a little about each of the three parts of this book. Finally, we will indicate why we feel these techniques are so important in telecommunications, and will grow more and more with time.

1.3 Modern Heuristic Techniques

There are a range of well-known methods in operations research, such as Dynamic Programming, Integer Programming, and so forth, which have traditionally been used to address various kinds of optimization problems. However, a large community of computer scientists and artificial intelligence researchers are devoting a lot of effort these days into more modern ideas called 'metaheuristics' or often just 'heuristics'. The key difference between the modern and the classical methods is that the modern methods are fundamentally easier to apply. That is, given a typical, complex real-world problem, it usually takes *much* less effort to develop a simulated annealing approach to solving that problem than it does to formulate the problem in such a way that integer linear programming can be applied to it.

That is not to say that the 'modern' method will outperform the 'classical' method. In fact, the likely and typical scenarios, when both kinds of method have been applied, are:

- A metaheuristics expert compares the two kinds of technique: the modern method outperforms the classical method.
- A classical operations research expert compares the two kinds of technique: the classical method outperforms the modern method.

Although tongue-in-cheek, this observation is based on an important aspect of solving optimization problems. The more you know about a particular technique you are applying, the better you are able to tailor and otherwise exploit it to get the best results.

In this section we only provide the basic details of a few modern optimization algorithms, and hence do not provide quite enough information for a reader to be able to tailor them appropriately to particular problems. However, although we do not tell you how to twiddle them, we do indicate where the knobs are. How to twiddle them depends very much on the problem, and later chapters will provide such information for particular cases. What should become clear from this chapter, however, is that these techniques are highly general in their applicability. In fact, whenever there is some decent way available to evaluate or score candidate solutions to your problem, then these techniques can be applied.

The techniques essentially fall into two groups: local search and population-based search. What this means is discussed in the following.

1.3.1 Local Search

Imagine you are trying to solve a problem P, and you have a huge set S of potential solutions to this problem. You don't really *have* the set S, since it is too large to ever enumerate fully. However, you have some way of generating solutions from it. For example, S could be a set of connection topologies for a network, and candidate solutions s, s', s'', and so on, are particular candidate topologies which you have come up with somehow. Also, imagine that you have a fitness function $f(s)$ which gives a score to a candidate solution. The better the score, the better the solution. For example, if we were trying to find the most reliable network topology, then $f(s)$ might calculate the probability of link failure between two particularly important nodes. Strictly, in this case we would want to use the reciprocal of this value if we really wanted to call it 'fitness'. In cases where the *lower* the score is, the better (such as in this failure probability example), it is often more appropriate to call $f(s)$ a *cost function*.

We need one more thing, which we will call a *neighbourhood operator*. This is a function which takes a candidate s, and produces a new candidate s' which is (usually) only slightly different to s. We will use the term 'mutation' to describe this operator. For example, if we mutate a network topology, the resulting mutant may include an extra link not present in the 'parent' topology, but otherwise be the same. Alternatively, mutation might remove, or move, a link.

Now we can basically describe local search. First, consider one of the simplest local search methods, called *hillclimbing*, which amounts to following the steps below:

1. Begin: generate an initial candidate solution (perhaps at random); call this the current solution, c. Evaluate it.

2. Mutate c to produce a mutant m, and evaluate m.

3. If $f(m)$ is better than or equal to $f(c)$, then replace c with m. (i.e. c is now a copy of m).

4. Until a termination criterion is reached, return to step 2.

The idea of hillclimbing should be clear from the algorithm above. At any stage, we have a *current* solution, and we take a look at a neighbour of this solution – something slightly different. If the neighbour is fitter (or equal), then it seems a good idea to *move to* the neighbour; that is, to start again but with the neighbour as the new current solution. The fundamental idea behind this, and behind local search in general, is that good solutions 'congregate'. You can't seriously expect a very reliable topology to emerge by, for example, adding a single extra link to a very unreliable topology. However, you *can* expect that such a change could turn a very reliable topology into a slightly more reliable one.

In local search, we exploit this idea by continually searching in the neighbourhood of a current solution. We then move to somewhere reasonably fit in that neighbourhood, and reiterate the process. The danger here is that we might get stuck in what is called a 'local optimum', i.e. the current solution is not quite good enough for our purposes, but all of its neighbours are even worse. This is bad news for the hillclimbing algorithm, since it will simply be trapped there. Other local search methods distinguish themselves from hillclimbing, however, precisely in terms of having some way to address this situation. We will look at just two such methods here, which are those in most common use, and indeed, are those used later on in the book. These are simulated annealing and tabu search.

Simulated Annealing
Simulated annealing is much like hillclimbing. The only differences are the introduction of a pair of parameters, an extra step which does some book-keeping with those parameters, and, which is the main point, step 3 is changed to make use of these parameters:

1. Begin: generate and evaluate an initial candidate solution (perhaps at random); call this the current solution, c. Initialize temperature parameter T and cooling rate r ($0 < r < 1$).
2. Mutate c to produce a mutant m, and evaluate m.
3. If test($f(m), f(c), T$) evaluates to true, then replace c with m. (ie: c is now a copy of m).
4. Update the temperature parameter (e.g. T becomes rT)
5. Until a termination criterion is reached, return to step 2.

What happens in simulated annealing is that we sometimes accept a mutant even if it is worse than the current solution. However, we don't do that very often, and we are much less likely to do so if the mutant is *much* worse. Also, we are less likely to do so as time goes on. The overall effect is that the algorithm has a good chance of escaping from local optima, hence possibly finding better regions of the space later on. However, the fundametal bias of moving towards better regions is maintained. All of this is encapsulated in the *test* function in step 3. An example of the kind of test used is first to work out:

$$e^{(f(m)-f(c))/T}$$

This assumes that we are trying to maximize a cost (otherwise we just switch $f(m)$ and $f(c)$). If the mutant is better or equal to the current solution, then the above expression will come to something greater or equal to one. If the mutant is worse, then it will result to something smaller than 1, and the worse the mutant is, the closer to zero it will be. Hence, the result of this expression is used as a probability. A random number is generated, *rand*, where $0 <$ *rand* < 1, and the test in step 3 simply checks whether or not the above expression is smaller than *rand*. If so (it will always be so if the mutant is better or equal), we accept the mutant. T is a 'temperature' parameter. It starts out large, and is gradually reduced (see step 4) with time. As you can tell from the above expression, this means that the probabilities of accepting worse mutants will also reduce with time.

Simulated annealing turns out to be a very powerful method, although it can be quite difficult to get the parameters right. For a good modern account, see Dowsland (1995).

Tabu Search
An alternative way to escape local optima is provided by tabu search (Glover 1989; 1989a; Glover and Laguna, 1997). There are many subtle aspects to tabu search; here we will just indicate certain essential points about the technique. A clear and full introduction is provided in Glover and Laguna (1995; 1997).

Tabu search, like some other local search methods which we do not discuss, considers several neighbours of a current solution and eventually chooses one to move to. The distinguishing feature of tabu search is how the choice is made. It is not simply a matter of choosing the fittest neighbour of those tested. Tabu search also takes into account the particular kind of mutation which would take you there. For example, if the best neighbour of your current solution is one which can be reached by changing the other end of a link emerging from node k, but we have quite recently made such a move in an earlier iteration, then a different neighbour may be chosen instead. Then again, even if the current best move is of a kind which has been used very recently, and would normally be disallowed (i.e. considered 'taboo'), tabu search provides a mechanism based on 'aspiration criteria', which would allow us to choose that more if the neighbour in question was sufficiently better than the (say) current solution.

Hence, any implementation of tabu search maintains some form of memory, which records certain attributes of the recently made moves. What these attributes are depends much on the problem, and this is part of the art of applying tabu search. For example, if we are trying to optimize a network topology, one kind of mutation operator would be to change the cabling associated with the link between nodes a and b. In our tabu search implementation, we would perhaps record only the fact that we made a 'change-cabling' move at iteration i, or we might otherwise, or additionally, simply record the fact that we made a change in association with node a and another in association with node b. If we only recorded the former type of attribute, then near-future possible 'change-cabling' moves would be disallowed, independent of what nodes were involved. If we recorded only the latter type of attribute, then near-future potential changes involving a and/or b might be disallowed, but 'change-cabling' moves *per se* would be acceptable.

Artful Local Search

Our brief notes on simulated and tabu search illustrate that the key aspect of a good local search method is to decide which neighbour to move to. All local search methods employ the fundamental idea that *local* moves are almost always a good idea, i.e. if you have a good current solution, there may be a better one nearby, and that may lead to even better ones, and so forth.

However, it is also clear that occasionally, perhaps often, we must accept the fact that we can only find improved current solutions by temporarily (we hope) moving to worse ones. Simulated annealing and tabu search are the two main approaches to dealing with this issue. Unfortunately, however, it is almost never clear, given a particular problem, what is the best way to design and implement the approach. There are many choices to make; the first is to decide how to *represent* a candidate solution in the first place. For example, a network topology can be represented as a list of links, where each link is a pair of nodes (a,b). Decoding such a list into a network topology simply amounts to drawing a point-to-point link for each node-pair in the list. Alternatively, we could represent a network topology as a binary string, containing as many bits as there are *possible* links. Each position in the bit string will refer to a particular potential point-to-point link, So, a candidate solution such as '10010...' indicates that there is a point-to-point link between nodes 1 and 2, none between nodes 1 and 3, or 1 and 4, but there is one between nodes 1 and 5, and so forth.

Generally, there are many ways of devising a method to represent solutions to a problem. The choice, of course, also affects the design of neighbourhood operators. In the above example, removing a link from a topology involves two different kinds of operation in the two representations. In the node-pair list case, we need to actually remove a pair from the list. In the binary case, we change a 1 to a 0 in a particular position in the string.

Devising good representations and operators is part of the art of effectively using local search to solve hard optimization problems. Another major part of this art, however, is to employ problem specific knowledge or existing heuristics where possible. For example, one problem with either of the two representations we have noted so far for network topology is that a typical randomly generated topology may well be unconnected. That is, a candidate solution may simply not contain paths between every pair of nodes. Typically, we would only be interested in connected networks, so any algorithmic search effort spent in connection with evaluating unconnected networks seems rather wasteful. This is where basic domain knowledge and existing heuristics will come in helpful. First, any good network designer will know about various graph theoretic concepts, such as spanning trees, shortest path algorithms, and so forth. It is not difficult to devise an alteration to the node-pair list representation, which ensures that every candidate solution s contains a spanning tree for the network, and is hence connected. One way to do this is involves interpreting the first few node-pairs *indirectly*. Instead of (a, b) indicating that this network contains a link between a and b, it will instead mean that the ath connected node will be linked to the bth unconnected node. In this way, every next node pair indicates how to join an as yet unused node to a growing spanning tree. When all nodes have been thus connected, remaining node-pairs can be interpreted directly.

Better still, the problem we address will probably involve cost issues, and hence costs of particular links will play a major role in the fitness function. Domain knowledge then tells us that several well known and fast algorithms exits which find a minmal-cost spanning tree (Kruskal, 1956; Prim, 1957). It may therefore make good sense, depending on various other details of the problem in hand, to initialize each candidate solution with a minimal cost tree, and all that we need to *represent* are the links we add onto this tree.

There are many, many more ways in which domain knowledge or existing heuristics can be employed to benefit a local search approach to an optimization problem. Something about the problem might tell us, for example, what kinds of mutation have a better chance of leading to good neighbours. An existing heuristic method for, say, quickly assigning channels in a mobile network, might be used to provide the initial point for a local search which attempts to find better solutions.

1.3.2 *Population-Based Search*

An alternative style of algorithm, now becoming very popular, builds on the idea of local search by using a *population* of 'current' solutions instead of just one. There are two ways in which this potentially enhances the chances of finding good solutions. First, since we have a population, we can effectively spend time searching in several different neighbourhoods at once. A population-based algorithm tends to share out the computational effort to different candidate solutions in a way biased by their relative fitness. That is, more time will be spent searching the neighbourhoods of good solutions than moderate solutions.

However, at least a little time will be spent searching in the region of moderate or poor solutions, and should this lead to finding a particularly good mutant along the way, then the load-balancing of computational effort will be suitably revised.

Another opportunity provided by population-based techniques is that we can try out *recombination* operators. These are ways of producing mutants, but this time from two or more 'parent' solutions, rather than just one. Hence the result can be called a *recombinant*, rather than a mutant. Recombination provides a relatively principled way to justify large neighbourhood moves. One of the difficulties of local search is that even advanced techniques such as simulated annealing and tabu search will still get stuck at local optima, the only 'escape' from which might be a rather drastic mutation. That is, the algorithm may have tried all of the possible local moves, and so must start to try non-local moves if it has any chance of getting anywhere. The real problem here is that there are so many potential non-local moves. Indeed, the 'non-local' neighbourhood is actually the entire space of possibilities!

Recombination is a method which provides a way of choosing good non-local moves from the huge space of possibilities. For example, if two parent solutions are each a vector of k elements, a recombination operator called uniform crossover (Syswerda, 1989) would build a child from these two parents by, for each element in turn, randomly taking its value from either point. The child could end up being 50% different from each of its parents, which is vastly greater than the typical difference between a single parent solution and something in its neighbourhood.

Below are the steps for a generic population based algorithm.

1. Begin: generate an initial population of candidate solutions. Evaluate each of them.
2. Select some of the population to be parents.
3. Apply recombination and mutation operators to the parents to produce some children.
4. Incorporate the children into the population.
5. Until a termination criterion is reached, return to step 2.

There are many ways to perform each of these steps, but the essential points are as follows. Step 2 usually employs a 'survival of the fittest' strategy; this is where the 'load sharing' discussed above comes into play. The fitter a candidate solution is, the more chance it has of being a parent, and therefore the more chances there are that the algorithm will explore its neighbourhood. There are several different selection techniques, most of which can be parameterised to alter the degree to which fitter parents are preferred (the *selection pressure*). Step 3 applies either recombination or mutation operators, or both. There are all sorts of standard recombination and mutation operators, but – as we hinted above – the real benefits come when some thought has been put into designing specific kinds of operator using domain knowledge. In step 4, bear in mind that we are (almost always) maintaining a fixed population size. So, if we have a population of 100, but 20 children to incorporate, then 20 of these 120 must be discarded. A common approach is to simply discard the 20 least fit of the combined group, but there are a few other approaches; we may use a technique called 'crowding', e.g. De Jong (1975), in which diversity plays a role in the decision about what candidate solutions to discard. For example, we would certainly prefer to discard solution *s* in favour of a less fit solution *t*, if it happens to be the case that *s* already has a duplicate in the population, but *t* is 'new'.

Finally, we should point out some terminological issues. There are many population based algorithms, and in fact they are usually called *evolutionary algorithms* (EAs). Another common term employed is *genetic algorithms*, which strictly refers to a family of such methods which always uses a recombination operator (Holland, 1965; Goldberg, 1989), while other families of such algorithms, called *evolutionary programming* (Fogel, 1995) and *evolution strategies* (Bäck, 1996), tend to use mutation alone, but are quite clever about how they use it. In all cases, a candidate solution tends to be called a *chromosome* and its elements are called *genes*.

1.4 Adaptive Computation Techniques

To address the optimization needs inherent in constantly changing telecommunications environments, direct use of local or population based optimization techniques may sometimes be quite inappropriate. This might be because it may take too long to converge to a good solution, and so by the time the solution arrives, the problem has changed!

What we need instead in this context is a way to make very fast, but very good, decisions. For example, to decide what the best 'next hop' is for a packet arriving at node *a* with destination *d*, we could perhaps run a simulation of the network given the current prevailing traffic patterns and estimate the likely arrival times at the destination for given possible next hops. The result of such a simulation would provide us with a well-informed choice, and we can then send the packet appropriately on its way.

Now, if we could the above in a few microseconds, we would have a suitable and very profitable packet routing strategy. However, since it will probably take several hours to do an accurate simulation using the sort of processing power and memory typically available at network switches, in reality it is a ridiculous idea!

Instead, we need some way of making good decisions quickly, but where the decision will somehow take into account the prevailing circumstances. Ideally, we are looking for a 'black box' which, when fed with a question and some environmental indicators, provides a sensible and immediate answer. One example of such a black box is a routing table of the sort typically found in packet-switched networks. The question asked by a communication packet is: 'eventually I want to get to *d*, so where should I go next?'. The routing table, via a simple lookup indexed by *d*, very quickly provides an answer and sends it on its way. The trouble with this, of course, is that it is fundamentally non-adaptive. Unless certain advanced network management protocols are in operation, the routing table will always give the same answer, even if the onward link from the suggested next hop to *d* is heavily congested at the moment. Our black boxes must therefore occasionally change somehow, and adapt to prevailing conditions. In fact, chapters eight and nine both discuss ways of doing this involving routing tables in packet switched networks.

So, adaptive techniques in the context of telecommunications tends to involve black boxes, or models, which somehow learn and adapt 'offline' but can react very quickly and appropriately when asked for a decision. In some of the chapters that involve adaptation, the techniques employed are those we have discussed already in section 1.3, but changed appropriately in the light of the need for fast decisions. Though, in others, certain other important modern techniques are employed, which we shall now briefly introduce. These are *neural computation*, *fuzzy logic* and *game theory*. The role of neural computation in this context is to develop a model offline, which learns (by example) how to make the right decisions in varied sets of circumstances. The resulting trained neural network is then employed online as the decision maker. The role of fuzzy logic is to provide a way of producing robust *rules* which provide the decisions. This essentially produces a black box decisionmaker, like a neural network, but with different internal workings. Finally, game theory provides another way of looking at complex, dynamic network scenarios. Essentially, if we view certain aspects of network management as a 'game', a certain set of equations and well known models come into play, which in turn provide good approximations to the dynamics of real networks. Hence, certain network management decisions can be supported by using these equations.

1.4.1 Neural Computation

Neural computation (Rumelhart and MacClelland, 1989; Haykin, 1998) is essentially a pattern classification technique, but with vastly wider application possibilities than that suggests at first sight. The real power of this method lies in the fact that we do not need to know how to distinguish one kind of pattern from another. To build a classical rule-based expert system for distinguishing patterns, for example, we would obviously need to acquire rules, and build them into the system. If we use neural computation, however, a special kind of black box called a 'neural network' will essentially learn the underlying rules by example. A classic application of the technique is in credit risk assessment. The rules which underlie a decision about who will or will not be a good credit risk, assuming that we disregard clear cases such as those with pre-existing high mortgages but low salaries (i.e. one of the co-editors of this book), are highly complex, if indeed they can be expressed at all. We could train a neural network to predict bad risks, however, simply by providing a suite of known examples, such as 'person p from region r with salary s and profession y ... defaulted on a loan of size m; person q with salary t ...' and so forth. With things like p, r, s, and so forth, as inputs, the neural network gradually adjusts itself internally in such a way that it eventually produces the correct output (indication the likelihood of defaulting on the loan) for each of the examples it is trained with. Remarkably, and very usefully, we can then expect the neural network to provide good (and extremely fast) guesses when provided with previously unseen inputs.

Internally, a neural network is a very simple structure; it is just a collection of nodes (sometimes called 'artificial neurons') with input links and output links, each of which does simple processing: it adds up the numbers arriving via its input links, each weighted by a strength value (called a weight) associated with the link it came through, it then processes this sum (usually just to transform it to a number between 0 and 1), and sends the result out across its output links. A so-called 'feed-forward' neural network is a collection of such nodes, organised into layers. The problem inputs arrive as inputs to each of the first layer of nodes, the outputs from this layer feed into the second layer, and so on, although usually there are just three layers. Obviously, the numbers that come out at the end depend on what arrives at the input layer, in a way intimately determined by the weights on the links. It is precisley these weights which are altered gradually in a principled fashion by the training process. Classically, this is a method called backpropagation (Rumelhart and MacClelland,

1989), but there are many modern variants. Indeed, the kind of network we have briefly described here is just one of many types available (Haykin, 1999).

1.4.2   Fuzzy Logic

In some cases we can think of decent rules for our problem domain. For example, 'if traffic is heavy, use node a', and 'if traffic is very heavy, use node b'. However, such rules are not very useful without a good way of deciding what 'heavy' or 'very heavy' actually mean in terms of link utilization, for example. In a classical expert system approach, we would adopt pre-determined thresholds for these so-called 'linguistic variables', and decide, for example, that 'traffic is heavy' means that the utilization of the link in question is between 70% and 85%. This might seem fine, but it is not difficult to see that 69.5% utilization might cause problems; in such a scenario, the rule whose condition is 'moderate traffic' (55% to 70%, perhaps) would be used, but it may have been more appropriate, and yield a better result, to use the 'heavy traffic'.

Fuzzy logic provides a way to use linguistic variables which deals with the thresholds issue in a very natural and robust way. In fact, it almost rids us of a need for thresholds, instead introducing things called 'membership functions'. We no longer have traffic which is *either* heavy *or* moderate. Instead, a certain traffic value is *to some extent* heavy, and *to some other extent* moderate. The extents depend upon the actual numerical values by way of the membership functions, which are often simple 'triangular functions'. For example, the extent to which traffic is heavy might be 0 between 0% and 35% utilization, it may then rise smoothly to 1 between 35% and 75%, and then drop smoothly to 0 again between 75% and 90%, being 0 from then on. The membership function for the linguistic variable 'very heavy' will overlap with this, so that a traffic value of 82.5% might be 'heavy' to the extent 0.5, and 'very heavy' to the extent 0.7.

Given certain environmental conditions, different rules will therefore apply to different extents. In particular, fuzzy logic provides ways in which to determine the degree to which different rules are applicable when the condition parts of the rules involve several linguistic variables. Chapter 8, which uses fuzzy logic, discusses several such details.

The main strength of fuzzy logic is that we only need to ensure that the membership functions are intuitively reasonable. The resulting system, in terms of the appropriateness of the decisions that eventually emerge, tends to be highly robust to variations in the membership functions, within reasonable bounds. We may have to do some work, however, in constructing the rules themselves. This is where the 'offline' learning comes in when we are using fuzzy logic in an adaptive environment. Sometimes, genetic algorithms may be employed for the task of constructing a good set of rules.

1.4.3   Game Theory

Finally, game theory provides another way to look at certain kinds of complex, dynamic commuications-related problems, especially as regards network management and service provision. Consider the complex decision processes involved in deciding what tariff to set for call connection or data provision service in a dynamic network environment, involving competition with many other service providers. Under certain sets of assumptions, the dynamics of the network, which includes, for example, the continual activity of users switching to different service providers occasionally based on new tariffs being advertised, will bear considerable resemblance to various models of competition which have been developed in theoretical biology, economics, and other areas. In particular, certain equations will apply which enable predictions as to whether a proposed new tariff would lead to an *unstable* situation (where, for example, more customers would switch to your new service than you could cope with).

Especially in future networks, many quite complex management decisions might need to be made quickly and often. For example, to maintain market share, new tariffs might need to be set hourly, and entirely automatically, on the basis of current activity in terms of new subscriptions, lapsed customers, and news of new tariffs set by rival providers. The game theoretic approaches being developed by some of the contributors to this book will have an ever larger role to play in such scenarios, perhaps becoming the central engine in a variety of adaptive optimization techniques aimed at several service provision issues.

1.5   The Book

Heuristic and adaptive techniques are applied to a variety of telecommunications related optimization problems in the remainder of this book. It is divided into three parts, which roughly consider three styles of problem. The first part is network design and planning. This is, in fact, where most of the activity seems to be in terms of using modern heuristic techniques. Network design is simple to get to grips with as an optimization problem, since the problems involved can be quite well defined, and there is usually no dynamicity involved. That is, we are able to invest a lot of time and effort in developing a good optimization based approach, perhaps using specialized hybrid algorithms and operators, which may take quite a while to come up with a solution. Once a solution is designed, it might be gradually implemented, built and installed over the succeeding weeks or months.

In contrast, routing and protocols, which are the subject of Part Two, involve optimization issues which are almost always dynamic (although two of the chapters in Part Two sidestep this point by providing novel uses for time-consuming optimization method – see below). The bulk of this part of the book considers various ways to implement and adapt the 'black box' models discussed earlier. In one case, the black box is a neural network; in another, it is a fuzzy logic ruleset and in another it is a specialized form of routing table, adapted and updated via a novel algorithm reminiscent of a genetic algorithm.

Part Three looks at a range of issues, covering software, strategy, and various types of traffic management. Software development is a massive and growing problem in the telecommunications industry; it is not a telecommunications problem *per se*, but good solutions to it will have a very beneficial impact on service providers. The 'problem' is essentially the fact that service provision, network equipment, and so forth, all need advanced and flexible software, but such software takes great time and expense to develop. The rising competition, service types and continuing influx of new technologies into the telecommunications arena all exacerbate this problem, and telecommunications firms are therefore in dire need of ways to speed up and economize the software development process. One of the chapters in Part Three addresses an important aspect of this problem. Part Three also addresses *strategy*, by looking at the complex dynamic network management issues involved in service provision, and in admission and flow control. In each case, an approach based on game theory is adopted. Finally, Part Three also addresses traffic management, in both static and mobile networks.

We therefore cover a broad range of telecommunications optimization issues in this book. This is, we feel, a representative set of issues, but far from complete. The same can be said in connection with the range of techniques which are covered. The story that emerges is that heuristic and adaptive computation techniques have gained a foothold in the telecommunications arena, and will surely build on that rapidly. The emerging technologies in telecommunications, not to mention increasing use of the Internet by the general public, serve only to expand the role that advanced heuristic and adaptive methods can play.

Part One

Network Planning and Design

Essentially, a telecommunications network is a collection of devices (computers, telephone handsets, bridges, and so forth) linked together by wires and cables. Cables range in length and type, from the short wires that connect your home PC to a modem, to gargantuan cables which span the Atlantic Ocean. The task that a network needs to perform – that is the type of traffic it is meant to carry – lays constraints upon various aspects of its design. These aspects include the type and length of cable employed, the topology of the network's interconnections, the communications protocols employed at different levels, and so forth.

The networks (or network) which we are most familiar with have not really been designed – rather, they have evolved, as more and more cables and equipment have been linked in with older installations to cope with increasing demand. However, there are increasing contexts and needs for which *ab initio* or *greenfield* network design is an urgent issue. For example, private data networks installed in large corporations.

In this part, various network design problems are addressed by various heuristic optimization techniques. The problems cover different types of network, including copper-based, optical mesh and leased-line, and different types of criteria, ranging through cost, reliability and capacity. The message from the chapters in this part is strongly positive in favour of modern optimization techniques, particularly evolutionary algorithms. They are found to work very well, and very flexibly, over an impressive range of quite distinct types of network design problem. In fact, one of the chapters discusses a sophisticated tool, involving sophisticated hybrids of modern heuristic techniques, which is now commonly used by a large telecommunications firm to plan and design networks.

2

Evolutionary Methods for the Design of Reliable Networks

Alice E. Smith and Berna Dengiz

2.1 Introduction to the Design Problem

The problem of how to design a network so that certain constraints are met and one or more objectives are optimized is relevant in many real world applications in telecommunications (Abuali *et al.*, 1994a; Jan *et al.*, 1993; Koh and Lee, 1995; Walters and Smith, 1995), computer networking (Chopra *et al.*, 1984; Pierre *et al.*, 1995), water systems (Savic and Walters, 1995) and oil and gas lines (Goldberg, 1989). This chapter focuses on design of minimum cost reliable communications networks when a set of nodes and their topology are given, along with a set of possible bi-directional arcs to connect them. A variety of approaches are cited, and the previous work of the authors using genetic algorithms is discussed in detail. It must be noted that the design problem solved by these methods is significantly simplified. A large number of components and considerations are not treated here. Instead, the approaches focus on the costs and reliabilities of the network links.

2.1.1 Costs

Costs can include material costs of the cabling, installation costs such as trenching or boring, land or right of way costs, and connection or terminal costs inherent with the cabling. Many of these are 'unit costs', i.e. they depend on the length of the arc. However, there can be fixed costs per arc and these are easily accommodated in the methods discussed. In many papers, a unit cost is not specifically mentioned; instead each arc is assigned a weight which is used as the complete cost of the arc (Aggarwal *et al.*, 1982; Atiqullah and Rao, 1993; Kumar *et al.*, 1995).

2.1.2 Reliability

Associated with each type of connection is a reliability (with an implicit mission time), or equivalently, a stationary availability. This reliability has a range from 0 (never operational) to 1 (perfectly reliable). It is assumed (with good justification) that reliability comes at a cost. Therefore, a more reliable connection type implies a greater unit cost. The trade-off between cost and reliability is not linear. An increase in reliability causes a greater than equivalent increase in cost; often a quadratic relationship is assumed. Other simplifying assumptions commonly made are that nodes are perfectly reliable and do not fail, and that arcs have two possible states – good or failed. Arcs fail independently and repair is not considered.

There are two main reliability measures used in network design, namely all-terminal (also called *uniform* or *overall* network reliability) and source-sink (also called *two terminal* reliability). Sections 2.4 and 2.5 in this chapter consider only all-terminal reliability, while section 2.6 includes a source-sink reliability problem. All-terminal network reliability is concerned with the ability of each and every network node to be able to communicate with all other network nodes through some (non-specified) path. This implies that the network forms at least a minimum spanning tree. Source-sink reliability is concerned with the ability of the source node (pre-specified) to communicate with the sink node (also pre-specified) through some (non-specified) path.

The problem of calculating or estimating the reliability of a network is an active area of research related to the network design problem. There are four main approaches – exact calculation through analytic methods, estimation through variations of Monte Carlo simulation, upper or lower bounds on reliability, and easily calculated, but crude, surrogates for reliability. The issue of calculating or estimating the reliability of the network is so important for optimal network design, section 2.3 covers it in detail.

2.1.3 Design Objectives and Constraints

The most common objective is to design a network by selecting a subset of the possible arcs so that network reliability is maximized, and a maximum cost constraint is met. However, in many situations, it makes more sense to minimize network cost subject to a minimum network reliability constraint. There may be side constraints, such as minimum node degree (a node's degree is simply the number of arcs emanating from it) or maximum arc length allowed in the network. In this chapter, the objective is to find the minimum cost network architecture that meets a pre-specified minimum network reliability. That is, a cost function $C(x)$ is minimized over network archiectures with the constraint that the reliability $R(x)$ exceeds some minimum required level, R_0.

2.1.4 Difficulty of the Problem

The network design problem, as described, is an NP-hard combinatorial optimization problem (Garey and Johnson, 1979) where the search space for a fully connected network with a set of nodes N and with k possible arc choices is:

$$k^{|N|(|N|-1)/2} \tag{2.1}$$

Evolutionary Methods for the Design of Reliable Networks

Compounding the exponential growth in number of possible network topologies is the fact that the exact calculation of network reliability is also an NP-hard problem, which grows exponentially with the number of arcs.

2.1.5 Notation

The notation adopted in the remainder of this chapter is as detailed in Table 2.1.

Table 2.1 Notation used in chapter 2.

Notation	Meaning
N	Set of given nodes.
L	Set of possible arcs.
l_{ij}	Option of each arc ($\in \{1,2,...,k\}$).
$p(l_k)$	Reliability of arc option.
$c(l_k)$	Unit cost of arc option.
x	Topology of a network design.
$C(x)$	Total cost of a network design.
C_0	Maximum cost constraint.
$R(x)$	Reliability of a network design.
R_0	Minimum network reliability constraint.
g	Generation number in a genetic algorithm.
s	Population size of the genetic algorithm.
$m\%$	Percentage of mutants created per generation in the genetic algorithm.
r_p	Penalty rate in the genetic algorithm.
r_m	Mutation rate in the genetic algorithm.
t	Number of Monte Carlo reliability simulation iterations.

2.2 A Sampling of Optimization Approaches

The optimal design problem, when considering reliability, has been studied in the literature using alternative methods of search and optimization. Jan *et al.* (1993) developed an algorithm using decomposition based on branch and bound to minimize arc costs with a minimum network reliability constraint; this is computationally tractable for fully connected networks up to 12 nodes. Using a greedy heuristic, Aggarwal *et al.* (1982) maximized reliability given a cost constraint for networks with differing arc reliabilities and an all-terminal reliability metric. Ventetsanopoulos and Singh (1986) used a two-step heuristic procedure for the problem of minimizing a network's cost subject to a reliability constraint. The algorithm first used a heuristic to develop an initial feasible network configuration, then a branch and bound approach was used to improve this configuration. A deterministic version of simulated annealing was used by Atiqullah and Rao (1993) to find the optimal design of very small networks (five nodes or less). Pierre *et al.* (1995) also used simulated annealing to find optimal designs for packet switch networks where delay and capacity were considered, but reliability was not. Tabu search was used by Glover *et al.* (1991) to choose network design when considering cost and capacity, but not reliability. Another tabu search approach by Beltran and Skorin-Kapov (1994) was used to design reliable networks by searching for the least cost spanning 2-tree, where the 2-tree objective was a crude surrogate for reliability. Koh and Lee (1995) also used tabu search to find telecommunication network designs that required some nodes (special offices) have more than one arc while others (regular offices) required only one arc, using this arc constraint as a surrogate for network reliability.

Genetic algorithms (GAs) have recently been used in combinatorial optimization approaches to reliable design, mainly for series and parallel systems (Coit and Smith, 1996; Ida *et al.*, 1994; Painton and Campbell, 1995). For network design, Kumar *et al.* (1995) developed a GA considering diameter, average distance, and computer network reliability and applied it to four test problems of up to nine nodes. They calculated all-terminal network reliability exactly and used a maximum network diameter (minimal number of arcs between any two nodes) as a constraint. The same authors used this GA to design the expansion of existing computer networks (Kumar *et al.*, 1995a). Their approach has two significant limitations. First, they require that all network designs considered throughout the search be feasible. While relatively easy to achieve using a cost constraint and a maximum reliability objective, this is not as easy when using a cost objective and a reliability constraint. The second limitation is their encoding, which is a list of all possible arcs from each node, arranged in an arbitrary node sequence. Presence (absence) of an arc is signaled by a 1 (0). For a ten node problem, the encoding grows to a string length of 90. However, the more serious drawback of the encoding is the difficulty in maintaining the agreement of the arcs present and absent after crossover and mutation. An elaborate repair operator must be used, which tends to disrupt the beneficial effects of crossover. Davis *et al.* (1993) approached a related problem considering arc capacities and rerouting upon arc failure using a problem-specific GA. Abuali *et al.* (1994) assigned terminal nodes to concentrator sites to minimize costs while considering capacities using a GA, but no reliability was considered. The same authors (Abuali *et al.*, 1994a) solved the probabilistic minimum spanning tree problem, where inclusion of the node in the network is stochastic and the objective is to minimize connection (arc) costs, again disregarding reliability. Walters and Smith (1995) used a GA for the design of a pipe network that connects all nodes to a root node using a non-linear cost function. Reliability and capacity were not considered.

2.3 The Network Reliability Calculation During Optimal Design

Iterative (improvement) optimization techniques depend on the calculation or estimation of the reliability of different network topologies throughout the search process. However, the calculation of all-terminal network reliability is itself an NP-hard problem (Provan and Ball, 1983). Much of the following discussion also applies to source-sink reliability, which is also NP-hard, but easier than all-terminal network reliability.

Assuming that the arcs (set L) fail independently, the number of the possible network states is 2^L. For large L, it is computationally impossible to calculate the exact network reliability using state enumeration even once, much less the numerous times required by iterative search techniques. Therefore, the main interest is in crude surrogates, simulation methods and bounding methods. Crude surrogates to network reliability include a constraint on minimum node degree or minimum path connectedness. These are easily calculated, but they are not precisely correlated with actual network reliability. For the all-terminal network reliability problem, efficient Monte Carlo simulation is difficult because simulation generally loses efficiency as a network approaches a fully connected state.

When considering bounds, both the tightness of the bound and its computational effort must be considered. Upper and lower bounds based on formulations from Kruskal (1963) and Katona (1968), as comprehensively discussed in Brecht and Colbourn (1988), are based on the reliability polynomial, and can be used for both source-sink and all-terminal network reliability. The importance of the reliability polynomial is that it transforms the reliability calculation into a counting of operational network states on a reduced set of arcs. Bounds on the coefficients lead directly to bounds on the reliability polynomial. The accuracy of the Kruskal-Katona bounds depends on both the number and the accuracy of the coefficients computed. Ball and Provan (1983) report tighter bounds by using a different reliability polynomial. Their bounds can be computed in polynomial (in L) time and are applicable for both source-sink and all-terminal reliability. Brecht and Colbourn (1988) improved the Kruskal-Katona bounds by efficiently computing two additional coefficients of the polynomial. Brown *et al.* (1993) used network transformations to efficiently compute the Ball-Provan bounds for all-terminal reliability. Nel and Colbourn (1990) developed a Monte Carlo method for estimating some additional coefficients in the reliability polynomial of Ball and Provan. These additional coefficients provide substantial improvements (i.e., tighter bounds). Another efficiently computable all-terminal reliability upper bound is defined by Jan (1993). Jan's method uses only the cut sets separating individual nodes from a network and can be calculated in polynomial (in N) time. Note the distinction between polynomial in N (nodes) and polynomial in L (arcs), where for highly reliable networks, L will far exceed N.

One of the important limitations of the bounding methods cited is that they requires all arcs to have the same reliability, which is an unrealistic assumption for many problems. In recent work by Konak and Smith (1998; 1998a), Jan's approach is extended to networks with unequal arc reliability. Also, a tighter upper bound is achieved, even for the case when all arc reliabilities are identical, at virtually no additional computational cost, i.e. the new bound is polynomial in N.

In solving the optimal design problem, it is likely that a combination of crude surrogates, bounding the network reliability along with accurately estimating it with Monte Carlo simulation, will be a good approach. Through much of the search, crude surrogates or bounds will be accurate enough, but as the final few candidate topologies are weighed, a very accurate method must be used (iterated simulation or exact calculation).

2.4 A Simple Genetic Algorithm Method When All Arcs have Identical Reliability

2.4.1 Encoding and GA Operators

In this section, a simple GA approach to optimal network design when all arcs have identical reliability is discussed. This approach was developed by Dengiz *et al.* (1997).

Each candidate network design is encoded as a binary string of length $|N|(|N|-1)/2$, the number of possible arcs in a fully connected network. This is reduced for networks where not all possible links are permitted. For example, Figure 2.1 shows a simple network that consists of 5 nodes and 10 possible arcs, but with only 7 arcs present.

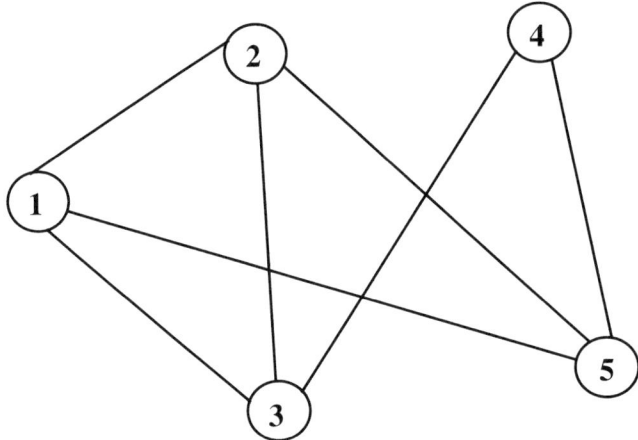

Figure 2.1 Five node network with arbitrarily numbered nodes.

The string representation of the network in Figure 2.1 is

$$[\quad 1 \quad 1 \quad 0 \quad 1 \quad 1 \quad 0 \quad 1 \quad 1 \quad 0 \quad 1 \quad]$$
$$[\quad x_{12} \quad x_{13} \quad x_{14} \quad x_{15} \quad x_{23} \quad x_{24} \quad x_{25} \quad x_{34} \quad x_{35} \quad x_{45} \quad]$$

In this GA, the initial population consists of randomly generated 2-connected networks (Roberts and Wessler, 1970). The 2-connectivity measure is used as a preliminary screening, since it is usually a property of highly reliable networks. A set of experiments determined the following GA parameter values: $s = 20$, $r_c = 0.95$, and $r_m = 0.05$.

The approach uses the conventional GA operators of roulette wheel selection, single point crossover and bit flip mutation. Each crossover operation yielded the two complementary children, and each child was mutated. Evolution continues until a preset number of generations, which varies according to the size of the network.

2.4.2 The Fitness Function

The objective function is the sum of the total cost for all arcs plus a quadratic penalty function for networks that fail to meet the minimum reliability requirement. The objective of the penalty function is to lead the GA to near-optimal feasible solutions. It is important to allow infeasible solutions into the population because good solutions are often the result of breeding between a feasible and an infeasible solution and the GA does not ensure feasible children, even if both parents are feasible (Smith and Tate, 1993; Coit et al., 1996). The fitness function considering possible infeasible solutions is:

$$Z(x) = \sum_{i=1}^{N-1} \sum_{j=i+1}^{N} c_{ij} x_{ij} + \delta(c_{max}(R(x) - R_0))^2 \qquad (2.2)$$

where $\delta = 1$ if the network is infeasible and 0 otherwise. c_{max} is the maximum arc cost possible in the network.

2.4.3 Dealing with the Reliability Calculation

This method uses three reliability estimations to trade off accuracy with computational effort:

- A connectivity check for a spanning tree is made on all new network designs using the method of Hopcroft and Ullman (1973).

- For networks that pass this check, the 2-connectivity measure (Roberts and Wessler, 1970) is made by counting the node degrees.

- For networks that pass both of these preliminary checks, Jan's upper bound (Jan, 1993) is used to compute the upper bound of reliability of a candidate network, $R_U(x)$.

This upper bound is used in the calculation of the objective function (equation 2.2) for all networks except those which are the best found so far (x_{BEST}). Networks which have $R_U(x) \geq R_o$ and the lowest cost so far are sent to the Monte Carlo subroutine for more precise estimation of network reliability using an efficient Monte Carlo technique by Yeh et al. (1994). The simulation is done for $t = 3000$ iterations for each candidate network.

2.4.4 Computational Experiences

Results compared to the branch and bound method of Jan et al. (1993) on the test problems are summarized in Table 2.2. These problems are both fully connected and non-fully connected networks (viz., only a subset of L is possible for selection). N of the networks ranges from 5 to 25. Each problem for the GA was run 10 times, each time with a different random number seed. As shown, the GA gives the optimal value for the all replications of problems 1–3 and finds optimal for all but two of the problems for at least one run of the 10. The two with suboptimal results (12 and 13) are very close to optimal. Table 2.3 lists the search space for each problem along with the proportion actually searched by the GA during a single run ($n \times g_{MAX}$). g_{MAX} ranged from 30 to 20000, depending on problem size. This proportion is an upper bound because GA's can (and often do) revisit solutions already considered earlier in the evolutionary search. It can be seen that the GA approach examines only a very tiny fraction of the possible solutions for the larger problems, yet still yields optimal or near-optimal solutions. Table 2.3 also compares the efficacy of the Monte Carlo estimation of network reliability. The exact network reliability is calculated using a backtracking algorithm, also used by Jan et al. (1993), and compared to the estimated counterpart for the final network for those problems where the GA found optimal. The reliability estimation of the Monte Carlo method is unbiased, and is always within 1% of the exact network reliability.

Table 2.2 Comparison of GA results from section 2.4.

Problem	N	L	p	R_o	Optimal Cost[2]	Results of Genetic Algorithm[1]		
						Best Cost	Mean Cost	Coeff. of Variation
FULLY CONNECTED NETWORKS								
1	5	10	0.80	0.90	255	255	255.0	0
2	5	10	0.90	0.95	201	201	201.0	0
3	7	21	0.90	0.90	720	720	720.0	0
4	7	21	0.90	0.95	845	845	857.0	0.0185
5	7	21	0.95	0.95	630	630	656.0	0.0344
6	8	28	0.90	0.90	203	203	205.4	0.0198
7	8	28	0.90	0.95	247	247	249.5	0.0183
8	8	28	0.95	0.95	179	179	180.3	0.0228
9	9	36	0.90	0.90	239	239	245.1	0.0497
10	9	36	0.90	0.95	286	286	298.2	0.0340
11	9	36	0.95	0.95	209	209	227.2	0.0839
12	10	45	0.90	0.90	154	156	169.8	0.0618
13	10	45	0.90	0.95	197	205	206.6	0.0095
14	10	45	0.95	0.95	136	136	150.4	0.0802
15	15	105	0.90	0.95	---	317	344.6	0.0703
16	20	190	0.95	0.95	---	926	956.0	0.0304
17	25	300	0.95	0.90	---	1606	1651.3	0.0243
NON FULLY CONNECTED NETWORKS								
18	14	21	0.90	0.90	1063	1063	1076.1	0.0129
19	16	24	0.90	0.95	1022	1022	1032.0	0.0204
20	20	30	0.95	0.90	596	596	598.6	0.0052

[1.] Over ten runs.
[2.] Found by the method of Jan *et al.* (1993).

2.5 A Problem-Specific Genetic Algorithm Method when All Arcs have Identical Reliability

The GA in the preceding section was effective, but there are greater computational efficiencies possible if the GA can exploit the particular structure of the optimal network design problem. This section presents such an approach as done in Dengiz *et al.* (1997a; 1997b). The encoding, crossover and mutation are modified to perform local search and repair during evolution and the initial population is seeded. These modifications improve both the efficiency and the effectiveness of the search process. The drawback, of course, is the work and testing necessary to develop and implement effective operators and structures.

2.5.1 *Encoding and Seeding*

A variable length integer string representation was used with every possible arc arbitrarily assigned an integer, and the presence of that arc in the topology is shown by the presence of that integer in the ordered string. The fully connected network in Figure 2.2(a), for example, uses the assignment of integer labels to arcs.

Table 2.3 Comparison of search effort and reliability estimation of the GA of section 2.4.

Problem	Search Space	Solutions Searched	Fraction Searched	R_o	Actual $R(x)$	Estimated $R(x)$	Percent Difference
1	1.02 E3	6.00 E2	5.86 E–1	0.90	0.9170	0.9170	0.000
2	1.02 E3	6.00 E2	5.86 E–1	0.95	0.9579	0.9604	0.261
3	2.10 E6	1.50 E4	7.14 E–3	0.90	0.9034	0.9031	–0.033
4	2.10 E6	1.50 E4	7.14 E–3	0.95	0.9513	0.9580	0.704
5	2.10 E6	1.50 E4	7.14 E–3	0.95	0.9556	0.9569	0.136
6	2.68 E8	2.00 E4	7.46 E–5	0.90	0.9078	0.9078	0.000
7	2.68 E8	2.00 E4	7.46 E–5	0.95	0.9614	0.9628	0.001
8	2.68 E8	2.00 E4	7.46 E–5	0.95	0.9637	0.9645	0.083
9	6.87 E10	4.00 E4	5.82 E–7	0.90	0.9066	0.9069	0.033
10	6.87 E10	4.00 E4	5.82 E–7	0.95	0.9567	0.9545	–0.230
11	6.87 E10	4.00 E4	5.82 E–7	0.95	0.9669	0.9668	–0.010
12	3.52 E13	8.00 E4	2.27 E–9	0.90	0.9050	*	
13	3.52 E13	8.00 E4	2.27 E–9	0.95	0.9516	*	
14	3.52 E13	8.00 E4	2.27 E–9	0.95	0.9611	0.9591	–0.208
15	4.06 E31	1.40 E5	3.45 E–27	0.95	@	0.9509	
16	1.57 E57	2.00 E5	1.27 E–52	0.95	@	0.9925	
17	2.04 E90	4.00 E5	1.96 E–85	0.90	@	0.9618	
18	2.10 E6	1.50 E4	7.14 E–3	0.90	0.9035	0.9035	0.000
19	1.68 E7	2.00 E4	1.19 E–3	0.95	0.9538	0.9550	0.126
20	1.07 E9	3.00 E4	2.80 E–5	0.90	0.9032	0.9027	–0.055

* Optimal not found by GA.
@ Network is too large to exactly calculate reliability.

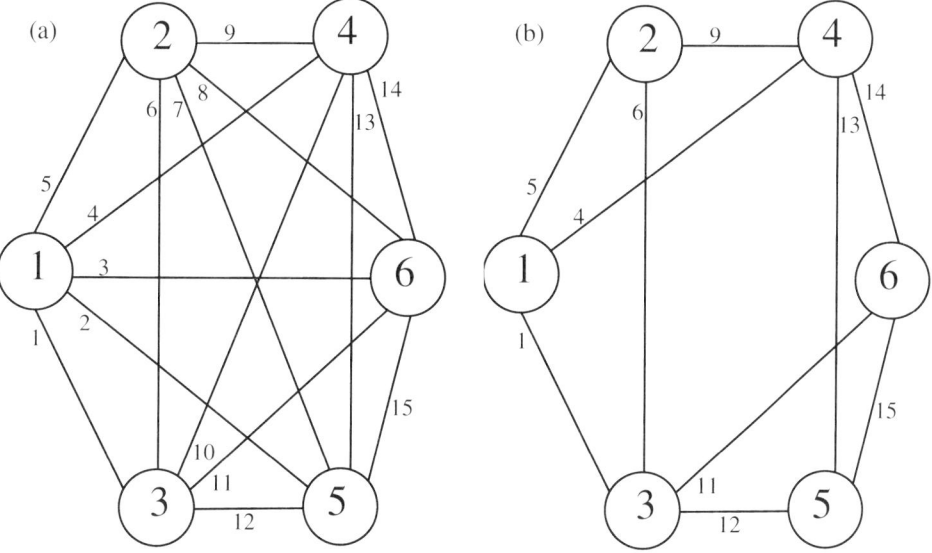

Figure 2.2 Two six-node networks: (a) fully connected, with arcs labeled arbitrarily from 1 to 15; (b) partially connected, with arcs labeled using the same scheme as in (a).

String representations of networks given in Figure 2.2 are [1 2 3 4 5 6 7 8 9 10 11 12 13 14 15] and [1 4 5 6 9 11 12 13 14 15], respectively. The first network includes all possible arcs using the labels above. The second contains ten arcs, using the same labeling scheme. The initial population consists of highly reliable networks, generated as follows.

1. A spanning tree is implemented through the depth-first search algorithm by Hopcroft and Ullman (1973), which grows a tree from a randomly chosen node.
2. Arcs selected randomly from the co-tree set (the set of arcs which are not yet used in the tree) are added to the spanning tree to increase connectivity.
3. If the network obtained by steps 1 and 2 does not have 2-connectivity (Roberts and Wessler, 1970), it is repaired by the algorithm explained in section 2.5.3.

2.5.2 The Genetic Algorithm

The flow of the algorithm is as follows:

1. Generate the initial population. Calculate the fitness of each candidate network in the population using equation 2.2 and Jan's upper bound (Jan, 1993) as $R(x)$, except for the lowest cost network with $R_U(x) \geq R_o$. For this network, x_{BEST}, use the Monte Carlo estimation of $R(x)$ in equation 2.2. Generation, $g, = 1$.
2. Select two candidate networks. An elitist ranking selection with stochastic remainder sampling without replacement is used (Goldberg, 1989).
3. To obtain two children, apply crossover to the selected networks and to the children.
4. Determine the 2-connectivity of each new child. Use the repair algorithm on any that do not satisfy 2-connectivity.
5. Calculate $R_U(x)$ for each child using Jan's upper bound and compute its fitness using equation 2.2.
6. If the number of new children is smaller than $s-1$ go to Step 2.
7. Replace parents with children, retaining the best solution from the previous generation.
8. Sort the new generation according to fitness. $i = 1$ to s.
   (a) If $Z(x_i) < Z(x_{BEST})$, then calculate the reliability of this network using Monte Carlo simulation, else go to Step 9.
   (b) $x_{BEST} = x_i$. Go to Step 9.
9. If $g = g_{MAX}$ stop, else go to Step 2 and $g = g+1$.

The parameters are $s = 50$, $r_c = 0.70$, $r_m = 0.30$ and $DR = 0.60$, which is used in mutation.

2.5.3 Repair, Crossover and Mutation

If a candidate network fails 2-connectivity, the network is repaired using three different alternatives according to how many nodes fail the test. The repair method is detailed below, where N_k refers to a set of nodes with degree k, N_{min} is the set of nodes with minimum degree, excepting nodes with degree 1, n_k is the number of nodes in the set N_k, m_{1j} are node labels in the the set N_1, m_{minj} are node labels in the set N_{min}, with $j = 1,2,..., |N_{min}|$.

1. Determine N_k, n_k; for k ranging from 1 to the maximum node degree in a network.
2. Rank all N_k and n_k, except N_1 and n_1, in increasing order from $k = 2$ to the maximum node degree; determine N_{min} and n_{min}.
   (a) If $n_1 = 1$, determine which arc between this node and the nodes in the set N_{min} has minimum cost and add it, stop.
   (b) If $n_1 = 2$,
      – Compute the connection cost of the two nodes ($c_{m_{11},m_{12}}$) in N_1.
      – Compute all $c_{m_{11},m_{min\,j}}$ and $c_{m_{12},m_{min\,j}}$ for $j = 1,2,...,n_{min}$.
      – If $c_{m_{11},m_{12}} < [\min(c_{m_{11},m_{min\,j}}) + \min(c_{m_{12},m_{min\,j}})]$ then connect the 2 nodes in N_1; else connect the nodes in N_1 to other nodes in N_{min}, through $\min(c_{m_{11},m_{min\,j}})$, $\min(c_{m_{12},m_{min\,j}})$.
   (c) If $n_1 > 2$,
      – Randomly select two nodes from N_1,
      – Apply (b) for these two nodes until $n_1 = 0$.

The crossover method, described next, is a form of uniform crossover with repair to ensure that each child is at least a spanning tree with 2-connectivity.

1. Select two candidate networks, called T1 and T2. Determine the common arcs = T1∩T2, other arcs are: $\overline{T1}$ = T1 - (T1∩T2); $\overline{T2}$ = T2 - (T1∩T2).
2. Assign common arcs to children, T1', T2'. T1' = T1∩T2; T2' = T1∩T2.
3. If T1' and T2' are spanning trees, go to step 5, else go to step 4.
4. Arcs from $\overline{T1}$, in cost order, are added to T1' until T1' is a spanning tree. Use the same procedure to obtain T2' from $\overline{T2}$.
5. Determine which arcs of T1 ∪ T2 do not exist in T1' and T2': CT1 = T1 \ T1'; CT2 = T2 \ T2'.
6. T1' = T1' ∪ CT2; T2' = T2' ∪ CT1.

Mutation, described next, takes the form of a randomized greedy local search operator. The mutation operator is applied differently according to node degrees of the network.

1. Determine node degrees deg(j) of the network for $j = 1,2,...,N$
   If deg(j) = 2 for all j; go to Step 2,
   If deg(j) > 2 for all j; go to Step 3,
   Else, deg(j) ≥ 2; for all j; go to Step 4.
2. Randomly select an allowable arc not in the network and add it; stop.
3. Rank arcs of the network in decreasing cost order. Drop the maximum cost arc from the network. If the network still has 2-connectivity, stop; otherwise cancel dropping

this arc, and retry the procedure for the remaining ranked arc until one is dropped or the list has been exhausted; stop.

4. Generate $u \sim U(0,1)$. If $u < (1-DR)$ (where DR is the drop rate) go to step 2, otherwise go to step 3.

2.5.4 Computational Results

This GA is compared with the branch and bound (B+B) technique by Jan *et al.* (1993) and the GA of section 2.4. 79 randomly generated test problems are used for the comparison. Each problem for the GAs was run ten times with different random number seeds to gauge the variability of the GA. Tables 2.4 and 2.5 list complete results of the three methods. The GAs find optimal solutions at a fraction of the computational cost of branch and bound for the larger problems. Both GA formulations found the optimal solution in at least one of the ten runs for all problems. The GA of this section is computationally more efficient and reduces variability among seeds when compared to the GA of section 2.4

2.6 A Genetic Algorithm when Links Can have Different Reliabilities

A simplistic and restrictive assumption of previous research is that all possible arcs must have identical reliability and unit cost. This is a limitation of the approaches to the problem. In real world design problems there are generally multiple choices for arcs, each with an associated reliability and unit cost, and other design attributes. When considering the economics of network design, it is important to allow designs with arcs of differing unit costs. The research presented in this section is from Deeter and Smith (1997, 1997a) and makes the significant relaxation that there are multiple choices of arc type for each possible arc, and the final network may have a heterogeneous combination of differing arc reliabilities and costs. While greatly improving the relevance of the problem to real world economic design, this complicates the network reliability calculation and exponentially increases the search space.

2.6.1 Encoding, Genetic Algorithm and Parameters

The following example for a problem with $|N| = 5$ and $k = 4$ levels of connection shows how a candidate network design is encoded. The chromosome: {0100203102} encodes the network illustrated in Figure 2.3. There are $(5 \times 4)/2 = 10$ possible arcs for this example but only five are present; the other five are at level of connection $l_{i,j} = 0$. This information is placed in a chromosome using the possible values of $0, 1, ..., k-1$.

The objective function of sections 2.4 and 2.5 (equation 2.2) is modified to

$$C_p(x) = C(x) + C(x^*) \times (1 + R_0 - R(x))^{r_p + \frac{s \times g}{50}} \quad (2.3)$$

where $C_p(x)$ is the penalized cost, $C(x)$ is the unpenalized cost and $C(x^*)$ is the cost of the best feasible solution in the population. This is a dynamic penalty that depends upon the length of search, g.

Table 2.4 Complete results comparing performance and CPU time on 79 test problems.

Problem					B+B		Section 2.4 GA		Section 2.5 GA	
No	N	L	p	R_o	Best Cost	CPU sec.	Coeff. Var.[1]	CPU sec.	Coeff. Var.[1]	CPU sec.
FULLY CONNECTED NETWORKS										
1	6	15	0.90 0.90		231	1.87	0.0245	57.50	0	11.97
2	6	15	0.90 0.90		239	0.01	0	41.05	0	8.28
3	6	15	0.90 0.90		227	0.04	0	38.90	0	12.30
4	6	15	0.90 0.90		212	0.17	0	46.32	0	12.60
5	6	15	0.90 0.90		184	0.28	0	52.39	0.0233	13.72
6	6	15	0.90 0.95		254	0.11	0	69.39	0.0217	19.48
7	6	15	0.90 0.95		286	0.00	0	50.17	0	13.04
8	6	15	0.90 0.95		275	0.06	0	48.37	0	12.40
9	6	15	0.90 0.95		255	0.06	0	59.32	0	14.36
10	6	15	0.90 0.95		198	0.01	0	53.65	0.0121	21.51
11	6	15	0.95 0.95		227	3.90	0.0357	57.98	0.0023	14.08
12	6	15	0.95 0.95		213	0.11	0.0235	47.83	0.0193	10.03
13	6	15	0.95 0.95		190	0.00	0.0280	42.32	0	10.09
14	6	15	0.95 0.95		200	0.44	0.0238	57.54	0.0173	13.04
15	6	15	0.95 0.95		179	0.66	0.0193	46.97	0.0256	11.36
16	7	21	0.90 0.90		189	11.26	0.0177	130.71	0.0175	21.77
17	7	21	0.90 0.90		184	0.17	0	76.74	0	18.80
18	7	21	0.90 0.90		243	0.50	0.0167	135.98	0.0202	26.93
19	7	21	0.90 0.90		129	1.21	0.0121	122.46	0.0195	28.91
20	7	21	0.90 0.90		124	0.05	0	83.45	0	23.77
21	7	21	0.90 0.95		205	0.83	0.0406	301.41	0.0337	71.40
22	7	21	0.90 0.95		209	0.06	0	71.4	0	37.06
23	7	21	0.90 0.95		268	0.06	0.0310	255.73	0.0187	56.39
24	7	21	0.90 0.95		143	0.17	0.0264	280.26	0.0193	78.72
25	7	21	0.90 0.95		153	0.01	0	160.43	0	52.93
26	7	21	0.95 0.95		185	22.85	0.0333	112.26	0.0111	28.89
27	7	21	0.95 0.95		182	1.27	0.0046	81.78	0.0035	16.99
28	7	21	0.95 0.95		230	1.76	0.0090	109.47	0.0072	26.64
29	7	21	0.95 0.95		122	2.31	0.0265	112.62	0.0259	27.82
30	7	21	0.95 0.95		124	0.39	0	74.49	0	19.64
31	8	28	0.90 0.90		208	21.9	0.0211	260.86	0.0161	79.55
32	8	28	0.90 0.90		203	20.37	0	175.06	0	75.37
33	8	28	0.90 0.90		211	140.66	0.0149	198.80	0.0119	79.67
34	8	28	0.90 0.90		291	173.01	0.0204	210.95	0.0108	83.66
35	8	28	0.90 0.90		178	159.34	0.0112	230.70	0	67.34
36	8	28	0.90 0.95		247	10162.53	0.0152	611.28	0.0140	168.79
37	8	28	0.90 0.95		247	15207.83	0.0274	808.94	0.0183	226.08
38	8	28	0.90 0.95		245	12712.21	0.0124	663.99	0.0034	184.31
39	8	28	0.90 0.95		336	9616.80	0.0169	743.39	0.0177	303.50
40	8	28	0.90 0.95		202	9242.10	0.0231	629.13	0.0235	266.47

[1] Over 10 runs.

Table 2.5 Complete results comparing performance and CPU time on 79 test problems.

Problem						B+B	Section 4 GA		Section 5 GA	
No	N	L	P	R_O	Best Cost	CPU sec.	Coeff. Var.[1]	CPU sec.	Coeff. Var.[1]	CPU sec.
FULLY CONNECTED NETWORKS										
41	8	28	0.95	0.95	179	0.11	0	133.32	0	43.81
42	8	28	0.95	0.95	194	2.69	0.0053	202.57	0.0033	40.56
43	8	28	0.95	0.95	197	26.97	0.0052	173.74	0.0080	58.04
44	8	28	0.95	0.95	276	20.76	0.0133	187.02	0.0100	50.64
45	8	28	0.95	0.95	173	72.78	0.0190	189.02	0.0206	53.51
46	9	36	0.90	0.90	239	8.02	0.0105	324.38	0.0066	98.19
47	9	36	0.90	0.90	191	23.78	0.0277	365.31	0.0081	153.77
48	9	36	0.90	0.90	257	702.05	0.0301	530.37	0.0171	176.79
49	9	36	0.90	0.90	171	0.82	0.0255	292.01	0	81.18
50	9	36	0.90	0.90	198	12.36	0.0228	378.91	0	90.49
51	9	36	0.90	0.95	286	8321.87	0.0821	1215.28	0.0325	404.93
52	9	36	0.90	0.95	220	14259.48	0.0330	998.79	0.0309	358.28
53	9	36	0.90	0.95	306	9900.87	0.0313	1256.82	0.0163	560.89
54	9	36	0.90	0.95	219	17000.04	0.0457	865.38	0.0226	340.13
55	9	36	0.90	0.95	237	7739.99	0.0760	1024.77	0.0778	391.52
56	9	36	0.95	0.95	209	4.95	0.0576	274.83	0	59.24
57	9	36	0.95	0.95	171	21.75	0.0137	293.43	0.0092	99.98
58	9	36	0.95	0.95	233	525.03	0.0375	372.18	0.0268	97.95
59	9	36	0.95	0.95	151	0.99	0.0471	252.71	0	65.78
60	9	36	0.95	0.95	185	25.92	0.0381	385.59	0	71.67
61	10	45	0.90	0.90	131	4623.19	0.0518	1047.60	0.0231	375.14
62	10	45	0.90	0.90	154	2118.75	0.0651	794.83	0.0223	214.63
63	10	45	0.90	0.90	267	1860.74	0.0142	999.01	0.0061	415.53
64	10	45	0.90	0.90	263	1466.73	0.0126	678.02	0	171.04
65	10	45	0.90	0.90	293	2212.70	0.0329	1093.36	0.0182	488.12
66	10	45	0.90	0.95	153	5712.97	0.0257	1718.45	0.0150	982.98
67	10	45	0.90	0.95	197	7728.21	0.0203	1689.51	0.0177	726.31
68	10	45	0.90	0.95	311	8248.16	0.0367	1967.61	0.0136	984.30
69	10	45	0.90	0.95	291	6802.16	0.0404	1529.61	0.0244	825.45
70	10	45	0.90	0.95	358	12221.39	0.0276	2662.34	0.0048	1071.99
71	10	45	0.95	0.95	121	3492.17	0.0563	793.22	0.0124	177.31
72	10	45	0.95	0.95	136	1125.89	0.0291	615.29	0.0185	81.87
73	10	45	0.95	0.95	236	987.64	0.0276	781.68	0.0160	139.53
74	10	45	0.95	0.95	245	2507.89	0.0369	632.11	0	98.31
75	10	45	0.95	0.95	268	1359.91	0.0513	630.37	0.0120	131.55
76	11	55	0.90	0.90	246	59575.49	0.0499	1532.34	0	472.11
NON FULLY CONNECTED NETWORKS										
77	14	21	0.90	0.90	1063	23950.01	0.0129	7293.97	0.0079	1672.75
78	16	24	0.90	0.95	1022	131756.43	0.0204	2699.38	0.0185	2334.15
79	20	30	0.95	0.95	596	[2]	0.0052	5983.24	0.0152	4458.81

[1] Over 10 runs.
[2] Optimum solution taken from Jan *et al.* (1993). CPU time unknown.

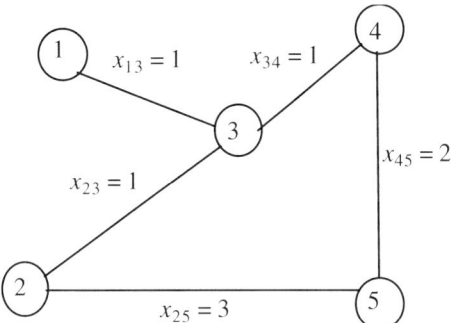

Figure 2.3 Example network design for chromosome 0100203102 (section 2.6).

Below is the GA algorithm, followed by a more detailed description of the key steps.

1. Randomly Generate Initial Population
    Send initial population to the reliability and cost calculation function and calculate fitness using equation 2.3
    Check for initial *Best Solution*
    if no solution is feasible the best infeasible solution is recorded
2. Begin Generational Loop
    Select and Breed Parents
        copy *Best Solution* to new population
        two distinct parents are chosen using the rank based procedure of Tate and Smith (1995)
        children are generated using uniform crossover
        children are mutated
    when enough children are created the parents are replaced by the children
    Send new population to the reliability and cost calculation functions, and calculate fitness using equation 2.3
    Check for new *Best Solution*
        if no solution is feasible the best infeasible solution is recorded
    Repeat until g_{max} generations have elapsed.

Crossover is uniform by randomly taking an allele from one of the parents to form the corresponding allele of the child. This is done for each allele of the chromosome. For example, a potential crossover of parents x_1 and x_2 is illustrated below.

x_1 {0120131011}
x_2 {1111012002}
———————
child {0110132001}

After a new child is created it goes through mutation. A solution undergoes mutation according to the percentage of population mutated. For example, if $m\% = 20\%$ and $s = 30$, then six members are randomly chosen and mutated. Once a solution is chosen to be mutated then the probability of mutation per allele is equal to the mutation rate, r_m. So if $r_m = 0.3$ then each allele will be mutated with probability 0.3. When an allele is mutated its value *must* change. If an arc was turned off, $l_{i,j} = 0$, then it will be turned on with an equal probability of being turned to any of the states 1 through $k-1$. If an allele is originally on, then it will either be turned off ($k = 0$) or it will be turned to one of the different on levels, with equal probability. An example is given below. The solution has been mutated by changing the seventh allele from a 2 to a 0 and changing the ninth allele from a 0 to a 1.

solution {0110132001}
mutated solution {0110130011}

2.6.2 Test Problem 1 – Ten Nodes

The ten node test problem was designed by randomly picking ten sets of (x,y) coordinates and using each of the points as nodes on an 100 by 100 grid. The Euclidean distances between the nodes were calculated, and the unit costs and reliabilities were taken from Table 2.6. The ten node problem was examined with a system reliability requirement of 0.95. Because of the network size, reliability could not be calculated exactly. The Monte Carlo estimator of reliability used both dynamic and static parameters. For the 'general' reliability check, which was used on every new population member, the total number of replications used was dynamic. At the first generation, the estimator replicated each network 1000 times ($t = 1000$). As the number of generations increased, the number of replications used in the general reliability check also increased. After every hundredth generation the number of replications used in the general reliability check was incremented by 1000 ($t = t + 1000$). This dynamic approach was used so that as search progressed the reliability estimates would improve. Whenever a network was created that met the reliability constraint using the general reliability estimator, and had better cost than the best found so far, a 'best check' reliability estimator was employed. This replicated a given system $t = 25000$ times. This was used to help ensure the feasibility and accuracy of the very best candidate designs.

From initial experimentation $s = 90$, $m\% = 25$, $r_c = 1.00$, $r_m = 0.25$, $r_p = 6$, and $g_{max} = 1200$. Since the problem has 1.24×10^{27} possible designs, it was impossible to enumerate. So, a random greedy search was used as a comparison. Ten runs of each algorithm using the same set of random number seeds were averaged and plotted as shown in Figure 2.4.

Table 2.6 Arc unit costs and reliabilities for problems in section 2.6.

Connection Type (k)	Reliability	Unit Cost
not connected, 0	0.00	0
1	0.70	8
2	0.80	10
3	0.90	14

Notice in Figure 2.4 that the GA best cost dips much more rapidly than does the best cost corresponding to the greedy algorithm, indicating that the GA will find good solutions much more efficiently than a myopic approach. Also, both lines appear to be asymptotically approaching a solution, however the line corresponding to the GA is approaching a much better solution than the line corresponding to the greedy search.

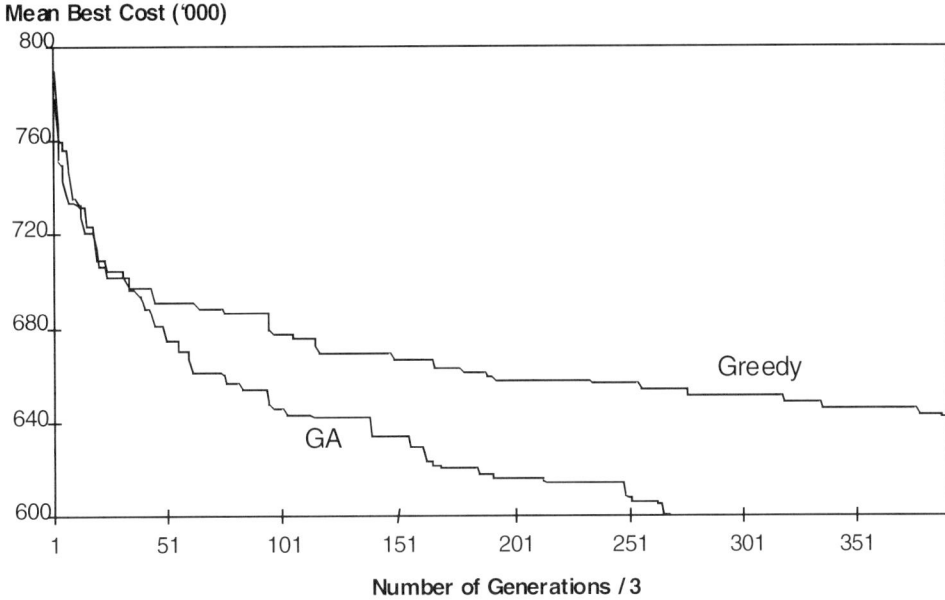

Figure 2.4 GA vs. greedy search averaged over 10 runs for the problem of section 2.6.2.

2.6.3 Test Problem 2 – Source-Sink Reliability

This problem demonstrates the flexibility of the GA approach in two respects. First, the calculation of reliability is different. Secondly, the architecture of arcs is restricted; 18 of 36 arcs are unavailable for the network design as shown in Figure 2.5. The GA easily accommodates these rather fundamental changes. The change in the reliability calculation is accomplished by simply modifying the backtracking algorithm of Ball and Van Slyke (1977) – this problem is small enough to calculate reliability exactly during search. The fact that not all possible arcs are allowed is accommodated by simply leaving these out of the chromosome string, as was done in some of the problems of sections 2.4 and 2.5.

This problem is taken from the literature (Kumamoto et al., 1977) and has 6.9×10^{10} possible topologies, thus precluding enumeration to identify the optimal design. A system reliability requirement $R_o(x) = 0.99$ is set. After some initial experimentation it was determined that $s = 40$, $m\% = 80$, $r_c = 1.00$, $r_m = 0.05$, $r_p = 6$ and $g_{max} = 2000$. Seven of the 10 runs found a best cost of 4680. The other three test runs found a best cost of 4726. Since the GA found only two distinct solutions over 10 runs, it is likely that both are near-optimal, if 4680 is not optimal.

2.7 Concluding Remarks

It can be seen that an evolutionary approach to optimal network design, when considering reliability, is effective and flexible. Differences in objective function, constraints and

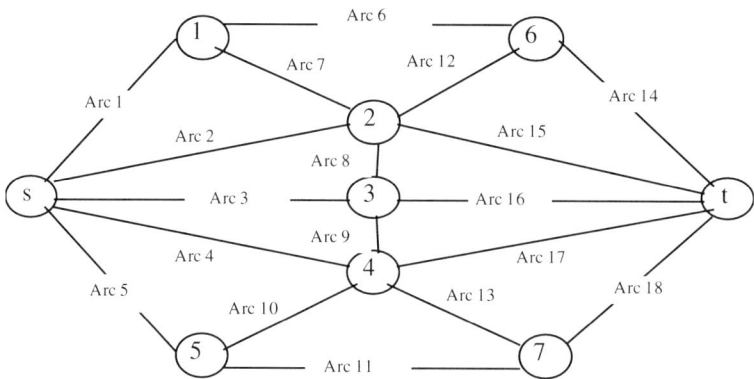

Figure 2.5 Source-sink problem topology of section 2.6.3.

reliability calculation are easily handled. One difficulty is the number of times that network reliability must be calculated or estimated. An effective search for problems of realistic size will use a combination of bounds, easily calculated reliability surrogates such as node degree, and Monte Carlo simulation estimators. Another emerging approach is to use artificial neural network approximators for network reliability (Srivaree-ratana and Smith, 1998; 1998a). One important attribute of evolutionary search that has yet to be exploited in the literature is the generation of multiple, superior network designs during the procedure. A human designer could more carefully examine and consider the few superior designs identified by the evolutionary algorithm. These are likely to be dissimilar, and thus show the designer the particularly promising regions of the design search space.

Acknowledgements

The authors gratefully acknowledge the support of U.S. National Science Foundation grant INT-9731207 and additional support from the Scientific and Technical Research Council of Turkey (Tubitak) for their collaboration.

3

Efficient Network Design using Heuristic and Genetic Algorithms

Jeffrey Blessing

3.1 Introduction

The efficient design of telecommunication networks has long been a challenging optimization problem. It is made difficult by the conflicting, interdependent requirements necessary to optimize the network's performance. The goal of the designer is to produce a minimum cost network that allows maximum flow of information (in the form of messages) between multiple source-sink pairs of nodes that simultaneously use the network. An optimum design method must also produce a network topology that efficiently routes these messages within an acceptable amount of time. The problem of designing minimum-cost, multi-commodity, maximum-flow networks with efficient routing of messages in a synthesized network topology, is *NP-complete* (Clementi and Di Ianni, 1996; Gavish, 1992; King-Tim et al., 1997). NP-complete problems are those for which no known algorithm can find the optimum solution besides brute force, exhaustive approaches. Thus, the challenge is to develop algorithms that run in polynomial time, which produce designs as near as possible optimum. Since lower bounds on the network design problem are only known for simple cases (involving only one type of communication link, for instance Dutta and Mitra (1993); Ng and Hoang (1983; 1987); Gerla and Kleinrock (1977)), the method of choice for selecting good algorithms is to compare their results on identical problems. The approach taken in this chapter is to implement several of the best known methods in order to objectively compare them on randomly generated instances of the network design problem.

Because there is keen interest in designing optimum networks, many approaches have been developed, including branch exchange (Maruyama, 1978; Steiglitz *et al.*, 1969), cut saturation (Gerla *et al.*, 1974), genetic algorithms (Elbaum and Sidi, 1995; King-Tim *et al.*, 1997; Pierre and Legault, 1996), MENTOR algorithms (Grover *et al.*, 1991; Kershenbaum, 1993), and simulated annealing. In this chapter, we will look at the results of several of these methods and perform an in-depth study of several heuristic and genetic techniques.

3.2 Problem Definition

The basic topology of a communication network can be modeled by an undirected graph of edges and nodes (vertices) and is referred to by the symbol $G(V, E)$. The network design problem is to synthesize a network topology that will satisfy all of the requirements set forth as follows.

3.2.1 Minimize Cost

Each edge represents a set of communication *lines* that connect the two nodes. Each *line type* has a *unit cost*, u_{ab}, which is the cost per unit distance of line type b on edge a (which links the nodes (i, j)). The unit cost of a line is a function of the line's capacity. For instance, a 6 Mbps line may cost $2 per mile; a 45 Mbps line, $5 per mile, etc. Communication line types are available in discrete capacities, set by the local telephone company or media carrier. Normally, telephone tariff charges follow an 'economy of scale' in which the unit cost decreases with increasing line capacity. Additionally, some tariffs may have a fixed charge per line type in addition to a cost per distance fee. In this case, both components of the line cost must be incorporated into one unit cost per line type, per edge (since each edge presumably has a unique length). The fixed cost is divided by the length of the line and added to the cost per unit distance to yield one unit cost per line type, per edge. If an undirected graph, $G(V, E)$, has m edges, and each edge (i, j) has a distance d_{ij} and represents l line types, each with a unit cost u_{ab}, then the objective function of the optimization problem is:

$$\min \sum_{a=1}^{m} \sum_{b=1}^{l} d_{ij} u_{ab} x_{ab} \quad (3.1)$$

where x_{ab} is the quantity of line type b's selected for edge a. Note that each edge a has two other representations: (i, j), and (j, i). So, d_{ij} could also be referred to as d_a.

3.2.2 Maximize Flow

Since minimizing cost while maximizing flow are diametrically opposed objectives, the resolution of this conflict is to specify a requirements matrix, **R**, where r_{st} represents the minimum amount of continuous capacity required between nodes s and t. The value r_{st} is often referred to as the *demand* between s and t. Similarly, if the total flow between every pair of nodes is given by a flow matrix, $\boldsymbol{F}$, and f_{st} is the flow in $G(V, E)$ from source s to sink t, then the maximum flow requirement could be written as:

$$f_{st} > r_{st} \tag{3.2}$$

Note that, since $\boldsymbol{R}$ and $\boldsymbol{F}$ are symmetric matrices, $f_{st} = f_{ts}$ and $r_{st} = r_{ts}$. Also, $f_{ii} = 0$ and $r_{ii} = 0$ along the major diagonal of $\boldsymbol{F}$ and $\boldsymbol{R}$. The notion of undirected graphs and symmetric matrices here is supported by the fact that carrier lines are inherently bi-directional (for instance, a T-2 line simultaneously carries two signals at 6 Mbps in both directions).

3.2.3 Multi-commodity Flow

Simultaneous use of the network by multiple source-sink pairs is modeled by assigning a unique *commodity* to the flow between every pair of nodes in $G(V, E)$. Thus, there are $n(n\text{-}1)\ni 2$ distinct flows, or commodities, in the network. Capacity restrictions limit the total flow on each edge to be at, or below, c_{ij}, the total capacity of edge (i, j). Let the term f_{ij}^k represent the flow of commodity k in edge (i, j). This constraint is expressed as:

$$\sum_{k=1}^{b} f_{ij}^k \leq c_{ij} \tag{3.3}$$

where b is the number of commodities in edge (i, j). Note that $f_{ij}^k = f_{ji}^k$ and $f_{ii}^k = 0$.

3.2.4 Efficient Routing

Whatever the final topology $G(V, E)$ may be, it is necessary for a design algorithm to provide a way to route all concurrent flows from sources to sinks. This may take the form of one path assigned to a commodity (*virtual path* routing), or a set of paths assigned to a single commodity (*bifurcated* routing). In either case, the assignment of flow to each edge must be made explicit by any network synthesis method. Most methods use some form of shortest path routing (Dijkstra, 1959), with the only difference being how the "length" of an edge is defined. In some models, the length may denote physical distance. In others, it may indicate delay (in seconds) or unit cost. When the length of each edge is one, then the path with the minimum number of 'hops' is selected. Unless otherwise stated, shortest distance routing will be used by the methods presented in this chapter.

3.2.5 Sufficient Redundancy

In some instances of the network design problem, the goal of achieving minimum cost is realized by a graph of minimum connectivity. For instance, a tree is a graph which connects all nodes with a minimum number of edges. However, any single node (or edge) failure will disconnect the network. Two routes are said to be *edge-disjoint* if they connect the same source and sink and have no common edges. Similarly, two routes are said to be *node-*

disjoint if the only nodes they share are the source and sink. Since single point failures are common in networks, an acceptable design method must provide enough redundancy to survive single node failures. However, any redundancy increases the cost of the network. To balance these conflicting goals, a minimum of two node-disjoint paths must exist between every pair of nodes. In the case of a single point network failure, traffic can be sent along an alternate path, albeit at a much slower rate (which, most likely, will not continue to meet the minimum required capacity constraints in R).

3.2.6 Acceptable Delay

The average branch delay, T, is a network-wide metric which measures the average amount of time (in seconds) that a packet will wait before being transmitted along an edge in the network. Kleinrock (1964) has developed a widely accepted model for delay in communication networks, in which each edge is modeled as an independent M/M/1 queue in a network of queues. Each queue has an exponentially distributed mean service time, and an average arrival rate of new packets which follows a Poisson distribution. The packet lengths are exponentially distributed, with an average packet length of $1/\Phi$. According to Kleinrock, the average delay on edge i is:

$$T_i = \frac{1}{\mu_i c_i - f_i} \tag{3.4}$$

where c_i is the total capacity and f_i is the total flow (of all commodities) on edge i. Since nothing specific can be known about the average packet length (it varies with application), set $\Phi = 1$. Kleinrock defines T, the average delay on any edge in the network, as:

$$T = \frac{1}{\gamma} \sum_{i=1}^{m} \frac{f_i}{c_i - f_i} \tag{3.5}$$

where γ is the total of all minimum flow requirements in the graph, and is defined as:

$$\gamma = \sum_{i \neq j} r_{ij} \tag{3.6}$$

Notice that each demand is counted twice, once for r_{ij} and again for r_{ji}. Unless otherwise specified, equation 3.5 will be used to estimate delay in the network. Still, some may consider this delay model to be too limiting, because it ignores propagation and nodal delay. Kleinrock (1970) defines a more comprehensive formula for delay as:

$$T = \frac{1}{\gamma} \sum_{i=1}^{m} \lambda_i [T_i + P_i + K_i] \tag{3.7}$$

where γ_i is the average packet rate on edge i, P_i is the propagation delay on edge i, and K_i is the nodal processing time at the node in which edge i terminates. The term T_i depends upon the nature of traffic on edge i and the packet length distribution. Since some of these values are application dependent, the more general delay model of equation 3.5 is used.

3.2.7 Conservation of Flow

At each node, the amount of flow into the node must equal the flow out of the node, unless the node is a source or sink of flow for a particular commodity. For a given commodity k, and a given node q, this requirement is expressed as:

$$\sum_{\forall(p,q)} f^k_{pq} - \sum_{\forall(q,r)} f^k_{qr} = \begin{cases} -f_{st} & \text{if } q = s \\ 0 & \text{if } q \neq s, t \\ f_{st} & \text{if } q = t \end{cases} \quad (3.8)$$

where p and r are neighbors of q. Notice that f_{st} represents the flow of one commodity in the network while f^k_{ij} represents the amount of flow of commodity k in edge (i, j).

3.3 Problem Complexity

To measure the complexity of the network design problem, it is necessary to employ a model of machine computation and data storage that fits the architecture of modern day computers. The computational model used is that of a *scalar machine*. A scalar machine is any computing device that executes unary or binary operations on *scalar* (i.e. single-valued) data stored in memory, in one machine cycle (or *clock tick*). Such a machine will be configured with a Central Processing Unit (CPU), an on-line, Randomly Accessible Memory component (RAM) which stores single-valued data in consecutively addressed memory locations, and a simple Input/Output device (I/O). The Arithmetic and Logic Unit (ALU) of the central processor must possess the circuitry to perform operations such as: addition (+), subtraction (−), assignment (=), comparisons (<, ≤, >, ≥), shifting (<<, and >>), reading, and writing of data, all in unit time. All vector operations on a scalar machine take place in time proportional to the vector.

Given the above definition of a scalar computing machine, the goal of any network design method is to produce an $n \times n$ adjacency matrix which contains the optimum capacity label for each edge in the final topology. A label of zero in cell (i, j) indicates that there is no edge between nodes i and j in the final design. Since edge capacities are discrete quantities, let k be the number of discrete values needed to be considered for each cell in an optimum design. An exhaustive search of all possible labelings would produce $k^{(n^2-n)/2}$ labeled networks in the solution space to be tested for optimality. Additionally, each network would have a large number of ways to route flow, only one of which is optimal. For even small instances of the network design problem, any approach on a scalar machine that has to consider every possible outcome, is intractable. For example, in the next section an instance of this problem, known as the Ten Chinese Cities network design problem, will be defined in which $k = 12$ and $n = 10$. In this problem, there are more than 10^{48} possible solutions to be considered. Clearly, problems of this level of complexity can not be addressed by exhaustive methods. Even traditional optimization algorithms, such as linear programming and integer programming methods, will face exponential execution times.

If the network design problem of section 3.2 is formulated as a non-linear optimization problem, the objective function is to *minimize cost*, subject to proper *flow assignment* and *capacity assignment*, which meets the *delay* requirement in the proposed *topology*. The constraints of topology selection, flow assignment, capacity assignment and delay, are inter-related with one another, and with the objective function. For instance, a change in topology would affect the cost of the network, as would a change in the flow assignment or capacity assignment. A decrease in the delay would increase the cost of the network, since a lower delay requires more excess capacity in the network. Flow and capacity assignment are inter-related, since the routing of the flow is dependent on finding an augmenting path for each commodity which fits within the capacity constraints of each edge in the topology. Also, one only needs to look at equation 3.5 to see that delay is a function of both the flow and the capacity assigned to each edge in the network. Figure 3.1 attempts to show how these sub-problems are related to one another in the network design problem. The arrowhead indicates the direction of the 'affects' relationship. Of all the factors affecting cost, only topology is unaffected by the other constraints. Also, the cycle formed by flow assignment, capacity assignment, and delay, indicates that any ordering may be used among these three sub-problems. Since any acceptable network design algorithm must deal with all the constraints in Figure 3.1, the implication is that topology should be addressed first, followed by any ordering of flow assignment, capacity assignment and delay, in order to address the objective of minimizing cost.

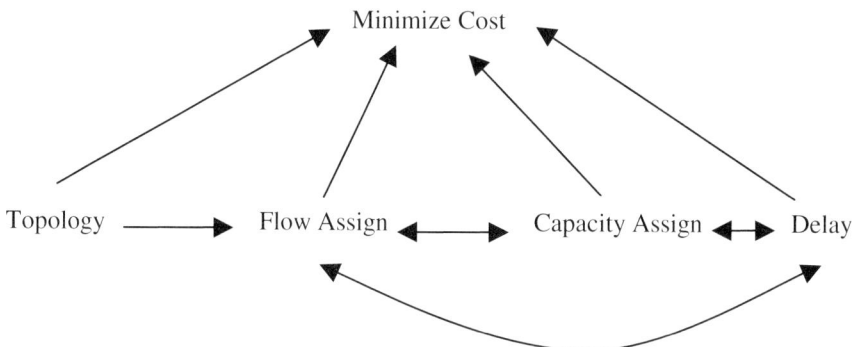

Figure 3.1 Sub-problems of the network design problem and their inter-relationships.

3.4 Heuristic Algorithms

Although any network design algorithm can be said to be heuristic (since none guarantee the optimal answer), the genetic algorithms all share a common strategy of imitating evolutionary systems by using the principle of 'survival of the fittest' to guide the search process. Thus, genetic algorithms present a unique and distinctly different approach to the problem, and are described in the next section. There are many heuristic algorithms that have been developed to address the general problem of network design. They include branch exchange (Maruyama, 1978; Steiglitz *et al.*, 1969), cut saturation (Gerla *et al.*, 1974), the MENTOR algorithm (Grover *et al.*, 1991; Kershenbaum, 1993), and a new method introduced in this section, called the Union of Rings algorithm (Blessing, 1998). Empirical studies suggest that cut saturation produces superior designs to branch exchange (Boorstyn and Frank, 1977; Gerla *et al.*, 1974). In several studies where cut saturation results are reported (King-Tim *et al.*, 1997; Pierre and Legault, 1996), the Union of Rings method produces a less costly design within the minimum delay requirement. Since the Union of Rings method seems to produce the most promising results (albeit, on a small number of test cases), it will be used as a representative method to compare with the genetic methods of the following section.

The cut saturation method, like the Union of Rings method, is based on the analysis of maximum flows and minimum cuts in a network. A *cut* is a set of edges which, when removed from the graph, disconnects the graph into two or more components. In a network flow problem, the labels on the edges of the graph denote the capacity of the edge. Thus, the capacity of a cut is simply the sum of the labels on each edge in the cut set. A cut whose edges are filled to capacity is called a *saturated cut*. The cut saturation algorithm begins by routing flow until a cut is saturated in the proposed graph. In order to continue sending flow through the network, an edge must be added to the saturated cut. An edge is added to the graph which spans the saturated cut, thus increasing its capacity, and allowing more flow through the network. Also, edges on either side of the saturated cut can be removed from the graph, thus reducing cost. The addition or deletion of edges causes flow to be re-routed, thus producing new saturated cuts in the graph. Edges are added and deleted from the network, as long as the overall cost of the network is improving (i.e. decreasing). The algorithm terminates when a *locally optimum* point is reached, where neither the addition or deletion of an edge improves the graph.

Let G refer to any connected, undirected graph of n nodes. In their seminal paper on network flows, Ford and Fulkerson (1962) show that the *maximum flow* between any source and sink in G equals the value of the *minimum cut* separating the source and sink – this is known as the *max-flow, min-cut* theorem. When considering all possible flows (i.e. commodities) in a network, Gomory and Hu (1961) show that there are only $n-1$ minimum cuts in G separating all possible source-sink pairs. Further, these $n-1$ *essential cuts* in G form a tree, called the Gomory-Hu *cut tree*, T. Since T preserves the maximum flows between all pairs of nodes in G, T is said to be *flow-equivalent* to G. Thus, each edge in T represents both a minimum cut and a maximum flow in G. The significance of Gomory and Hu's *multi-terminal, maximal flows* result is that only $n-1$ maximum flow problems need be done to find the maximum flow between $n(n-1)/2$ pairs of nodes in G.

Two problems exist with using the Gomory-Hu cut tree as a network design algorithm for the problem defined in section 3.2. First, a tree cannot survive a single edge or node failure, and second, multi-terminal, maximal flows allow for only one commodity at a time to use the network.

Also, it may be surprising to know that the Gomory-Hu cut tree is *not* a minimum weight flow-equivalent graph which connects all the nodes of G (the weight of a graph is simply the sum of its edge capacities). Nor is the minimum spanning tree of Prim (1957)! The significance of a *minimum weight flow-equivalent graph* is that, if the cost to send one unit of flow over any edge is one (a *unit cost condition*), then the minimum weight graph is also a minimum cost graph, and is *optimal* under unit cost conditions. Another appealing property of the Minimum Weight Flow-Equivalent Graph (MWFEG) is that it is, at least, bi-connected (thus solving problem 1 above). If the MWFEG can be modified to allow for concurrent use of the graph by multiple commodities, and accommodate a more robust cost function, it may produce near-optimum results. This is the main idea behind the Union of Rings algorithm.

Frank and Frisch (1971) describe a synthesis process by which the minimum weight flow-equivalent graph of G can be constructed. However, it involves the complicated and time consuming tasks of computing the *principally partitioned* and *semi-principally partitioned* matrices of the adjacency matrix of G. As shown below, these steps are unnecessary in order to compute the MWFEG of G. The Union of Rings algorithm, which makes use of the MWFEG, is outlined as follows:

1. Draw the requirements matrix, R, as a complete graph, labeling each edge (i, j) with the minimum flow requirement r_{ij}.
2. Compute T, the maximum spanning tree of R.
3. Convert T into a *linear flow-equivalent graph*, L, by using Algorithm 1 below.
4. Factor L into a set of *uniform capacity rings*, as described in Frank and Frisch (1971).
5. Superimpose the set of rings from step 3 to form a network topology, N.
6. Remove any short cut edges in N which are not *cost efficient*, and re-route their flow on other edges in N. This step produces the final network topology, NN.

Steps 1 and 2 determine the dominant requirements of the problem. Hu has shown that, if the dominant requirements are satisfied between all nodes, then all requirements can be satisfied (Hu, 1982). Obviously, the maximum spanning tree of a complete requirements graph satisfies all dominant requirements (taken one requirement at a time). Steps 3, 4 and 5 transform the tree into a biconnected, flow-equivalent graph. Step 3 is best explained by Example 1, below. Step 6 is a process of local optimization, where only some of the edges in N are considered for removal. For the sake of completeness, Algorithm 1 is as follows:

1. Initially, all nodes are unmarked. Arbitrarily pick a starting node in T and mark it. This node is the start of the linear graph L.
2. Select the maximum capacity edge incident to one marked, and one unmarked, node in T as the next edge-node pair to append to L. Break ties arbitrarily.
3. Mark the unmarked node that is incident to the edge selected in step 2.
4. If all nodes are marked, then stop. Otherwise, go to step 2.

Example 1

Figure 3.2 shows a maximum spanning tree, T, of a typical requirements matrix. Algorithm 1 is used to convert T into the flow-equivalent linear graph, L, shown below T in Figure 3.2. To demonstrate that any tree can be converted into a flow-equivalent linear graph, a proof of Algorithm 1 is contained in the Appendix to this chapter.

The Union of Rings algorithm is best described by an example. The particular design problem in question is published in King-Tim *et al.* (1997). The problem is to link ten Chinese cities into one network, subject to redundancy requirements (at least two disjoint paths between every pair of nodes), minimum flow requirements (expressed as the adjacency matrix in Table 3.1), and a maximum branch delay of 0.1 second.

There are three types of communication lines available, which have the following capacities and costs:

- 6 Mbps lines cost 1 unit per kilometer
- 45 Mbps lines cost 4 units per kilometer
- 150 Mbps lines cost 9 units per kilometer

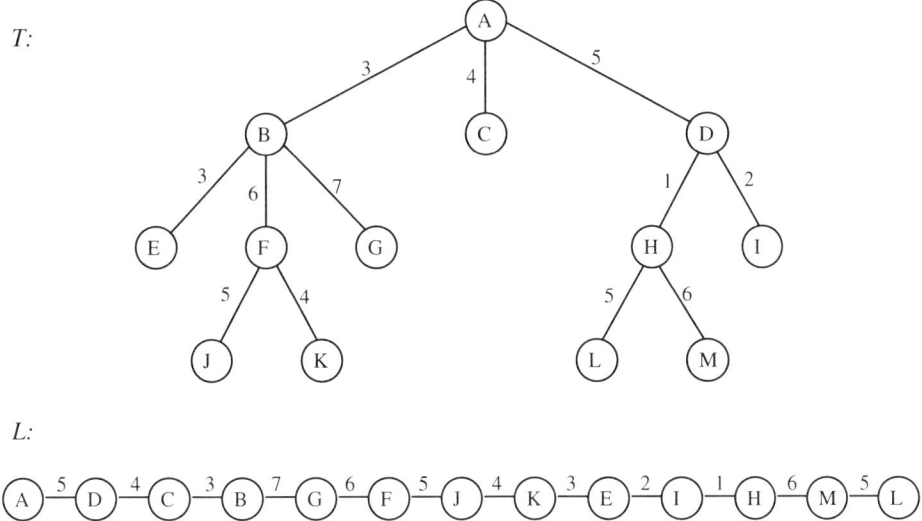

Figure 3.2 A maximum spanning tree and its flow-equivalent linear graph.

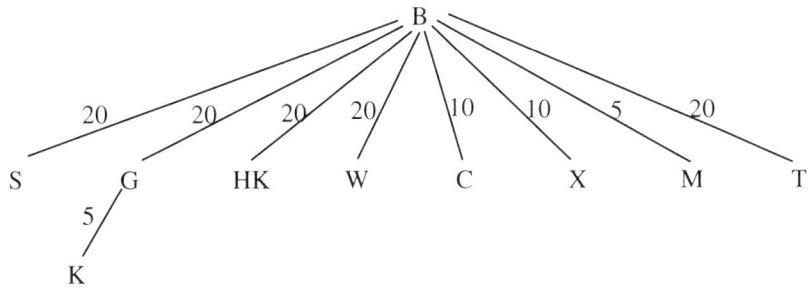

Figure 3.3 Maximum spanning tree of the complete requirements graph of Table 3.1.

Table 3.1 Minimum flow requirements

Cities	Beijing	Shanghai	Guangzhou	Hong Kong	Wuhan	Chengdu	Xi'an	Kunming	Harbin	Tianjin
Beijing	0	20	20	20	20	10	10	2	5	20
Shanghai	20	0	20	20	20	5	5	2	1	20
Guangzho	20	20	0	20	10	5	5	5	1	5
Hong Kong	20	20	20	0	10	5	2	2	1	5
Wuhan	20	20	10	10	0	5	5	0	1	5
Chengdu	10	5	5	5	5	0	5	2	0	2
Xi'an	10	5	5	2	5	5	0	0	0	2
Kunming	2	2	5	2	0	2	0	0	0	0
Harbin	5	1	1	1	1	0	0	0	0	5
Tianjin	20	20	5	5	5	2	2	0	5	0

3.4.1 Topology

To make the figures more readable, each node will be labeled with the first letter of the corresponding city it represents in the problem. The maximum spanning tree of the complete requirements graph is given in Figure 3.3. The *linear flow-equivalent graph* which corresponds to the maximum spanning tree is shown in Figure 3.4. The process by which *uniform capacity rings* are extracted from the linear flow-equivalent graph is described in Frank and Frisch (1971), and is illustrated in Figure 3.5. Basically, the minimum capacity edge is factored out of each edge in the linear graph. This is repeated until all edges from the original graph are reduced to zero. Once all the uniform capacity linear graphs have been determined, each linear graph is made into a uniform capacity ring by connecting the first and last nodes with a 'wrap around' edge. When a ring has been formed, the capacity of each edge in the ring can be reduced by half and still preserve the flow-equivalence property. This is possible since each linear flow can now be divided evenly into two flows, one which goes 'clockwise' around the ring and the other in the 'counter-clockwise' direction. The uniform capacity rings are shown in Figure 3.6.

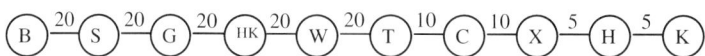

Figure 3.4 Linear flow-equivalent graph corresponding to the spanning tree of Figure 3.3.

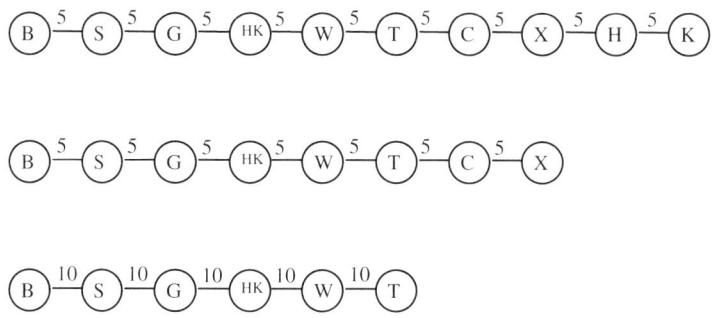

Figure 3.5 Extracting uniform capacity rings from the linear flow-equivalent graph.

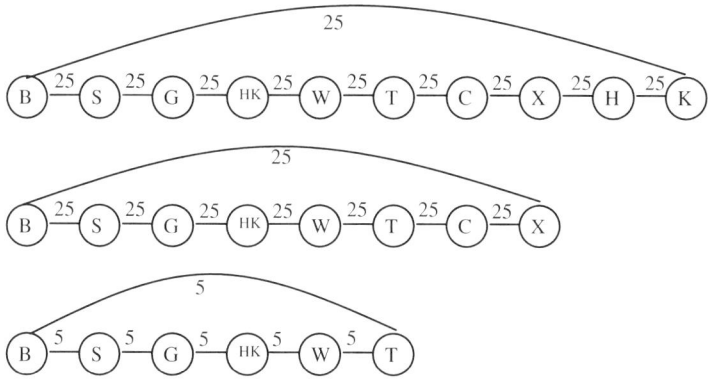

Figure 3.6 Uniform capacity rings extracted from the graph of Figure 3.4.

Notice that, since each ring is of a uniform capacity, the order of the nodes around the ring can be modified without upsetting the flow-equivalence property. This property is employed to order the nodes of the first ring (the one which contains all n nodes) in a minimum distance tour or cycle. Recall from equation 3.1 that distance is a factor in the total network cost. Heuristically, we desire to find the minimum distance ring in order to minimize the cost of connecting the nodes. The minimum distance tour is used to superimpose rings of uniform capacity in a cost-effective manner, to form an initial network topology N. Connecting n nodes in a minimum distance tour is commonly known as the Traveling Salesman Problem (TSP) (Kruskal, 1956). Although TSP is an NP-complete problem (Garey et al., 1976), there are many good heuristics which come close to the optimum TSP solution, such as 2-opt, 3-opt, and methods based on the Lin-Kernighan algorithm (Johnson and McGeoch, 1997). In practice, one needs to choose which TSP method to employ based on the size of the problem. If the problem is small, the algorithms based on Lin and Kernighan (abbreviated often to LK –which are of greater time complexity) may be the best choice. If the problem is large, then methods such as 3-opt or even 2-opt may be all that is practical. Whatever method is used, it is generally the case that the Union of Rings method produces better results with better solutions to the TSP sub-problem. Once a tour has been determined for the first ring, all nodes in subsequent rings are sequenced in the same order. For our example, the tour selected is

$$B–H–T–S–W–HK–G–K–C–X–B$$

The second ring drops out nodes H and K, so the tour for that ring is

$$B–T–S–W–HK–G–C–X–B$$

Lastly, the third ring drops out nodes C and X, so the tour for the third ring is

$$B–T–S–W–HK–G–B$$

The resulting network topology, N, is shown in Figure 3.7.

In the process of forming the initial topology N, several 'short-cut' edges (edges which 'cut short' the original tour) were added to the graph, namely B–T, C–G, and B–G. These short-cuts are analyzed for 'cost effectiveness' after the flow and capacity assignments are made. Beginning with the most costly edge and continuing onward in this manner, if the network cost is reduced by eliminating the short-cut edge, its traffic is rerouted along remaining edges and the short-cut edge is removed from the topology. Removal of the cost inefficient short-cut edges comprise the only locally improving moves made to the network topology. This step results in the final topology NN, which for the Ten Chinese Cities problem, is shown in Figure 3.8.

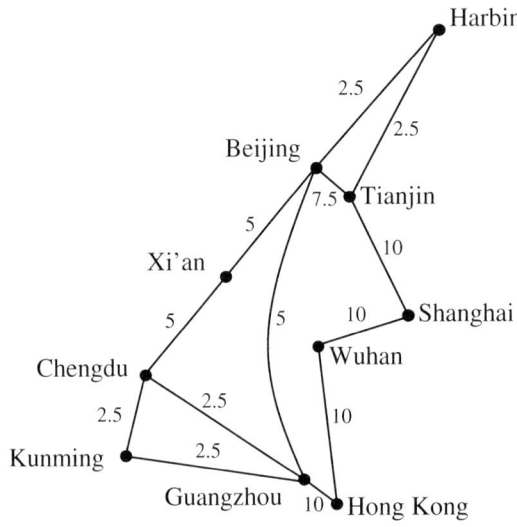

Figure 3.7 The resulting network topology after applying the union of rings algorithm.

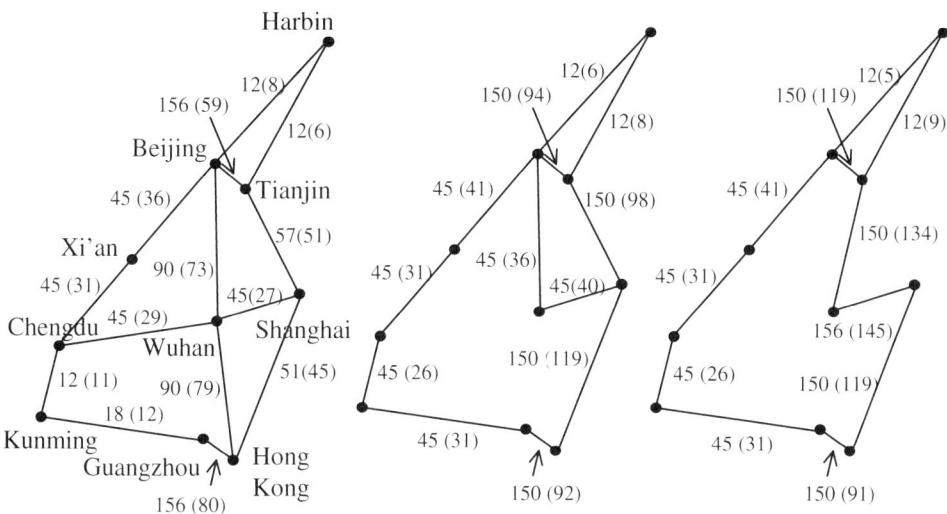

Figure 3.8 The resulting network topologies from each of the three methods compared on the Ten Chinese Cities problem. The leftmost network is the result of the Branch Exchange method, the network in the middle is the result of the genetic algorithm, and the network on the right is the result from the union of rings algorithm.

3.4.2 Flow Assignment

Now that the initial topology has been determined, the next problem from Figure 3.1 is the flow assignment problem. In many network design algorithms, the routing of flow through a network is done by using shortest path routes (Fratta *et al.*, 1973; Hoang and Ng, 1983; Ng and Hoang, 1987; Pierre and Legault, 1996). In fact, when the cost function is simplified to a single line type (having a discrete capacity and cost), the set of multi-commodity flows satisfying the requirements matrix R forms a convex polyhedron (Ng and Hoang, 1987). The extreme points of such a polyhedron will all correspond to shortest route policies (Gerla and Kleinrock, 1977). Since a shortest path routing policy is optimal for a simplified version of our problem, it seems wise to use it for any variation of the network design problem. This leads to another simplification made by the Union of Rings method: instead of trying to assign flow to fit the capacity of every edge, route all commodities first, along the shortest distance paths in N, and then fit the capacity to the sum of the flows on each of the edges.

3.4.3 Capacity Assignment

With the flows for all the commodities routed along shortest distance paths in N, the next task is to assign capacity to each edge such that the sum of the flows on each edge is 'covered'. A flow is said to be covered when the capacity of the edge exceeds the sum of the flows on the edge. The capacity of an edge is determined by selecting the quantity of each line type, from the set of line types offered by the local telephone company or media carrier. This is an integer programming problem, since the optimum way to cover a given flow value will not necessarily use the highest density line available. The *density* of a line is simply the capacity divided by the cost per unit distance. For our example design problem, the line densities are 6/1 = 6.0, 45/4 = 11.25 and 150/9 = 16.66. Notice that, if it wasn't for the integer restrictions, the problem would be trivial. The optimal solution would always be to take as much of the highest density line as necessary, paying a pro-rated amount for any fractional part of a line that may be needed to exactly fit the flow. Integer restrictions forbid using half a line and paying half the rate! Fortunately, the optimum cover to any flow value can be found by formulating the capacity assignment problem as a *knapsack* problem, which can be solved by using the methods of *dynamic programming*. The dynamic programming formulation of the knapsack problem is given in Hu (1982). The dynamic programming formulation of the integer capacity assignment problem is given as follows.

Let n be the number of line types available, with each line type having a discrete capacity (c_i) and unit cost (u_i). Let x be the solution vector, with x_i being the quantity of line type i in the optimum solution. Finally, let b be the maximum flow in any edge of the graph. The optimization problem is:

$$\min \sum_{i=1}^{n} u_i x_i \qquad (3.9)$$

subject to:

$$\sum_{i=1}^{n} c_i x_i \geq b \qquad (3.10)$$

Dynamic programming methods solve this problem iteratively by using the following recursive relation:

$$F_i(y) = m\{F_{i-1}(y), F_i(y - c_i) + u_i\} \qquad (3.11)$$

$F_i(y)$ is the minimum cost solution using only the first i line types on edges with flow y. In other words:

$$F_i(y) = \min \sum_{j=1}^{i} u_j x_j$$

(where $0 \neq i \neq n$) with the condition that:

$$\sum_{j=1}^{i} c_j x_j \geq y$$

(where $0 \neq y \neq b$). To compute the table of $F_i(y)$ values, the following set of boundary conditions are needed:

$$F_0(y) = \infty$$
$$F_i(0) = 0$$
$$F_i(\text{negative number}) = 0$$

For dynamic programming to work, the flows and capacities must be integers. However, the costs can be real numbers. Equation 3.11 works by deciding first how to best cover all flow values using only one line type. Then, when a second line type is considered, it looks at all possible ways of dividing the flow between the two line types. When a third line type is added, equation 3.11 is simply choosing the best amount of flow to cover with the third line type, leaving the rest of the flow to be covered optimally among the first two line types (it had decided those questions optimally after the first two iterations of the recursion). The term $F_{i-1}(y)$ means "Don't take any more of the ith line type" while the term $F_i(y-c_i) + u_i$ means "Take at least one more of the ith line type" in the final decision. Since all the previous decisions have been made optimally, the only decision to make in equation 3.11 is whether one more instance of line type i is necessary to cover the flow optimally.

Since the values of $F_i(y)$ are costs of optimal solutions, the problem remains as to how to recover the values of x in these solutions. Here, the same method used in Hu (1982) to recover the solutions to the knapsack problem is adopted to recover the values of x. In parallel with recording new values for $F_i(y)$, another array of values $I_i(y)$ is recorded, using the following definition:

$$I_i(y) = \begin{cases} I_{i-1}(y), & F_{i-1}(y) < F_i(y-c_i) + u_i \\ i, & F_i(y-c_i) + u_i \leq F_{i-1}(y) \end{cases} \quad (3.12)$$

The capital I stands for *index*, since $I_i(y)$ is the *maximum index* for the line types used in arriving at the corresponding value for $F_i(y)$ in equation 3.11. Thus, if $I_i(y) = k$, then increment x_k in x (initially $x = 0$), and recursively recover the remaining x values from $I_i(y-c_k)$. Here, the boundary conditions are $I_1(y) = 0$ if $F_1(y) = 0$, and $I_1(y) = 1$ otherwise.

In the Ten Chinese Cities instance of our network design problem, the unit cost of each line type is the same for each edge in the graph. Thus, only one instance of the capacity assignment problem (equations 3.8 and 3.9) needs to be solved, with b being the maximum flow value of any edge in the network. If the local tariffs add a fixed charge per line (based on line type) in addition to the distance cost, then that fixed cost must be divided by the length of the line before being added to the cost per unit distance. The new unit cost would be computed as follows:

$$u_k(i,j) = \text{Cost}_k + \frac{\text{fixed}_k}{d_{ij}} \quad (3.13)$$

In equation 3.13, $u_k(i, j)$ is the unit cost of line type k on edge (i, j); Cost_k is the cost per unit distance of line type k; fixed_k is the fixed cost for line type k; and d_{ij} is the distance from node i to node j. When both cost per unit distance and fixed costs appear in our cost function, the capacity assignment problem must be recalculated for every edge in the topology (since unit cost is now a function of the distance of each edge). Still, the solution given by the dynamic programming method outlined above would be optimum.

3.4.4 Delay

The use of dynamic programming in the capacity assignment problem minimizes the amount of excess (or unused) capacity in the network. However, according to equation 3.5, the lower the amount of excess capacity, the higher the average delay. In fact, some excess capacity is required in every edge to keep the delay from approaching infinity! If the average branch delay is above the threshold set in the design requirements, either the flows must be re-routed in the hopes of finding a flow assignment that yields a lower delay, or more capacity could be added to the network. The Union of Rings method addresses the problem of delay by repeatedly allocating more capacity to the most congested edge in the network, until the delay threshold is met. Note that this greedy approach may not find the least expensive modification to N which brings delay below the required threshold.

3.4.5 Results

The same instance of the Ten Chinese Cities network design problem was solved using three different network design algorithms: Branch Exchange, a Genetic Algorithm due to King-Tim et al. (1997), and the Union of Rings method. Table 3.2 summarizes the results using the three approaches. The results show that the Union of Rings algorithm produced the least expensive solution to the problem. Figure 3.8 shows the final topologies produced by the three methods, along with the link capacities (which label each edge) and the assigned flows (in parentheses).

Table 3.2 Summary of final designs.

Algorithm	Total Capacity (Mbps)	Total Cost (units)	Delay (sec.)
Branch Exchange	834	55,310	0.0999
Genetic Algorithm	894	50,590	0.0644
Union of Rings	954	47,820	0.0672

3.5 Genetic Algorithms

The most common formulation of the network design problem as a genetic algorithm is to use 0 and 1 bits as the genes of a chromosome that is $n(n-1)/2$ bits long. Each bit corresponds to the presence or absence of an edge in the complete graph of n nodes. In this manner, the chromosome only represents the topology of the candidate network. In King-Tim et al. (1997), the authors also use a separate chromosome representation for the flow assignment and capacity assignment sub-problems, solving each sub-problem genetically. In Pierre and Legault (1996) the authors use shortest distance path routing to solve the flow assignment problem and a simple selection method to pick the first line type which covers the flow in each edge to solve the capacity assignment problem. They address the connectivity question by simply testing if the minimum degree of the graph is at least 3. Note that, since node connectivity can be less than the minimum degree of the graph, a bi-connected topology is not guaranteed by specifying a degree of 3 in the fitness function. However, the authors also 'seed' the population with an initial graph in which every node is at least of degree 3. All other members of the initial population are randomly generated from the first individual. Single point crossover with a mutation rate of 0.005 is used, along with a proportional selection operator (also known as the *roulette wheel* method of selection), to evolve the fittest 100 members from each generation, for a total of 1000 generations. Since the objective is to minimize cost, the fitness function, f, is the reciprocal of the network cost, and is defined as follows:

$$f = \begin{cases} \dfrac{1}{C}, & 0 \leq T_n \leq 1 \\ \dfrac{1}{CT_n}, & \text{otherwise} \end{cases} \tag{3.14}$$

In equation 3.14, C is the total cost of the network, and $T_n = T/T_{max}$, is the ratio of the network's delay to the maximum acceptable delay in the problem statement. Multiplying the denominator by T_n is meant to penalize designs which do not meet the acceptable delay threshold, but does not entirely eliminate them from contributing to the next generation. The results in Pierre and Legault (1996) claim improved designs over Cut Saturation as the number of nodes in the problem increases. In every case of 15 nodes or more, the genetic algorithm of Pierre and Legault (1996) produced a superior solution to the results from Cut Saturation.

As a first point of comparison, the Union of Rings method was run on the sample problem of Pierre and Legault, with the results shown in Table 3.3. The Cut Saturation method produced a result 35% better in cost and 21% better in delay than the genetic algorithm of Pierre and Legault (1996). The Union of Rings algorithm, on the other hand, produced a result that is 5.25% better than the Cut Saturation solution.

Table 3.3 Final results.

Algorithm	Cost ($'s)	Delay (sec.)
GA of Pierre/Legault	33,972.67	0.077
Cut Saturation	22,221.30	0.061
Union of Rings	21,048.70	0.062

To more thoroughly compare the relative performance of heuristic and genetic algorithms with respect to the problem of network design, the Union of Rings algorithm and the genetic algorithm according to Pierre and Legault were implemented according to the descriptions of their authors. A random problem generator was also implemented, which selects a problem size ($4 \leq n \leq 25$), and a sequence of $n(n-1)/2$ demands for flow ($0 \leq d_i \leq 25$). The maximum acceptable delay was set to 0.1 seconds and the following line types were available in integer quantities (the three line capacities correspond to T-2, T-3 and OC-3 category digital leased lines):

- 6 Mbps lines cost 1 unit per kilometer
- 45 Mbps lines cost 4 units per kilometer
- 150 Mbps lines cost 9 units per kilometer

In 100 trial design problems, the Union of Rings algorithm outperformed the genetic algorithm in 54% of the cases, with an average improvement of 2% over the genetic algorithm results for all test cases. However, the variance was wide, with the best Union of Rings result being 24.75% better than the GA result, and the best GA result being 24% better than the corresponding Union of Rings result.

In order to improve upon the genetic algorithm approach, the dynamic programming method used by the Union of Rings algorithm (to solve the capacity assignment problem) was adopted for use in the genetic algorithm. With this modification, the only difference between the GA and the Union of Rings approach is in how they establish a solution topology. The same genetic algorithm parameters were used from the first comparison. In 100 randomly generated problems, the enhanced GA outperformed the Union of Rings result in 85% of the cases, with an average improvement of 10.5% over all cases tested. Still, the variance is wide, with the best GA result being 42.5% better than the Union of Rings solution. Conversely, there is a 25% improvement in the best Union of Rings result, compared to the corresponding GA solution. When the sample problem of Pierre and Legault was attempted, the GA with dynamic programming produced a design costing $19,797.25 with a delay of 0.065 seconds.

3.6   Conclusions

From the results above, it appears that the use of dynamic programming principles to address the capacity assignment sub-problem of the network design problem offers a significant reduction in the cost of the resulting designs. However, it is also apparent that, for any particular instance of the network design problem, a variety of solution methods must be attempted, since it is possible that any one method may dominate the others on any individual problem. With regard to problem size, genetic algorithm runs typically required 30 minutes to 1 hour (on a dedicated Pentium 233 MHz processor), where the Union of Rings algorithm would complete in 3 to 5 seconds. Thus, for very large scale network design problems, the Union of Rings method may be the only viable method to provide quality designs in reasonable time.

In terms of flexibility, genetic algorithms seem vastly more flexible to changing conditions than do the heuristics. Since the network design problem can be defined in so many different ways, it is important to be able to change the design algorithm to fit new problem requirements. For instance, when considering the topology of the graph, biconnectivity requirements necessitate that every node have at least two incident edges (this is a *necessary*, but not a *sufficient* condition). However, in terms of reliability, if a node has too many incident edges, its likely to be involved in many primary or secondary paths for the commodities of the problem, thus reducing path redundancy. If such a node fails, the amount of traffic re-routed would overwhelm the remaining cuts in the network. Thus, it may be desirable to keep the maximum degree of each node below some threshold value. Such a change to the design requirements might present difficulties for heuristic methods. Indeed, every new requirement many force a re-thinking of the reasoning in support of the heuristic method. However, such changes are easily handled by genetic algorithms. The fitness function only needs to be modified to return poor fitness values for graphs which do not meet all requirements precisely. This forces the search towards more acceptable topologies. Such changes were easily accommodated by the genetic algorithm approach implemented in this study.

Finally, when addressing large scale optimization problems such as network design, it may well be the case that the general search method used (genetic algorithms, simulated annealing, best-first search, hill climbing, etc.) is less important than the quality of the heuristics used to guide the search process. In the case of the genetic algorithms of this study, it was the decision making in the fitness function that lead to better solutions to the capacity assignment sub-problem, which in turn produced results that were 12% better than the GA results using a simple capacity assignment solution. Thus, when faced with a difficult optimization problem, one is better off spending time developing good heuristics to guide the search method of choice, than to experiment with other general purpose search algorithms.

3.8   APPENDIX

3.8.1   Proof of Algorithm 1

Algorithm 1 is a *marking* algorithm which deals with marked and unmarked nodes. Initially, all nodes are unmarked. One way to see how the algorithm works is to view all marked nodes as one *super-node*. When nodes become marked in step 3, the unmarked node u is merged with the super-node $V_{s..t}$ to form a slightly larger super-node $V_{s..u}$. The tree formed by merging nodes in T forms a slightly smaller tree TN. The algorithm, therefore, only needs to check for the maximum edge incident to the super-node to determine the next edge-node pair to append to L.

To prove that the algorithm is correct, one needs to show that merging two neighboring nodes, i and j, to form TN will not affect the maximum flow between either i or j and the rest of T. Let $\{N\}$ be the set of nodes in tree T. Let $\{N-i\}$ be the set of nodes in T, omitting node i. Similarly, $\{N-i-j\}$ would be the set of nodes in $\{N\}$, omitting nodes i and j. Let $V_{i..j}$ be the super-node formed by merging i and j in T. Finally, let TN be the tree that remains after merging i and j in T.

Theorem 1
Merging two neighboring nodes, i and j, in tree T will not affect the edges in the cut separating $V_{i..j}$ and $\{N-i-j\}$.

Proof of Theorem 1
For two neighbors, i and j, in tree T, the edge (i, j) is unique in T. $V_{i..j}$ will have the same incident edges in TN as those incident to i in T, plus those incident to j in T, minus the edge (i, j). Other edges which are not incident to i or j are unaffected by the merge. Since merging i and j will only omit edge (i, j) from T, the edges incident to $\{i, j\}$ in TN will preserve the capacities between $V_{i..j}$ and $\{N-i-j\}$ in T. Once the edge (i, j) is present in L, it can be removed from T (via merging) without affecting other cuts in T.

Theorem 2
In Algorithm 1, L is flow equivalent to T.

Consider the following inductive argument:

Base case
Starting at any node i, let (i, j) be the maximum capacity edge incident to i. This is the first edge in L. Since there is only one path in T between i and j, and that path consists of a single edge (i, j), the flow between i and j in T equals the flow between i and j in L.

Induction step
Assume, after k iterations, L_k is flow equivalent to the subset of T formed by the nodes in $V_{i..k}$ and the edges from T which connect nodes $i..k$ (i.e. the marked nodes in the super-node).

To prove Theorem 2, it must be shown that L_{k+1} is flow equivalent to the subset of T formed by the nodes in the expanded super-node $V_{i..k+1}$.

Proof of Theorem 2
In constructing L_k, the original tree, T, has been modified to state $T_k N$, which consists of one super-node, $V_{i..k}$, and $n-k-1$ edges from T connecting it to $\{N-V_{i..k}\}$. The algorithm selects the next edge in L_k to be the maximum edge incident to $V_{i..k}$. Call this edge (k, l) where $k \in V_{i..k}$ and $l \in \{N-V_{i..k}\}$. Notice that k is a marked node and l an unmarked node. Since marked nodes in T are adjacent to one another, any unmarked node can have, at most, one marked neighbor. The only marked neighbor of l is k. Clearly, the capacity of (k, l) is the upper bound on flow between k and l. If l tries to send flow to any other marked node, the flow must go through edge (k, l). So the capacity of edge (k, l) is an upper bound on the amount of flow l can send to any node in $V_{i..k}$. By the inductive hypothesis, L_k is flow equivalent to the subset of T formed by nodes in $V_{i..k}$ and their connecting edges. Since L_{k+1} is L_k with edge (k, l) and node l appended to the end, L_{k+1} is flow equivalent to the subset of T formed by the nodes in $V_{i..k+1}$.

Theorem 3
Every connected, undirected graph is flow equivalent to a linear graph.

Proof of Theorem 3
Gomory and Hu (1961) have shown that every connected, undirected graph, G, is flow equivalent to a tree, T. Theorem 2 proves that every tree is flow equivalent to a linear graph. Let T be the Gomory-Hu cut tree. Applying the above algorithm to T will always produce a corresponding linear graph, L, which is flow equivalent to G. Therefore, every connected,

undirected graph, G, is flow equivalent to a linear graph, L.

3.8.2   Time Complexity

Every tree of n nodes contains exactly $n-1$ edges (Sedgewick, 1988). The worst case for producing L from T is a cut tree where all $n-1$ edges are incident to the root. This will cause the algorithm to consider all the edges in T at once. Step 2 in the algorithm is the most time consuming operation in the loop and is thus the upper bound on calculating time complexity. Since the root is the starting node in the algorithm, $n-2$ comparisons must be made to determine the maximum edge incident to the root. The algorithm then marks the node neighboring the root and the edge between them is removed from future consideration. This, in effect, expands the super-node. The second pass makes $n-3$ comparisons to find the maximum incident edge, the third makes $n-4$ comparisons, and so forth, until all $n-1$ edges have been added to L. The total number of comparisons is thus $O(n^2)$.

A simple improvement to the algorithm is to store each edge, considered in step 2, in a priority queue or heap. Using a *max_heap*, the time needed to produce L from T is bounded by the time to do $n-1$ inserts and $n-1$ *delete_max* operations. The insert and delete_max operations on heaps each require $O(\log n)$ time (Sedgewick, 1988). Thus, the total run time is $2 \cdot (n-1) \cdot \log n$, which is $O(n \log n)$.

4

Tabu Search and Evolutionary Scatter Search for 'Tree-Star' Network Problems, with Applications to Leased-Line Network Design

Jiefeng Xu, Steve Y. Chiu and Fred Glover

4.1 Introduction

Digital Data Service (DDS) is widely used for providing private high quality digital transport service in the telecommunications industry. The network connections of DSS are permanent and its transmission facilities are dedicated, enabling it to transfer digital data with less interference and greater security than switched service. DSS also proves to be appropriate for linking sites that have applications which require a permanent connection and a demonstrated need for frequent data transfer. For example, it can be used for remote Local Area Network (LAN) access, entry into frame relay networks, support for transaction-based systems, and can be incorporated in IBM's System Network Architecture (SNA) and other networks. With optimal DSS network design and sufficient use, DSS becomes economically competitive with frame relay service in the higher transmission speed ranges, and with analog private line service in the lower transmission speed ranges.

In this chapter, we address a fundamental DDS network design problem that arises in practical applications of a telecommunications company in the United States. The decision elements of the problem consist of a finite set of inter-offices (hubs) and a finite set of customer locations that are geographically distributed on a plane. A subset of hubs are chosen to be active subject to the restriction of forming a network in which every two active hubs to communicate with each other, hence constituting a spanning tree. Each hub has a fixed cost for being chosen active and each link (edge) has a connection cost for being included in the associated spanning tree. Each customer location must be connected directly to its own designated end office which in turn needs to be connected with exactly one active hub, thereby permitting every two customers to communicate with each other via the hub network. This also incurs a connection cost on the edge between the customer location and its associated hub. The objective is to design such a network at minimum cost.

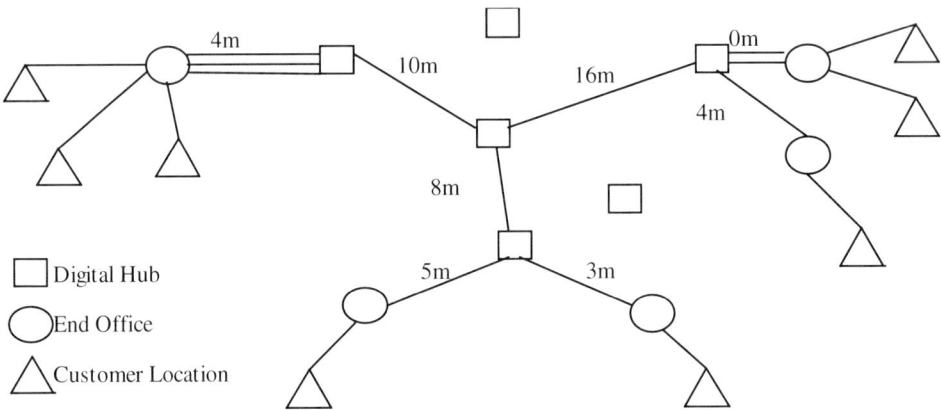

Figure 4.1 A DDS network.

Figure 4.1 shows a practical scenario of a small DDS network. The number of dedicated lines required for the link between an end office and its assigned hub is equal to the number of customer locations connected to the end office. Since the links between customer locations and end offices are always fixed, the costs of these links are constant and thereby can be ignored from the network design.

In practice, the line connection cost is distance sensitive and is calculated according to the tariff charges established by the Federal Communications Commission (FCC). These charges include a fixed cost for use and a variable cost that is related to the distance. For each active hub, in addition to the fixed bridging cost, a charge is also accessed for each incoming and outgoing line connected to this hub. To illustrate how these costs are associated with the DSS network, suppose the monthly cost data are given as in Table 4.1. Then, the monthly costs for the network given in Figure 4.1 are as detailed in Table 4.2.

The foregoing representation of the DDS network design problem can be simplified by reference to a Steiner Tree framework. Since the linking cost per line between an end office and a potential hub is known and the bridging cost per line for that hub is also available, we can pre-calculate the cost of connecting a customer location to a hub by adding up these two terms. Thus, the intermediate end offices can be eliminated and the DDS network problem can be converted into an extension of the Steiner Tree Problem. This extended problem was first investigated by Lee *et al.* (1996), who denote the hubs as 'Steiner nodes' and the customer locations as 'target nodes', thus giving this problem the name *Steiner tree-star (STS)* problem.

Table 4.1 Example monthly cost data for leased line networks.

Fixed bridging cost	$82.00		
Bridging cost per line	$41.00		
Line connecting cost:	Mileage	Fixed Cost	Variable Cost
	<1	$30.00	$0.00
	1—15	$125.00	$1.20
	≥16	$130.00	$1.50

Table 4.2 Monthly costs for network of Figure 4.1 based on Table 4.1

Bridging Cost	
fixed cost:	$82.00 × 4 = $328.00
variable cost	$41.00 × 14 = $574.00
Line Connecting Cost	
fixed cost:	$30.00 × 2 + $125 ×8 + $130 × 1 = $1190.00
variable cost:	$1.20 × (3 + 4 × 3 + 4 + 5 + 8 + 10) + $1.5 × 16 = $74.40
Total monthly cost:	**$2166.40**

Literature on the STS problem is limited. Lee *et al.* (1994) show that the STS problem is strongly NP-hard and identify two mixed zero-one integer programming formulations for this problem. Lee *et al.* (1996) further investigate valid inequalities and facets of the underlying polytope of the STS problem, and implement them in a branch and cut scheme. More recently, Xu *et al.* (1996a; 1996b) have developed a Tabu Search (TS) based algorithm. Their computational tests demonstrated that the TS algorithm is able to find optimal solutions for all problem instances up to 100 nodes. Applied to larger problems that the branch and cut procedure (Lee *et al.*, 1996) could not solve, the TS algorithm consistently outperforms the construction heuristic described in Lee *et al.* (1996).

In this chapter, we explore an implementation of Scatter Search (SS) for the STS problem. Scatter search, and its generalized form called path relinking, are evolutionary methods that have recently been shown to yield promising outcomes for solving combinatorial and nonlinear optimization problems. Based on formulations originally proposed in the 1960s (Glover, 1963; 1965) for combining decision rules and problem constraints, these methods use strategies for combining solution vectors that have proved effective for scheduling, routing, financial product design, neural network training, optimizing simulation and a variety of other problem areas (see, e.g., Glover (1999)).

Our chapter is organized as follows. The problem formulation is presented in the next section. Section 4.3 briefly describes the tabu search algorithm for the STS problem. We further describe the SS based heuristic for the STS problem in section 4.4 and examine several relevant issues, such as the diversification generator, the reference set update method, the subset generation method, the solution combination method and the improvement method. In section 4.5, we report computational results on a set of carefully designed test problems, accompanied by comparisons with the solutions obtained by the TS algorithm (Xu et al., 1996a; 1996b) which has been documented as the best heuristic available prior to this research. In the concluding section, we summarize our methodology and findings.

4.2   Mathematical Formulation

We formulate the STS problem as a 0-1 integer programming problem as follows. First we define:

M set of target nodes;
N set of Steiner nodes;
c_{ij} cost of connecting target node i to Steiner node j;
d_{jk} cost of connecting Steiner nodes j and k;
b_j cost of activating Steiner node j.

The decision variables of this formulation are:

x_i a binary variable equal to 1 if and only if Steiner node j is selected to be *active*.
y_{jk} a binary variable equal to 1 if and only if Steiner node j is linked to Steiner node k.
z_{ij} a binary variable equal to 1 if and only if target node i is linked to Steiner node j.

The model is then to minimize:

$$\sum_{i \in N} b_i x_i + \sum_{j \in N} \sum_{k \in N, k > j} d_{jk} y_{jk} + \sum_{i \in M} \sum_{j \in N} c_{ij} z_{ij} \qquad (4.1)$$

subject to:

$$\sum_{j \in N} z_{ij} = 1, \qquad \text{for } i \in M \qquad (4.2)$$

$$z_{ij} \leq x_j, \qquad \text{for } i \in M, j \in N \qquad (4.3)$$

$$y_{jk} \leq (x_j + x_k)/2, \qquad \text{for } j < k, \ j, k \in N \qquad (4.4)$$

$$\sum_{j \in N} \sum_{k > j, k \in N} y_{jk} \leq \sum_{j \in N} x_j - 1, \qquad \text{for } w \in S, S \subset N \qquad (4.5)$$

$$\sum_{j \in N} \sum_{k > j, k \in N} y_{jk} \leq \sum_{j \in (S-w)} x_j, \qquad \text{for } |S| \geq 3 \qquad (4.6)$$

$$x_j \in \{0,1\}, \qquad \text{for } j \in N \qquad (4.7)$$

$$y_{jk} \in \{0,1\}, \qquad \text{for } j < k, \ j,k \in N \qquad (4.8)$$

$$z_{jk} \in \{0,1\}, \qquad \text{for } i \in M, j \in N \qquad (4.9)$$

In this formulation, the objective function (equation 4.1) seeks to minimize the sums of the connection costs between target nodes and Steiner nodes, the connection costs between Steiner nodes, and the setup costs for activating Steiner nodes. The constraint of equation 4.2 specifies the star topology that requires each target node to be connected to exactly one Steiner node. Constraint 4.3 indicates that the target node can only be connected to the active Steiner node. Constraint 4.4 stipulates that two Steiner nodes can be connected if and only if both nodes are active. Constraints 4.5 and 4.6 express the spanning tree structure over the active Steiner nodes. In particular, equation 4.5 specifies the condition that the number of edges in any spanning tree must be equal to one fewer than the number of nodes, while equation 4.6 is an *anti-cycle* constraint that also ensures that connectivity will be established for each active Steiner node via the spanning tree. Constraints 4.7–4.9 express the non-negativity and discrete requirements. All of the decision variables are binary.

Clearly, the decision variable vector x is the critical one for the STS problem. Once this n-vector is determined, we can trivially determine the y_{jk} values by building the minimal spanning tree over the selected Steiner nodes (those for which $x_j = 1$), and then determine the z_{ij} values for each target node i by connecting it to its nearest active Steiner node, i.e. we have $z_{ij} = 1$ if and only if $c_{ij} = \min \{c_{ik} \mid x_k = 1\}$.

4.3 The Tabu Search Algorithm

In this section, we provide an overview of the tabu search algorithm for this problem, which was first proposed in Xu et al. (1996b). Although we do not describe the method in minute detail, we are careful to describe enough of its form to permit readers to understand both the similarities and differences between this method and the scatter search method that is the focus of our current investigation. The tabu search algorithm starts at a trivial initial solution and proceeds iteratively. At each iteration, a set of candidate moves is extracted from the neighborhood for evaluation, and a 'best' (highest evaluation) move is selected. The selected move is applied to the current solution, thereby generating a new solution. During each iteration, certain neighborhood moves are considered *tabu* moves and excluded from the candidate list. The best non-tabu move can be determined either deterministically or probabilistically. An aspiration criterion can over-ride the choice of a best non-tabu move by selecting a highly attractive tabu move. The algorithm proceeds in this way, until a pre-defined number of iterations has elapsed, and then terminates. At termination, the algorithm outputs the all-time best feasible solution. In subsequent subsections, we describe the major components of the algorithm.

4.3.1 Neighborhood Structure

Once the set of active Steiner nodes is determined, a feasible solution can easily be constructed by connecting the active Steiner nodes using a spanning tree and by linking the target nodes to their nearest active Steiner nodes. Based on this observation, we consider three types of moves: constructive moves which add a currently inactive Steiner node to the current solution, destructive moves which remove a active Steiner node from the current solution, and swap moves which exchange an active Steiner node with an inactive Steiner node. The swap moves induce a more significant change in the current solution and hence require a more complex evaluation. For efficiency, swap moves are executed less frequently. More specifically, we execute the swap move once for every certain number of iterations (for perturbation) and consecutively several times when the search fails to improve the current solution for a pre-specified number of iterations (for intensification). Outside the swap move phase, constructive and destructive moves are executed, selecting the best candidate move based on the evaluation and aspiration criteria applied to a subset of these two types of moves. In addition, since destructive moves that remove nodes deform the current spanning tree, we restrict the nodes removed to consist only of those active Steiner nodes whose degree does not exceed three. This restriction has the purpose of facilitating the move evaluation, as described next.

4.3.2 Move Evaluation and Error Correction

To quickly evaluate a potential move, we provide methods to estimate the cost of the resulting new solution according to the various move types. For a constructive move, we calculate the new cost by summing the fixed cost of adding the new Steiner node with the connection cost for linking the new node to its closest active Steiner node. For a destructive move, since we only consider those active Steiner nodes with degree less than or equal to three in the current solution, we can reconstruct the spanning tree as follows. If the degree of the node to be dropped is equal to one, we simply remove this node; If the degree is equal to two, we add the link that joins the two neighboring nodes after removing the node; If the degree is equal to three, we choose the least cost pair of links which will connect the three nodes previously adjacent to node removed. The cost of the new solution can be calculated by adjusting the connection cost for the new spanning tree and the fixed cost for the node removed. The swap can be treated as a combination of the constructive and destructive moves by first removing a tree node and then adding a non-tree node.

The error introduced by the preceding estimates can be corrected by executing a minimum spanning tree algorithm. We apply this error correction procedure every few iterations and also whenever a new best solution is found. Throughout the algorithm, we maintain a set of elite solutions that represent the best solutions found so far. The error correction procedure is also applied to these solutions periodically.

4.3.3 TS Memory

Our TS approach uses both a short-term memory and a long-term memory to prevent the search from being trapped in a local minimum and to intensify and diversify the search. The short term memory operates by imposing restrictions on the set of solution attributes that are permitted to be incorporated in (or changed by) candidate moves. More precisely, a node added to the solution by a constructive move is prevented from being deleted for a certain number of iterations, and likewise a node dropped from the solution by a destructive move is prevented from being added for a certain (different) number of iterations. For constructive and destructive moves, therefore, these restrictions ensure that the changes caused by each move will not be 'reversed' for the next few iterations. For each swap move, we impose tabu restrictions that affect both added and dropped nodes. The number of iterations during which a node remains subject to a tabu restriction is called the *tabu tenure* of the node. We establish a relatively small range for the tabu tenure, which depends on the type of move considered, and each time a move is executed, we select a specific tenure randomly from the associated range. We also use an aspiration criterion to over-ride the tabu classification whenever the move will lead to a new solution which is among the best two solutions found so far.

The long-term memory is a frequency based memory that depends on the number of times each particular node has been added or dropped from the solution. We use this to discourage the types of changes that have already occurred frequently (thus encouraging changes that have occurred less frequently). This represents a particular form of frequency memory based on attribute transitions (changes). Another type of frequency memory is based on residence, i.e. the number of iterations that nodes remain in or out of solution.

4.3.4   Probabilistic Choice

As stated above, a best candidate move can be selected at each iteration according to either probabilistic or deterministic rules. We find that a probabilistic choice of candidate move is appropriate in this application since the move evaluation contains 'noise' due to the estimate errors. The selection of the candidate move can be summarized as follows. First, all neighborhood moves (including tabu moves) are evaluated. If the move with the highest evaluation satisfies the aspiration criterion, it will be selected. Otherwise, we consider the list of moves ordered by their evaluations. For this purpose, tabu moves are considered to be moves with highly penalized evaluations.

We select the top move with a probability p and reject the move with probability $1-p$. If the move is rejected, then we consider the next move on the list in the same fashion. If it turns out that no move has been selected at the end of this process, we select the top move.

We also make the selection probability vary with the quality of the move by changing it to $p\beta_1^{r-\beta_2}$, where r is the ratio of the current move evaluation to the value of the best solution found so far, and β_1 and β_2 are two positive parameters. This new fine-tuned probability will increase the chance of selecting 'good' moves.

4.3.5   Solution Recovery for Intensification

We implement a variant of the restarting and recovery strategy in which the recovery of the elite solution is postponed until the last stage of the search. The elite solutions, which are the best *K* distinct solutions found so far, are recovered in reverse order, from the worst solution to the best solution. The list of elite solutions is updated whenever a new solution is found better than the worst solution in the list. Then the new solution is added to the list and the worst is dropped. During each solution recovery, the designated elite solution taken from the list becomes the current solution, and all tabu restrictions are removed and reinitialized. A new search is then launched that is permitted to constitute a fixed number of iterations until the next recovery starts. Once the recovery process reaches the best solution in the list, it moves circularly back to the worst solution and restarts the above process again. (Note that our probabilistic move selection induces the process to avoid repeating the previous search trajectory.)

4.4 The SS Algorithm

Our SS algorithm is specifically designed for the STS problem and consists of the following components, based on Glover (1997):

1. A Diversification Generator: to generate a collection of diverse trial solutions, using an arbitrary trial solution (or seed solution) as an input.
2. An Improvement Method: to transform a trial solution into one or more enhanced trial solutions. (Neither the input nor output solutions are required to be feasible, though the output solutions will more usually be expected to be so. If no improvement of the input trial solution results, the 'enhanced' solution is considered to be the same as the input solution.)
3. A Reference Set Update Method: to build and maintain a Reference Set consisting of the b best solutions found (where the value of b is typically small, e.g. between 20 and 40), organized to provide efficient accessing by other parts of the method.
4. A Subset Generation Method: to operate on the Reference Set, to produce a subset of its solutions as a basis for creating combined solutions.
5. A Solution Combination Method: to transform a given subset of solutions produced by the Subset Generation Method into one or more combined solution vectors.

In the following subsections, we first describe the framework of our SS algorithm, and then describe each component which is specifically designed for the STS problem.

4.4.1 *Framework of SS*

We specify the general template in outline form as follows. This template reflects the type of design often used in scatter search and path relinking.

Initial Phase
1. (*Seed Solution Step.*) Create one or more seed solutions, which are arbitrary trial solutions used to initiate the remainder of the method.
2. (*Diversification Generator.*) Use the Diversification Generator to generate diverse trial

solutions from the seed solution(s).

3. (*Improvement and Reference Set Update Methods.*) For each trial solution produced in Step 2, use the Improvement Method to create one or more enhanced trial solutions. During successive applications of this step, maintain and update a Reference Set consisting of the b best solutions found.

4. (*Repeat.*) Execute Steps 2 and 3 until producing some designated total number of enhanced trial solutions as a source of candidates for the Reference Set.

Scatter Search Phase

5. (*Subset Generation Method.*) Generate subsets of the Reference Set as a basis for creating combined solutions.

6. (*Solution Combination Method.*) For each subset X produced in Step 5, use the Solution Combination Method to produce a set C(X) that consists of one or more combined solutions. Treat each member of the set C(X) as a trial solution for the following step.

7. (*Improvement and Reference Set Update Methods.*) For each trial solution produced in Step 6, use the Improvement Method to create one or more enhanced trial solutions, while continuing to maintain and update the Reference Set.

8. (*Repeat.*) Execute Steps 5–7 in repeated sequence, until reaching a specified cut-off limit on the total number of iterations.

We follow the foregoing template and describe in detail each of the components in the subsequent subsections.

4.4.2 Diversification Generators for Zero-One Vectors

Let x denote an 0-1 n-vector in the solution representation. (In our STS problem, x represents a vector of the decision variables which determines if the corresponding Steiner node is active or not.) The first type of diversification generator we consider takes such a vector x as its seed solution, and generates a collection of solutions associated with an integer $h = 1, 2,..., h^*$, where $h^* \leq n - 1$ (recommended is $h^* \leq n/5$).

We generate two types of solutions, x' and x'', for each h, by the following pair of solution generating rules:

> Type 1 Solution: Let the first component x'_1 of x' be $1 - x_1$, and let $x'_{1+kh} = 1 - x_{1+kh}$ for $k = 1, 2, 3,..., k^*$, where k^* is the largest integer satisfying $k^* \leq n/h$. Remaining components of x' equal 0.

To illustrate for $x = (0,0,...,0)$: the values $h = 1, 2$ and 3 respectively yield $x' = (1,1,...,1)$, $x' = (1,0,1,0,1 ...)$ and $x' = (1,0,0,1,0,0,1,0,0,1,.....)$. This progression suggests the reason for preferring $h^* \leq n/5$. As h becomes larger, the solutions x' for two adjacent values of h differ from each other proportionately less than when h is smaller. An option to exploit this is to allow h to increase by an increasing increment for larger values of h.

Type 2 Solution: Let x'' be the complement of x'.

Again to illustrate for $x = (0,0,...,0)$: the values $h = 1$, 2 and 3 respectively yield $x'' = (0,0,...,0)$, $x'' = (0,1,0,1,.....)$ and $x'' = (0,1,1,0,1,1,0,...)$. Since x'' duplicates x for $h = 1$, the value $h = 1$ can be skipped when generating x''.

We extend the preceding design to generate additional solutions as follows. For values of $h \geq 3$ the solution vector is shifted so that the index 1 is instead represented as a variable index q, which can take the values 1, 2, 3, ..., h. Continuing the illustration for $x = (0,0,...,0)$, suppose $h = 3$. Then, in addition to $x' = (1,0,0,1,0,0,1,...)$, the method also generates the solutions given by $x' = (0,1,0,0,1,0,0,1,...)$ and $x' = (0,0,1,0,0,1,0,0,1....)$, as q takes the values 2 and 3.

The following pseudo-code indicates how the resulting diversification generator can be structured, where the parameter MaxSolutions indicates the maximum number of solutions desired to be generated. (In our implementation, we set MaxSolutions equal to the number of 'empty slots' in the reference set, so the procedure terminates either once the reference set is filled, or after all of the indicated solutions are produced.) Comments within the code appear in italics, enclosed within parentheses.

```
NumSolutions = 0
For h = 1 to h*
    Let q* = 1 if h < 3, and otherwise let q* = h
```
(q^* denotes the value such that q will range from 1 to q^*. We set $q^* = 1$ instead of $q^* = h$ for $h < 3$ because otherwise the solutions produced for the special case of $h < 3$ will duplicate other solutions or their complements.)
```
        For q = 1 to q*
            let k* = (n-q)/h <rounded down>
            For k = 1 to k*
```
$$x'_{q+kh} = 1 - x_{q+kh}$$
```
            End k
            If h > 1, generate x" as the complement of x'
                (x' and x" are the current output solutions.)
            NumSolutions = NumSolutions + 2 (or + 1 if h = 1)
            If NumSolutions ≥ MaxSolutions, then stop
            generating solutions.
        End q
End h
```

The number of solutions x' and x'' produced by the preceding generator is approximately $q^*(q^*+1)$. Thus if $n = 50$ and $h^* = n/5 = 10$, the method will generate about 110 different output solutions, while if $n = 100$ and $h^* = n/5 = 20$, the method will generate about 420 different output solutions.

Since the number of output solutions grows fairly rapidly as n increases, this number can be limited, while creating a relatively diverse subset of solutions, by allowing q to skip over various values between 1 and q^*. The greater the number of values skipped, the less 'similar' the successive solutions (for a given h) will be. Also, as previously noted, h itself can be incremented by a value that differs from 1.

If further variety is sought, the preceding approach can be augmented as follows. Let $h = 3, 4, ..., h^*$, for $h \leq n-2$ (preferably $h^* \leq n/3$). Then for each value of h, generate the following solutions.

Type 1A Solution: Let $x'_1 = 1 - x_1$ and $x'_2 = 1 - x_2$. Thereafter, let $x'_{1+kh} = 1 - x_{1+kh}$ and let $x'_{2+kh} = 1 - x_{2+kh}$ for $k = 1, 2, ..., k^*$, where k^* is the largest integer such that $2 + kp \leq n$. All other components of x' are the same as in x.

Type 2A Solution: Create x'' as the complement of x', as before.

Related variants are evident. The index 1 can also be shifted (using a parameter q) in a manner similar to that indicated for solutions of type 1 and 2.

4.4.3 Maintaining and Updating the Reference Set

The Reference Set Update method is a very important component in the SS template. Basically, it employs the update operation which consists of maintaining a record of the b all-time best solutions found. Several issues are relevant. First, since the Reference Set is a collection of the top-ranked solutions, it can be implemented as a sorted list. Initially, the list is empty. Then, solutions are added into the list and the list is kept sorted on solution evaluations. Once the list is full (i.e. the number of elite solutions in the list reaches its pre-defined limit, of b), the solution currently under consideration is added to the list only if it is better than the current worst solution and does not duplicate any of the other solutions on the list. In this case it replaces the worst solution, is be inserted into the proper position based on its evaluation.

The check-for-duplication procedure is expedited by using a hash function. If two solutions have the same objective function value and the same hash value, they are compared against each other for full duplication check.

Finally, it is useful to collect some types of statistics throughout the execution of the Reference Set Update method. These statistics include the number of times the Update method is called, as well as the number of times a new solution is added, which we use to control the progress of the SS method. Other auxiliary statistics include a count of the number of partial duplication checks, full duplication checks, and the number of occurrences where duplications were found.

4.4.4 Choosing Subsets of the Reference Solutions

We now describe the method for creating different subsets X of the reference set (denoted as RefSet), as a basis for implementing Step 5 of the SS Template. It is important to note the SS Template prescribes that the set $C(X)$ of combined solutions (i.e. the set of all combined solutions we intend to generate) is produced in its entirety at the point where X is created. Therefore, once a given subset X is created, there is no merit in creating it again. Therefore, we seek a procedure that generates subsets X of *RefSet* that have useful properties, while avoiding the duplication of subsets previously generated. Our approach for doing this is organized to generate the following four different collections of subsets of *RefSet*, which we refer to as *SubSetType* = 1, 2, 3 and 4. Let *bNow* denote the number of solutions currently recorded on *RefSet*, where *bNow* is not permitted to grow beyond a value *bMax*.

SubsetType = 1: all 2-element subsets.

SubsetType = 2: 3-element subsets derived from the 2-element subsets by augmenting each 2-element subset to include the best solution not in this subset.

SubsetType = 3: 4-element subsets derived from the 3-element subsets by augmenting each 3-element subset to include the best solutions not in this subset.

SubsetType = 4: the subsets consisting of the best i elements, for $i = 5$ to *bNow*.

The reason for choosing the four indicated types of subsets of *RefSet* is as follows. First, 2-element subsets are the foundation of the first 'provably optimal' procedures for generating constraint vector combinations in the surrogate constraint setting, whose ideas are the precursors of the ideas that became embodied in scatter search (see, e.g., Glover (1965), Greenberg and Pierskalla (1970)). Also, conspicuously, 2-element combinations have for many years dominated the genetic algorithm literature (in '2-parent' combinations for crossover).

We extend the 2-element subsets since we anticipate the 3-element subsets will have an influence that likewise is somewhat different than that of the 2-element subsets. However, since the 3-element subsets are much more numerous than the 2-element subsets, we restrict consideration to those that always contains the best current solution in each such subset. Likewise, we extend the 3-element subsets to 4-element subsets for the same reason, and similarly restrict attention to a sub-collection of these that always includes the two best solutions in each such subset. In addition, to obtain a limited sampling of subsets that contain larger numbers of solutions we include the special subsets designated as *SubsetType* = 4, which include the b best solutions as b ranges from 5 to *bMax*.

The methods to create the four types of subsets where RefSet is entirely static (i.e. where *bNow=bMax* and the set of bMax best solutions never changes) are trivial. However, these algorithms have the deficiency of potentially generating massive numbers of duplications if applied in the dynamic setting (where they must be re-initiated when RefSet becomes modified). Thus we create somewhat more elaborate processes to handle a dynamically changing reference set.

A basic part of the Subset Generation Method is the iterative process which supervises the method and calls other subroutines to execute each subset generation method for a given SubsetType (for SubsetType = 1 to 4, then circularly return to 1). Inside each individual subset generation method, once a subset is formed, the solution combination method $C(X)$ (Step 6 of the SS template) is immediately executed to create one or more trial solutions, followed by the execution of the improvement method (Step 7 of the SS template) which undertakes to improve these trial solutions. When these steps find new solutions, not previously generated, that are better than the last (worse) solution in RefSet, RefSet must be updated. Since the solution combination method and the improvement method are deterministic, there is no need to generate the same subset X produced at some earlier time. To avoid such duplications, we organize the procedure to make sure that X contains at least one new solution not contained in any subset previously generated. At the beginning of each iteration, we sort the new solutions in RefSet. Any combination of solutions that contains at least one new solution will be generated as a legal subset of RefSet for a given SubsetType. The iterative process terminates either when there is no new solution in RefSet (RefSet remains unchanged from the last iteration), or when the cumulative number of executions of the Improvement Method, as it is applied following the solution combination step, exceeds a chosen limit.

4.4.5 Solution Combination Method

Once a subset of the reference set is determined, we apply a solution combination method to produce a series of trial solutions. Let S^* denote the subset we consider which contains k distinct vectors $(x(1), ..., x(k))$. Then the trial points are produced by the following steps:

1. Generate the centers of gravity for each $k-1$ subset of S^*, denoted by $y(i)$, that is:

$$y(i) = \sum_{j \neq i} x(j)/(k-1), \quad \text{for } i = 1,...,k$$

2. For each pair $(x(i), y(i))$, consider the line connecting $x(i)$ and $y(i)$ by the representation $z(w) = x(i) + w(y(i) - x(i))$. We restrict the attention to the four points $z(1/3)$, $z(-1/3)$, $z(2/3)$ and $z(4/3)$ (two of them are interior points and the other two are exterior points).

3. Convert each of the above four points to a 0-1 vector by applying the threshold rule, that is, set an element to 1 if it exceeds a pre-defined threshold u, set it to 0 otherwise.

We observe that the lines generated in step 2 all pass through the center of gravity y^* for all k points, and therefore it is not necessary to calculate the k points $y(i)$ explicitly, but only to identify equivalent values of w for lines through y^*. However, for small values of k, it is just as easy to refer to the $y(i)$ points as indicated.

Since the trial points are 'rounded' by the simple threshold in step 3, it is entirely possible to produce the same trial vector for different S^*. These trial vectors are first transformed to trial solutions (e.g. by building a minimum spanning tree on the active Steiner nodes and calculating the total cost) and then fed as the inputs to the Improvement Method (described next). An important aspect here is to avoid the effort of transforming and improving solutions already generated. Avoidance of duplications by controlling the combined solutions, which includes submitting them to constructive and improving heuristics, can be a significant factor in producing an effective overall procedure. To do this, we store only the $r = rNow$ most recent solutions generated (allowing $rNow$ to grow to a maximum of $rMax$ different solutions recorded), following a scheme reminiscent of a simple short-term recency memory approach in tabu search. In particular, we keep these solutions in an array $xsave[r]$, $r = 1$ to $rNow$, and also keep track of a pointer $rNext$, which indicates where the next solution x' will be recorded once the array is full.

Let $E0$ and $Hash0$ be the evaluation and hash function value for solution x', and denote associated values for the $xsave[r]$ array by $Esave(r)$ and $Hashsave(r)$. These are accompanied by a 'depth' value, which is 0 if no duplication occurs, and otherwise tells how deep in the list – how far back from the last solution recorded – a duplication has been found. For example, depth = 3 indicates that the current solution duplicates a solution that was recorded 3 iterations ago. (This is not entirely accurate since, for example, depth = 3 could mean the solution was recorded five iterations ago and then two other duplications occurred, which still results in recording only three solutions.)

The pseudo code for checking the duplication is shown as follows:

Initialization Step:

    $rNow = 0$; $rNext = 0$; $CountDup(depth) = 0$, for $depth = 1$ to $rMax$

Duplication Check Subroutine.

```
Begin Subroutine.
      depth = 0
      If rNow = 0 then:
            rNow = 1; rNext = 1;
            xsave[1] = x' (record x' in xsave[1]),
            Esave(1) = E0; Firstsave(1) = FirstIndex0
            End the Subroutine
      Elseif rNow > 0 then:
```

(Go through in 'depth order', from most recently to least recently stored. When a duplicate is found, index r (below) gives the value of rMax that would have been large enough to identify the duplication.)

```
                  i = rNext
                  For r = 1 to rNow
                        If Esave(i) = E0 then:
                              If Hash0 = Hashsave(i) then:
                                    If x' = x[i] then:
                                          (x' is a duplicate)
                                          depth = r
                                          End Duplication Check Subroutine
                                    Endif
                              Endif
                        Endif
                  i = i-1
                  if i < 1 then i = rNow
                  End r
```

(Here, no solutions were duplicated by x'. Add x' to the list in position rNext, which will replace the solution previously in rNext if the list is full.)

```
                  rNext = rNext + 1
                  If rNext > rMax then rNext = 1
                  If rNow < rMax then rNow = rNow + 1
                  xsave[rNext] = x'
                  Esave(rNext) = E0
                  Hashsave(rNext) = Hash0
      Endif
```

End of Duplication Check Subroutine

4.4.6 Improvement Method

We apply a local search heuristic to improve the initial solution and the trial solution produced by the combination method. The heuristic employs the same neighborhood of moves as used for the tabu search algorithm, i.e. constructive moves, destructive moves and swap moves. We also apply the same move evaluation for each type of neighborhood moves. The candidate moves of each move type are evaluated and the best moves for each move type are identified. Then the error correction method is applied for the best outcome obtained from destructive moves and swap moves to obtain the true cost. (Note that our evaluation method for the constructive moves is exact.) If the true cost of the best move for all three types is lower (better) than the cost of the current solution, that move is executed and the search proceeds. Otherwise, the local search heuristic terminates with the current solution.

Since the local search improvement method always ends with a local optimum, it is very likely to terminate with the same solution for different starting solutions. This further accentuates the importance of the method to avoid duplicating solutions in the reference set, as proposed in section 4.5.

4.5 Computational Results

In this section, we report our computational outcomes for a sets of randomly generated test problems. In this set, the locations of target nodes and Steiner nodes are randomly generated in Euclidean space with coordinates from the interval [0, 1000]. Euclidean distances are used because they are documented to provide the most difficult instances of classical randomly generated Steiner Tree problems (Chopra and Rao, 1994). The fixed cost of selecting a Steiner node is generated randomly from the interval [10,1000], which provides the most difficult tradeoff with the other parameters selected (Lee *et al.*, 1996).

We generate 21 test problems whose dimensions are as follows. We first set the value of n (the number of Steiner nodes) to 100, 125, 150, 175 and 200 respectively. For each fixed n, we generate three problems by setting m (the number of target nodes) equal to n, $n+50$ and $n+100$ respectively. Therefore, 15 test problems with the above dimensions are generated. Furthermore, we generate six additional problems which are designed to be particularly hard. These problems have dimensions (denoted by $m \times n$) as 250×250, 300×250, 350×250, 100×300, 200×300 and 300×300. As established in our previous research (Xu *et al.*, 1996b), these 21 problems are unable to be handled by the exact method (i.e. the branch and cut method by Lee *et al.* (1996) due to the computing times and memory limitations, and our advanced tabu search algorithm described in section 4.3 is the best heuristic available among the various construction heuristics.

4.5.1 Parameter Settings

Our TS method requires a few parameters to be set at the appropriate values. These values are initialized based on our computational experience and common sense, and then fine-tuned using a systematic approach (Xu *et al.*, 1998). First we select an initial solution produced simply by connecting every target node to its cheapest-link Steiner node, and then constructing a minimum spanning tree on the set of selected Steiner nodes. Then we randomly generate tabu tenures for the three types of moves in the TS procedure from an relatively small interval each time a move is executed. The interval [1,3] is used for constructive moves and the interval [2,5] is used for destructive moves. In the case of swap moves, an interval of [1,3] is used for each of the two elementary moves composing the swap. We execute swaps either once every seven iterations or in a block of five consecutive iterations when no 'new best' solution is found during the most recent 200 iterations. The termination condition is effective when $min\ \{20000,\ max\ \{3000,\ n^2\}/2\}$, where n is the number of Steiner nodes. The error correction procedure is executed each time a new best solution is found, and applied to the current solution after every three accumulated moves, not counting destructive moves that drop nodes of degree one. Error correction is also applied every 200 iterations to the priority queue that stores the twenty best solutions.

The other parameters for our TS approach include the iteration counter which triggers the long term memory. It is set to 500. The long term memory penalty is calculated as $300*f/F$ for constructive and destructive moves, where f denotes the frequency of the move under consideration and F denotes the maximum of all such frequencies. For a swap move, the penalty is calculated as $150*(f_1+f_2)/F$ where f_1 and f_2 are the respective frequencies for the two constituent constructive and destructive moves. In probabilistic move selection, we choose the probability of acceptance $p = 0.3$. In addition, we restrict the candidate list for the probabilistic rule to contain the ten best moves (adjusted for tabu penalties). We also pair up the ten best destructive moves and the ten best constructive moves to construct a candidate list for swap moves. The fine-tuned selection probability function as mentioned in section 4.3, is defined as $0.3^{r-0.15}$. Finally, we assign the following parameters for implementing the restart/recovery strategy. We recover 40 solutions in total. For each recovered solution, a block of 30 iterations is executed. Thus, the first recovery occurs at iteration 1200, and is executed every 30 iterations thereafter.

Now we describe the parameter setting for our SS method. Unlike the TS algorithm, our SS method contains very few parameters. First, we use an extremely trivial solution which sets all Steiner nodes active as the initial solution. The maximum number of solutions in the reference set, $bMax$, is set to be 30. The value of the threshold, u, which is used to map the points of the trial solution to binary variables, is set at 0.75. In addition, we set the parameter $h*$ in the diversification generators to 5. The maximum iteration in SS is set to 10. Finally, to speed up SS for our current preliminary testing, we skip subsets type 2 and 3, thus only subsets type 1 and 4 are evaluated.

4.5.2 Numerical Test Results for SS

We test both our TS and SS methods on the 21 benchmark problems. The TS method is coded in C and the SS approach in C++. The computational results on the 21 benchmark problems are provided in Table 4.3 as follows. For ease of comparison, we mark the SS solutions which are better than their TS counterparts by (+). Similarly, the SS solutions which are worse than their TS counterparts are marked by (-). The CPU times reported represent the execution time on a HP D380 machine with SPECint_rate95 of 111–211.

Table 4.3 Test results for TS and SS.

Problem	TS		SS	
(m×n)	Cost	CPU	COST	CPU
100×100	16166	01:39	16166	01:35
150×100	19359	02:32	19359	02:43
200×100	25102	03:27	25102	02:52
125×125	16307	04:52	16307	03:32
175×125	21046	06:36	21046	06:00
225×125	26213	08:25	26213	07:11
150×150	19329	10:43	19329	10:59
200×150	24358	14:48	24378(-)	15:00
250×150	28248	20:21	28248	15:02
175×175	20907	23:00	20918(-)	28:25
225×175	25003	30:33	25017(-)	34:11
275×175	27672	41:11	27672	30:23
200×200	22876	33:18	22880(-)	47:20
250×200	26122	45:43	26122	45:13
300×200	29879	1:01:42	29879	1:11:47
250×250	25568	1:06:52	25566(+)	2:52:50
300×250	29310	1:29:28	29310	6:22:06
350×250	32290	1:57:19	32290	8:08:12
100×300	13122	35:03	13119(+)	1:59:37
200×300	21238	1:12:56	21238	5:04:01
300×300	28727	2:00:27	28707(+)	20:25:20

From Table 4.3, it is clear that our SS method is highly competitive with the TS approach in term of solution quality. In the 21 benchmark problems, SS produces three better solutions than TS. It also ties 14 problems with TS, and produces four worse solutions, but the differences are truly marginal (less than 0.1%). Given the fact that our TS approach has been documented as the best heuristic available for the STS problem, and that it has produced optimal solutions for all test problems with up to 100 Steiner nodes (Xu *et al.*, 1996b), the quality of our SS method is quite high.

We observe from Table 4.3 that our SS approach takes the same order of CPU time as TS for problems with up to 200 Steiner nodes. Since in practice most problems do not contain more than 200 Steiner nodes, this indicates that the SS algorithm can be employed as an effective decision making tool. However, for problems whose sizes are over 200 Steiner nodes, SS requires significantly greater CPU time than TS. This can be imputed to several factors. First, SS uses simple local search to improve the trial solutions. Statistics show that more than 90% of the CPU time is spent on executing the local search method. Unlike our TS method, the local search method does not employ a candidate list strategy, and does not take long term memory into consideration. More specifically, our local search pays the same attention to the constructive moves, destructive moves and swap moves.

However, statistics show that the constructive and swap moves are more time consuming and therefore should be executed less frequently to achieve greater speed. The use of a candidate list strategy and long term memory, as is characteristically done in tabu search, appears effective for reducing the number of non-productive moves. Secondly, we employ relatively primitive types of subsets to generate trial points. There are a variety of ways to speed up the process by improving the subsets and solution combination method. As we show in the next subsection, a customary speedup can obtain significant savings on execution times. Thirdly, our solution combination method ignores the existence of 'strongly determined' or 'consistent' variables in the elite solutions. Again, the long term memory is useful to isolate these variables. Finally, our SS approach is not fine-tuned. Most parameters are set arbitrarily.

4.5.3 Improving the speed of SS

We realize that our foregoing subset generation method and solution combination method are not customized for the STS problem, so they may produce some wasted effort. More specifically, while the solution combination method described in section 4.5 is appropriate for general integer programming where the decision variables are not necessarily 0 and 1, it is less suitably designed for highly discrete 0-1 problem such as STS, where the decision to set variables equal to 0 or 1 is not based on meaningful rounding penalties derived from a fractional relaxed solution. For a 2-element subset (SubSet I), it is often not necessary to generate four trial points. For the other subset (SubSets II, III and IV), our previously identified linear combination will generate trial points fairly close to the overall center of gravity, which is likely to create many duplicate solutions after rounding.

For the 0-1 case, a highly relevant (though not exhaustive) set of combinations and roundings of r reference points consists simply of those equivalent to creating a positive integer threshold $t \leq r$, and stipulating that the offspring will have its ith component $x_i = 1$ if and only if at least t of the r parents have $x_i = 1$. (Different thresholds can be chosen for different variables, to expand the range of options considered.) In particular, for two parents, a setting of $t = 1$ gives the offspring that is the union of the 1's in the parents and $t = 2$ gives the offspring that is the intersection of the 1's in the parents. The inclusion of negative weights can give offspring that exclude $x_i = 1$ if both parents receive this assignment.

To compare with our preceding approach, we tested the following three simple rules that result by using trivial linear combinations and rounding decisions (variables not indicated to receive a value of 1 automatically receive a value of 0):

1. $x_i = 1$ if the ith component of both parents is equal to 1;
2. $x_i = 1$ if the ith component of the first parent, but not of the second, is equal to 1;
3. $x_i = 1$ if the ith component of the second parent, but not of the first, is equal to 1.

The rule that generates the union of 1's in the parents is excluded because its exploitation in the current setting requires the use of destructive moves to recover the tree structure, and such a recovery process has not been incorporated in this preliminary study.

We report the results from generating the offspring from rules (1), (2) and (3) independently, i.e. producing exactly one offspring for each pair of parents. The three different resulting approaches are labeled SS 1, SS 2 and SS 3 in Table 4.4, and we also provide a comparison with our SS results in the same table. Additionally, we test the strategy which generates all three offspring simultaneously, labeled 'SS 4' in Table 4.4.

Table 4.4 Comparisons of results with the simplified solution combination rules.

Problem (mxn)	SS COST	SS CPU	SS 1 COST	SS 1 CPU	SS 2 COST	SS 2 CPU	SS 3 COST	SS 3 CPU	SS 4 COST	SS 4 CPU
100×100	16166	01:35	16166	43	16166	25	16166	25	16166	2:00
150×100	19359	02:43	19359	1:12	16359	30	16359	30	16359	4:45
200×100	25102	02:52	25102	1:10	25102	36	25102	35	25102	6:16
125×125	16307	03:32	16307	1:34	16307	1:10	16307	1:09	16307	3:22
175×125	21046	06:00	21046	2:37	21046	1:24	21046	1:23	21046	11:26
225×125	26213	07:11	26213	3:17	26213	1:43	26213	1:40	26213	12:41
150×150	19329	10:59	19329	5:07	19329	2:47	19329	2:44	19329	18:45
200×150	24378	15:00	24378	7:01	24378	3:18	24378	3:14	24378	27:38
250×150	28248	15:02	28248	6:35	28248	3:49	28248	3:41	28248	31:20
175×175	20918	28:25	20918	13:13	20918	5:44	20918	5:35	20918	47:02
225×175	25017	34:11	25017	16:37	25017	6:42	25017	6:30	25017	1:02:52
275×175	27672	30:23	27672	14:00	27672	7:48	27672	7:35	27672	55:11
200×200	22880	47:20	22880	23:53	22880	10:57	22880	10:37	22880	1:28:50
250×200	26122	45:13	26122	21:32	26122	13:03	26122	12:39	26122	1:46:10
300×200	29879	1:11:47	29879	31:32	29900(-)	14:18	29900(-)	13:55	29879	2:37:49
250×250	25566	2:52:50	25566	1:30:37	25566	33:49	25566	33:58	25566	5:33:53
300×250	29310	6:22:06	29310	2:23:54	29343(-)	40:30	29343(-)	39:20	29310	10:57:22
350×250	32290	8:08:12	32290	3:49:50	32290	1:50:14	32290	1:41:30	32290	18:34:25
100×300	13119	1:59:37	13119	1:03:52	13119	39:02	13119	37:58	13119	3:33:32
200×300	21238	5:04:01	21238	2:39:34	21238	1:05:52	21238	1:06:21	21238	9:29:04
300×300	28707	20:25:20	28707	9:37:19	28707	3:41:03	28707	4:05:45	28707	36:11:25

Table 4.4 clearly shows that the three simplified rules can effectively reduce the execution time by comparison to the SS method. In particular, SS 1 obtains the same high quality solutions as SS does for all 21 test problems, while improving the CPU times by percentages that range from 46% to 62%. SS 2 and SS 3 further dramatically increase the speed relative to SS, each reducing the CPU times by an amount that ranges from approximately 67% to 89%, at the cost of producing two marginally inferior solutions by both rules. The notable efficiency improvement by SS 2 and SS 3 can be attributed to the fact that the simplified rules (2) and (3) often create more $x_j = 1$ assignments than the simplified rule (1) does, which requires fewer subsequent constructive moves to generate a complete solution. The SS 4 approach precisely matches the solutions produced by SS, but

takes more time (approximately twice as much). This suggests that the offspring produced by SS are quite likely a subset of those produced by SS 4, but a subset that is sufficient to produce the same best solutions. (This outcome could conceivably change for a different set of test problems.) Unexpectedly, SS 4 requires more CPU time than the sum of the CPU times of SS 1, SS 2 and SS 3. Perhaps the most surprising outcome is that the simple 'intersection rule' underlying SS 1 is able to produce solutions whose quality matches that of SS – a quality that is nearly as good overall as that obtained by the tabu search approach.

It is to be emphasized that we have only explored the very simplest rules for the solution combination method. The vectors generated as offspring are 'stripped down' by comparison with those typically generated by GA combination rules, though it is easy to see that different choices of rounding, as by thresholds that vary for different variables, can produce many more types of offspring than those available by 'genetic crossover' – including the variants provided by uniform and Bernoulli crossover (these GA crossovers are incapable of producing even the simple SS 2 and SS 3 types of offspring, for example).

It is novel that the SS 1 rule gives an offspring whose 1's are components of all the usual GA crossovers, although none of these GA crossovers will produce the outcome of SS 1. Note that an exception may occur within Bernoulli crossover under those rare circumstances where, by random selection, all 1's not shared by both parents happen to be excluded from the offspring. Also, from a historical perspective, it may be noted that the types of offspring examined here are special instances of those from the SS framework proposed in the 1970s, and that the 'refinements' of GA crossover such as embodied in Bernoulli crossover – which give only a subset of relevant possibilities – did not emerge until nearly a decade later

More refined rules than those we have tested are conspicuously possible, and afford an opportunity to further improve the SS performance, especially in application to subset types that contain more than two reference solutions. Furthermore, as previously observed, there are issues other than the solution combination method, such as the use of candidate lists, strategic oscillation and long term memory, that can also effectively improve the SS approach. These issues will be explored in our future research.

4.6 Conclusions

In this chapter, we have described and studied the Steiner Tree-Star (STS) telecommunications network problem, which has application to leased-line network design. The problem is to select a subset of hubs to form a backbone network, and then to connect each 'client' site to one of the selected hubs, to meet the objective of minimizing total network cost.

The main contribution of this chapter is to develop and test a Scatter Search (SS) algorithm for the STS problem. The components of the SS method, as detailed in preceding sections, include a diversification generator, an improvement method, a reference set update method, a subset generation method, and a solution combination method. The results on 21 benchmark problems show that our SS algorithm is very competitive with the Tabu Search (TS) method, previously documented as the best method available for the STS problem. Compared with TS, the SS approach produces new best solutions in three cases, ties in 14 cases, and gives worse solutions (by a very small margin) in four cases.

We recognize that our SS approach requires significantly more CPU time than the TS method for 'extra-large' problems. Since our SS approach uses strategies that are highly compatible with tabu search, efficiency for larger problems may similarly be improved by incorporating candidate list strategies and frequency-based long-term memory to enhance the method's performance. In addition, the independent success of both TS and SS suggests that the integration of TS and SS will be a promising avenue for future research.

5

Addressing Optimization Issues in Network Planning with Evolutionary Computation

John Tindle and K. F. Poon

5.1 Introduction

In line with the rapid growth of telecommunications networks in recent years, there has been a corresponding increase in the level of network complexity. Consequently, it is now generally accepted that advanced computer aided simulation and analysis methods are essential aids to the management of large networks.

The 1990s will be recalled as the decade of business process re-engineering. Most large conventional manufacturing organisations have already applied some re-engineering to improve their overall process efficiency, resulting in the establishment of streamlined management structures, optimised production schedules and facilities, reduced staffing levels and much closer control over working budgets and expenditure for major projects.

A similar process of re-engineering is now taking place in the telecommunications sector. In the future, information will assume a much greater level of importance. It is becoming increasingly evident that many businesses will achieve and actively maintain their competitive edge by continuously gathering, analysing and exploiting operational data in new and novel ways. Large-scale systems models can now be created and intelligently evolved to produce near-optimal topological structures. Macro-level network system analysis can now be attempted because of developments in heuristic, mathematical and intelligent methods, and the availability of low cost high performance desktop computers. Personal workstations can now solve complex problems by processing vast amounts of data

---

Telecommunications Optimization: Heuristic and Adaptive Techniques, edited by D. Corne, M.J. Oates and G.D. Smith
© 2000 John Wiley & Sons, Ltd

using new smart algorithms, thereby providing better solutions and deeper insight into the complex structure of network problems. The primary goals for telecommunications network operators are to develop methodologies for the following:

Optimal planning methods are required to minimise the cost of building new network structures.

Impact analysis is required to predict the impact that new components and systems will have on the network and business.

Business modelling is required to evaluate the effects that selected, individual and grouped changes will ultimately have upon the whole business and organisation.

Fault analysis is required to accurately identify and repair faults with the minimum disruption to network performance.

This chapter considers network planning and impact analysis for the Plain Old Telephone System (POTS). In the past, new network designs were predominately undertaken by human planners using simple manual or partially automated methods. The methods employed were characteristically labour-intensive with relatively low levels of productivity. A planner, for example, would usually be expected to design a typical network layout within three working days. Generally, this work would entail two stages or more of design iteration, each stage taking twelve hours. In most cases, the solutions generated were sub-optimal, and therefore the resultant capital costs were often higher than necessary. It became clear that there was an urgent need for new automated planning tools to handle complexity and aid network designers.

Henderson (1997) provides a very interesting and illustrative case study. This project involved the installation of a national SDH network for Energis UK. He states that, "Building of the network commenced before the design was complete. The truth is, the design was not complete until some time after it was built and working! In practice parallel design and build is becoming more common as the time scale pressures of our competitive environment increase". In addition, "Unfortunately, build never occurs as planned. Unforeseen difficulties delay sections, whilst others prove easy. With a virtual workforce of 1500 people, working to very tight time scales, logistics control was a major problem".

The inability to produce good plans sufficiently rapidly can sometimes result in poorly assigned capacity levels in a network. It is also likely that sub-optimal designs could be implemented which incur higher than necessary capital costs. In most cases, the additional costs of rework to solve these problems will be very high.

5.2 The Access Network

In the UK access, 90% of new systems are copper based and only 10% optical fibre. Incumbent service operators are finding it difficult to justify the installation of fibre as a replacement technology to support narrow band services carried by the existing copper network. At present, the cost of optical network components is considered too high, especially the electronic Line Termination Unit (LTU). In addition, the introduction of Asymmetric Digital Subscriber Line (ADSL) technology has the potential to greatly increase the available bandwidth in the copper access network, thus potentially extending its working lifetime.

Consequently, copper bearer platforms still attract significant levels of investment from many telecommunications operators. Although it was previously considered that fibre would be deployed more aggressively in the access sector, copper remains the dominant technology option for the narrow band service sector. The continuous use of copper into the near future has justified the development of a new smart planning tool to automate the planning process for greenfield developments. Although in this instance, the methods described in this chapter have been applied to solve the Copper Planning Problem (CPP), it is believed the approach in general can be successfully tailored to suit a variety of different network problems.

In the UK, an ongoing development programme is in place to provide telephony to new housing estates. It is essential that an estimate of the cost of work is produced, a bill of materials and a resource schedule. This information is passed on to construction team for installation. To produce a satisfactory project plan for the construction team by the manual method proved very difficult, time consuming and tedious because of the many complex factors planners were required to consider.

The smart tool, GenOSys, which has been developed primarily to aid network planners, automates the design of the secondary cable and duct networks necessary to connect new customers to the exchange. GenOSys employs Evolutionary Computation (EC) techniques to facilitate the rapid solution of large and complex network problems.

In order to develop an efficient search strategy yielding optimal or near-optimal solutions, it is essential to understand both the nature of the problem and the structure of the search space. The various merits of using EC methods to address the CPP are discussed in the following section.

5.2.1 An overview of the Greenfield CPP

The copper network provides customer access to a range of narrow and mid-band services. Most copper architectures comply with a three tier model consisting of the primary, secondary and distribution networks, all implemented in a tree structure.

The primary network connects the secondary network to the exchange via a Primary Connection Point (PCP), as shown in Figure 5.1. The distribution network connects customers to Distribution Points (DPs), which are connected via the secondary network to the PCP.

Access networks are normally accommodated within an underground duct network. The duct network forms the links between the access nodes, such as joint boxes, which are used to accommodate cable joints and DPs. The duct network is normally highly interconnected to provide flexible schemes routing between the exchange and customer. A typical secondary network structure connecting customers to PCPs, is shown in Figure 5.2,

In this context, local access network planning is defined in the following manner. For a predefined geographical area, duct structure and particular demand profile determine the following:

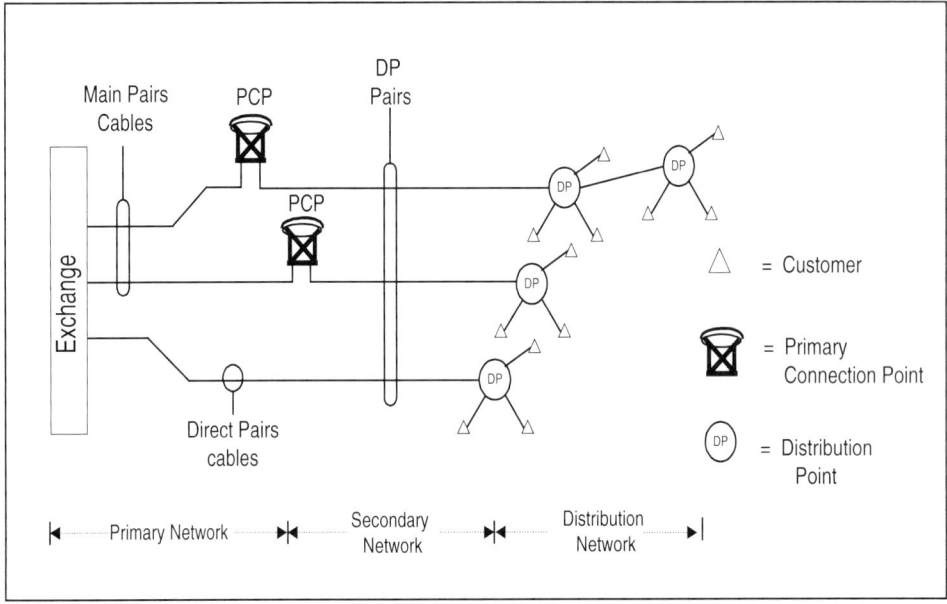

Figure 5.1 Copper access network.

1. location of all access nodes (footway boxes),
2. location of DPs,
3. assignment of customers to DPs,
4. aggregation of DPs into sub-networks,
5. assignment of DPs to cable paths, and
6. route of all cables to satisfy customer demand at the lowest cost.

The distribution network architecture is built from a number of different types of sleeves, joint boxes, cables and ducts. Customers are connected to the network at connection points called sleeves, which are normally accommodated in underground joint boxes. Ducts are used to house the secondary and distribution multicore cables, which range in size from 5 to 100 pairs.

5.2.2 Network Object Model

A model of the network was created using object-oriented development methods, Object oriented analysis OOA and design OOD enable development of flexible, intuitive models.

Object-oriented models are based upon the underlying domain structure. A system built around domain entity structures is normally more stable than one based around functionality. It has been recognised that most software modifications are associated with system functionality rather than the underlying domain (Paul *et al.*, 1994).

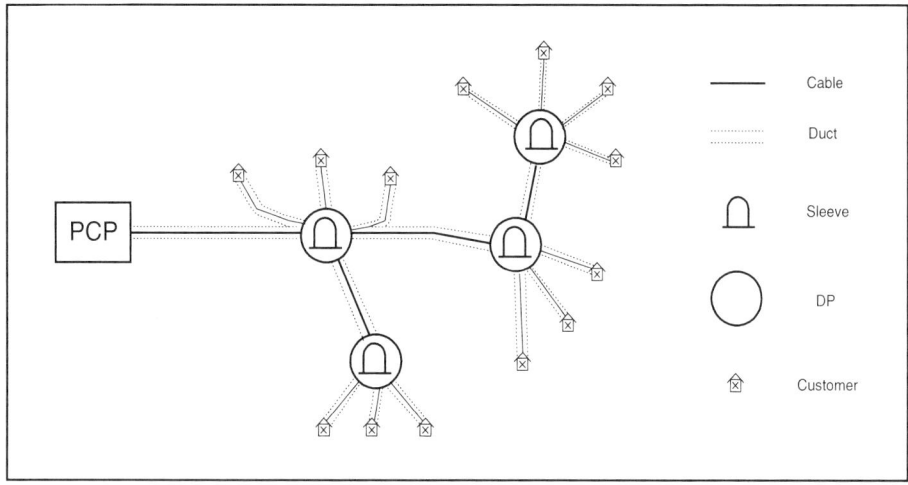

Figure 5.2 Secondary copper network.

These principles are then applied to create a model (Paul *et al.*, 1994).

Classification (abstraction) is the operation of grouping together objects with similar features.

Inheritance (generalisation) if there are specialised types of a class which exhibit common features, the principle of inheritance can be used to organise them.

Association represents a relationship between two different classes of objects.

Aggregation (whole/part structure) expresses a special relationship in which one class contains another.

As object modelling is based upon the organisation principles of human thought processes, domain specialists are able to contribute to the design of an object model without requiring specialist computer skills. In an object-oriented model, the objects and their associations correspond directly to real world entities and their relationships. Consequently, the object model is modular and closely resembles the real network. Objects can represent both physical items (copper cables) and abstract concepts (time slots in a frame). These characteristics make object models easy to understand, maintain and extend.

A network object model is required to capture the network topology and its connection rules. Persistent computer based objects capture the functionality of network components and state attributes. The use of object-oriented methods proved successful because they allowed the complex rules relating to network component connectivity to be represented in a flexible and concise manner. This scheme allows the seamless integration of engineering rules into the computer based objects. The majority of constraint checking for engineering rules is therefore carried out implicitly, through the mechanisms provided by the model.

However, practical experience has shown that another level of data abstraction is required to improve optimisation efficiency. The direct manipulation (creation and deletion) of network objects by the solver module during an optimisation run proved to be highly inefficient. To improve the efficiency of the process, relevant data from the object model is now held in the form of tables that can be accessed very rapidly. The design of the object schema employed the Object Modelling Technology (OMT) method developed by Rambaugh. This object model uses the class *Nodes* to represent access points, equipment cabinets and network components and the class *Links* to represent the cables and ducts. The high level view of the class schema is shown in Figure 5.3. The general-purpose object model has also been employed to support network performance and reliability analysis.

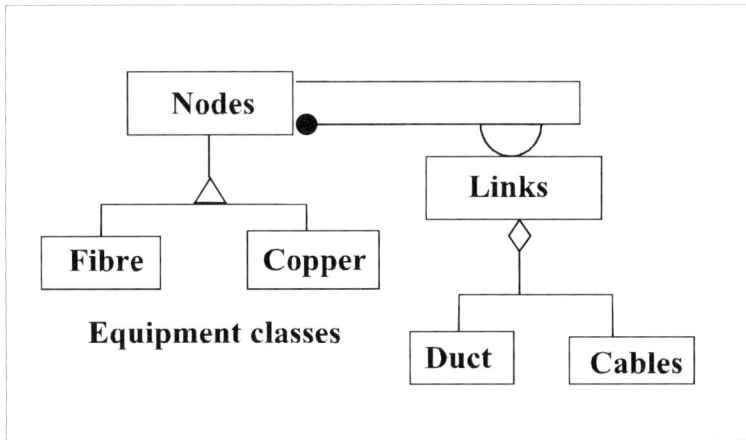

Figure 5.3 General telecommunications network object model.

5.2.3   *Structural Analysis Module*

The function of this module is to analyse network structure before an optimisation run. This pre-process generates data held in a distance matrix, tabulating the distance and optimal route between any two nodes. Graph theory is used to calculate the shortest distance between any two nodes and form a modified minimum spanning tree. Floyd's Algorithm is applied to generate the distance and path matrices. Distance information is required to initialise the search algorithm.

5.3   Copper Planning Tool

An integrated network planning environment based upon a Windows Graphical User Interface (GUI) has been created. This environment provides interfaces to existing external databases so that planning data can be easily transferred via a local area network.

The GUI allows the user to view networks in a number of different ways. A network is normally viewed against a geographical map background. The display is layered so that alternative layers can be switched on and off as required. A colour scheme and a standard set of icons have also been created so that the components shown on different layers can be readily identified. A pan and zoom feature is also available to facilitate the inspection of complex network structures. In addition, network designs can also be shown in a simplified schematic format.

The copper optimisation planning system has three modes of operation: manual, partially automated and fully automated. The manual option allows the planner to specify the placement of cables and duct components and to identify the location of DPs at network nodes. By using the partially automated option, the system determines the DP layout scheme and locations where cables can be inserted into the network. The fully automated option further increases the copper planning productivity, selects the network component types, provides routing information for ducting and cabling, optimises the cost and fully determines the connection arrangement of sleeves.

A high level schematic diagram of the planning system is in Figure 5.4. The core of the fully automated option involves the application of evolutionary computing methods and graph theory to produce optimal or near-optimal solutions. A map representing the area under consideration is digitised and sent to the input of the system. This scanned map provides the Ordinance Survey (OS) co-ordinates needed to accurately locate network component nodes and customers' premises. A network of possible duct routes along trenches is shown on the civil layer and overlaid on the scanned OS map. The subsequent optimisation process uses the duct network data. Users may interactively modify the network data stored in memory to customise network designs.

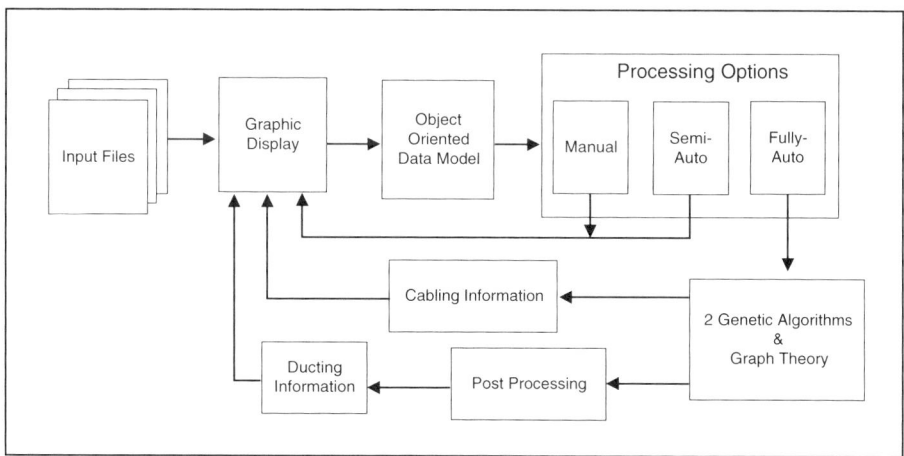

Figure 5.4 Copper planning system overview.

The objective of the fully automated option is to determine the best locations for DPs and form geographically optimal sub-networks. The duct network specified by the input data file for the optimiser can be configured as either tree or mesh networks. A tree structure is required for each solution within a sub-network to aggregate cables from customers to DPs and DPs to a PCP. The whole process is performed rapidly and cost-effectively without violation of technical constraints.

Figure 5.5 shows a network represented on the civil layer. Figure 5.6 shows DP locations and tree formations within each sub-network identified by the optimiser. Cable joints are not required within the nodes shown as dotted circles.

5.4 Problem Formulation for Network Optimisation

A specific planning problem is defined by an access network structure in terms of nodes and links and by the customer demand associated with the termination nodes. The principal aim of access network expansion planning is to determine the component and cable layouts for a given network, so that all customer demand is satisfied, at the minimum cost. A solution to the problem details information relating to all network components and cables. Demand may be satisfied by single or multiple sub-networks.

More formally, an optimisation problem is defined by a set of decision variables, an objective function to be minimised and a set of constraints associated with the decision variables. The set of all decision variables defines a solution to the problem. Decision variables are the location, quantity, type and connection arrangements for all copper system components. The objective function, incorporating the decision variables, allows the evaluation and comparison of alternative solutions. In this case, the objective function determines the actual cost of copper access network installations. There are two types of constraints on solutions. The first set of constraints ensures that all customer demand is satisfied by a solution, whereas the second set of constraints ensures the technical feasibility of a solution.

Expansion planning can be visualised as a mapping process in which one or more sub-networks are mapped onto a given access network duct structure, as depicted in Figure 5.6. During this process, components must be assigned to access nodes and cables to ducts. The total number of different mappings to evaluate is enormous. This is due to the vast range of possible configurations arising from alternative component combinations and the potentially large number of network nodes.

Access network expansion belongs to the class of constrained non-linear combinatorial optimisation problems. It has been recognised that network planning problems are generally difficult to solve because of the size and structure of the search space. The size of the search space for an access network expansion problem of moderate dimensions is already so large that even the fastest currently available computer would need far longer than the estimated age of the universe to search it exhaustively. The objective function and imposed constraints of the optimisation problem are non-linear and the decision variables have a discrete range. In addition, the problem cannot easily be converted into a form where existing solution procedures can be applied and guaranteed to succeed.

All of these factors, considered as a whole, make the successful application of standard optimisation methods very difficult to achieve. In some cases, conventional solution methods cannot be applied successfully. Furthermore, in such situations, a near optimal solution is the best that can be expected and consequently heuristic methods are often the only viable option.

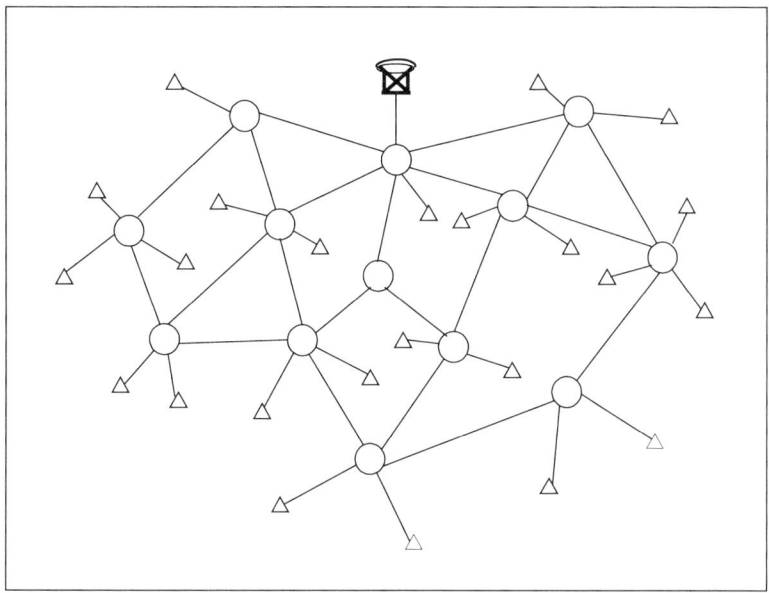

Figure 5.5 Network showing a set of potential DP locations.

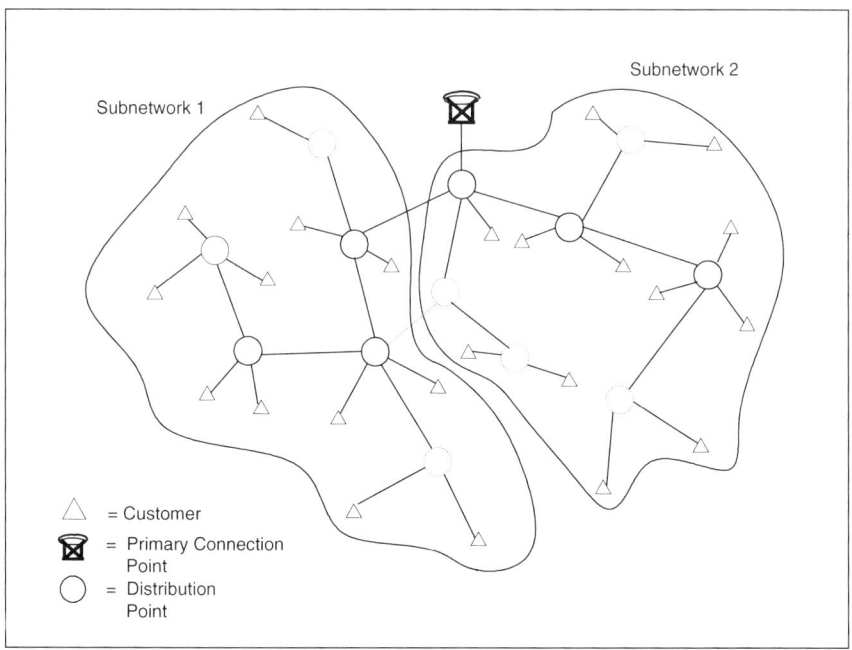

Figure 5.6 Optimised network showing DP locations and tree formation in each sub-network.

A realistic formulation of the problem at the outset is critically important because it directly affects the chances of success. If the problem formulation is oversimplified then solutions will be invalid because they do not address the full or real problem. However, if too many factors are taken into account the problem can be made over complex, to the extent that it may prove to be impossible to comprehend and solve. These conflicting factors must be considered during the algorithm design phase.

5.5   Evolutionary Algorithms

Evolutionary Algorithms (EAs), introduced in chapter one, are part of a family of heuristic search algorithms (Holland, 1975; Goldberg, 1989). EAs have already been used successfully to solve other difficult combinatorial optimisation problems. An important consideration is that EAs do not require special features such as linear or differentiable objective functions and are therefore applicable to problems with arbitrary structures.

Evolutionary methods seek to imitate the biological phenomenon of evolutionary reproduction. The principles of evolutionary computing are based upon Darwin's theory of evolution, where a population of individuals, in this case potential solutions, compete with one another over successive generations, as in 'survival of the fittest'. After a number of generations the best solutions survive and the less fit are gradually eliminated from the population. Instead of working with a single solution, as is the case for most optimisation methods, EAs effectively manipulate a population of solutions in parallel.

As solutions evolve during an optimisation run, the mechanism 'survival of the fittest' is employed to determine which solutions to retain in the population and which to discard. Each solution is evaluated and promoted into the next generation according to its fitness for purpose. The selection is performed stochastically and the probability of a solution surviving by moving into the next generation is proportional to its fitness. This process allows less fit solutions to be selected occasionally, and therefore helps to maintain genetic diversity in the population.

In an EA, the decision variables are normally represented in the form of an encoded string. The individual variables are called genes and the sum of all genes is called a chromosome. A fitness value for each solution is assigned according to a fitness function. In this particular case, the fitness function evaluator module is closely associated with the object model.

The Genetic Algorithm (GA) is a particular type of EC algorithm. In a GA, new solutions are produced after the application of genetic operators. For a canonical GA the fundamental operators are crossover and mutation. Crossover is based upon the principle of genetic mixing during which genes from one chromosome are mixed with those of another. Mutation, on the other hand, is an operator which works on a single chromosome, normally only making small changes to the genetic information. Mutation contributes to the maintenance of genetic diversity, which in turn helps to avoid premature convergence.

Recently, EC methods have attracted a considerable amount of attention because they have been successfully applied to solve a wide range of complex and difficult problems from various disciplines, including telecommunications, logistics, finance, planning, transportation and production.

As far as the authors can ascertain, Paul and Tindle (1996) and Paul *et al.* (1996) developed the first prototype smart Passive Optical Network (PON) planning tool. This tool employs evolutionary computing concepts within the search engine and conclusively demonstrates the viability of EC methods to solve large practical telecommunications network planning problems. The methods initially developed by Paul have been adapted and improved by Poon *et al.* (1998; 1998a) to solve the Copper Planning Problem (CPP).

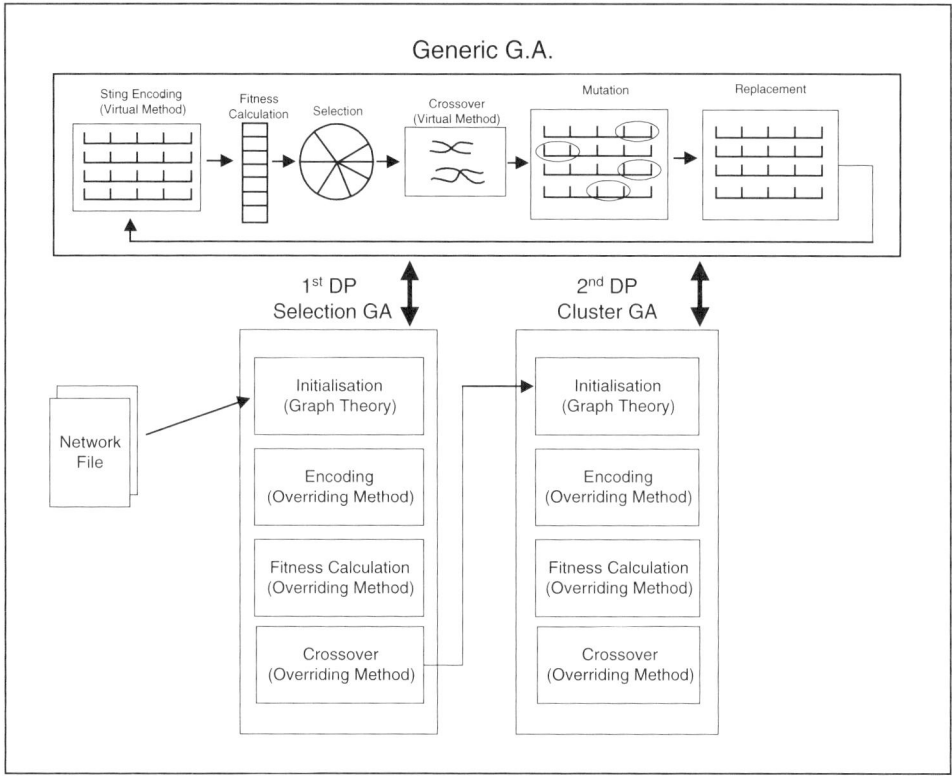

Figure 5.7 Generic GA class structure.

5.5.1 EC Search Engine

As previously outlined, to solve the CPP by an exhaustive search is impractical because of the combinatorial complexity of the problem, non-linear cost function and the number of constraints imposed. Consequently, EC methods have been applied to solve the copper problem. A generic object-oriented GA class has been developed so that code associated with the optimiser can be modified and reused to solve other network problems, as shown in Figure 5.7. A number of basic GA functions are depicted in Figure 5.7 which consist of string encoding, fitness calculation, selection, crossover, mutation and replacement operators. Standard crossover, mutation and replacement strategies are provided by default.

Alternative string encoding schemes and problem specific GA operators can be added by over-riding the existing functions. The GA class also allows the programmer to create a Multiple Objective Genetic Algorithm, (MOGA). In this context, a MOGA is useful because it allows more than one weighted design goal to be built into the cost function.

5.5.2   EA Initialisation

The genes placed in the initial population are not randomly generated as is commonly the case in other GAs. Initial values given to genes are predetermined taking into account geographical information. Furthermore, the initial population is seeded with good solutions by also including various domain specific information. This dramatically reduces the time required to find an acceptable solution. Normally, this produces initial solution costs within 20% of the best known value. The relationship between the object model and search algorithm is shown in Figure 5.8.

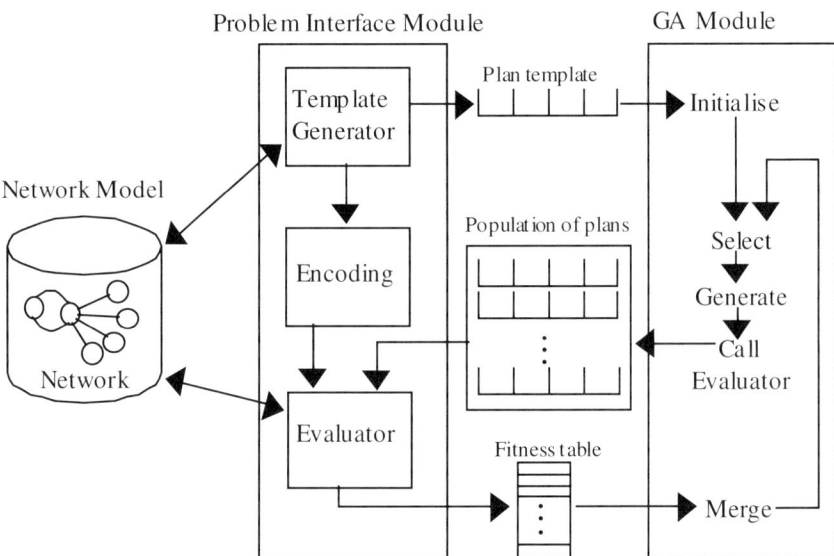

Figure 5.8 Basic structure of smart planning tool.

5.5.3   Main Search Algorithm – Optimiser Kernel

To make the problem solvable by computational optimisation methods, the overall problem has been sub-divided into two sub-problems which are solved sequentially. The kernel of the optimiser consists of two nested genetic algorithms, as shown in Figure 5.7. The general operation of the CPP optimiser is outlined below.

```
REPEAT
    GA1  The first GA identifies the location of DPs and forms
         clusters of customers around DPs.

    GA2  The second GA identifies sub-networks and forms clusters
         of DPs, taking into account the locations and
         maximum allowable demand constraints of DPs.

UNTIL termination criteria is true
```

A more detailed description of GA1 and GA2 is shown below. Solutions generated by the optimiser are encoded numerically. This data is converted into a list detailing the required number and location of network components. An associated cable schedule is also created defining the routes and connections of all multicore cables. From this detailed information, component inventories are produced for ordering and subsequent installation.

GA 1 – Select Customer Distribution Point (DP)

Steps 1 to 5 are only required during initialisation.

1. Employ Floyd's algorithm to create distance and path matrices to describe network connectivity.
2. Specify the population size and number of generation.
3. Define the encoded string length, using the total number of DPs.
4. Encode a binary string to represent selected DPs.
5. Create an initial population by randomly generating 0s and 1s.
6. Form a tree network to connect customers, select both DPs and Primary Connection Points (PCPs).
7. Evaluate the fitness of each string using the cable distances and equipment costs.
8. Define a sub-population for mating using roulette wheel selection.
9. Apply crossover to the binary encoded strings.
10. Perform mutation according to the mutation probability.
11. Replace the old population with the new population.
12. Repeat steps 6 to 12 until the number of generations reaches a predefined terminal value.

GA2 – Select Sub-Network DP Cluster

Steps 1 to 4 are only required during initialisation.

1. Estimate the number of sub-networks required by considering the total demand of the customers.
2. Define the encoded string length, using the estimated number of sub-networks required.

3. Encode an integer string with elements representing the centre of each sub-network.
4. Create an initial population by randomly generating a set of integers between 1 and the number of selected DPs determined by the first DP selection GA.
5. Evaluate the fitness of each string by considering the equipment cost and the cable distance of each customer from the nearest DP specified in the string.
6. Impose a penalty cost if the total demand of a sub-network violates the maximum allowable demand and consider the required distribution of spare capacity.
7. Define a population for mating using roulette wheel selection.
8. Apply crossover on the (integer) encoded string.
9. Perform mutation according to the mutation probability.
10. Replace the old population with the new population.
11. Repeat steps 5 to 11 until the number of generations reaches a predefined value.

The tool also allows the user to enter and manipulate data via the graphical interface. Data is automatically validated and checked within the system before being passed to the optimiser. An important advantage of this tool is that it produces accurate and consistent computer based network records.

5.6 Practical Results

A section of a real solution for a practical network is shown in the smart copper planning tool environment (Figure 5.9). The complete layout of this practical example and the optimised solution can be found in Figures 5.10 and 5.11, respectively. In this case, the tool provided a satisfactory solution with acceptable DP distributions and customer clusters. A total time of less than 13 minutes was taken to solve this 240-node problem. This example clearly demonstrates the effectiveness of EC methods when used to solve practical network problems. Table 5.1 gives a summary of the results and algorithm settings.

A special feature of the GenOSys tool is the ability to control the degree of spare capacity distribution. The distribution of the demand capacity for each sub-network can be adjusted by a special optimiser control variable, the *demand distribution bias factor*.

The optimised solution, given in Figure 5.11, shows the DP layout without a penalty imposed to control the spare capacity distribution. The total service demand for sub-networks 1 and 2 is 70 and 98, respectively.

Table 5.1 Summary of results and EA settings.

	Network results
Number of nodes; a network of realistic size	240
Number of runs	20
Number of generations	3000
Population size	50
Average elapsed time in minutes for each run (in P166 machine)	12
Average deviation from the best known minimum cost	< 1%

Addressing Optimization Issues in Network Planning with Evolutionary Computation 93

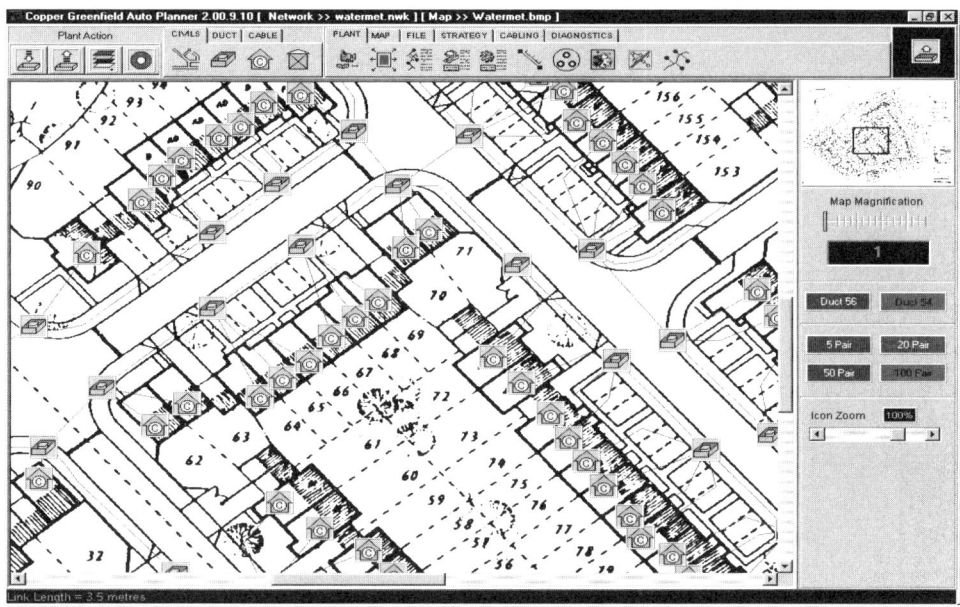

Figure 5.9 Copper planning tool GUI.

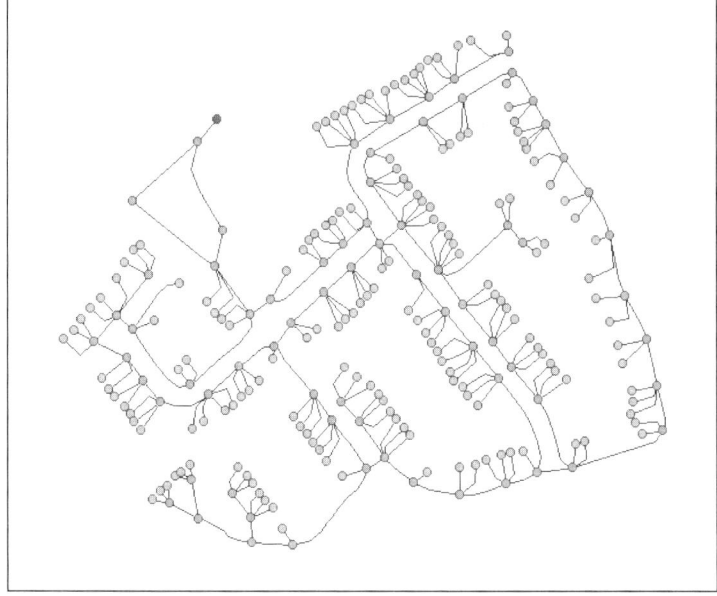

Figure 5.10 A complete layout for a practical network.

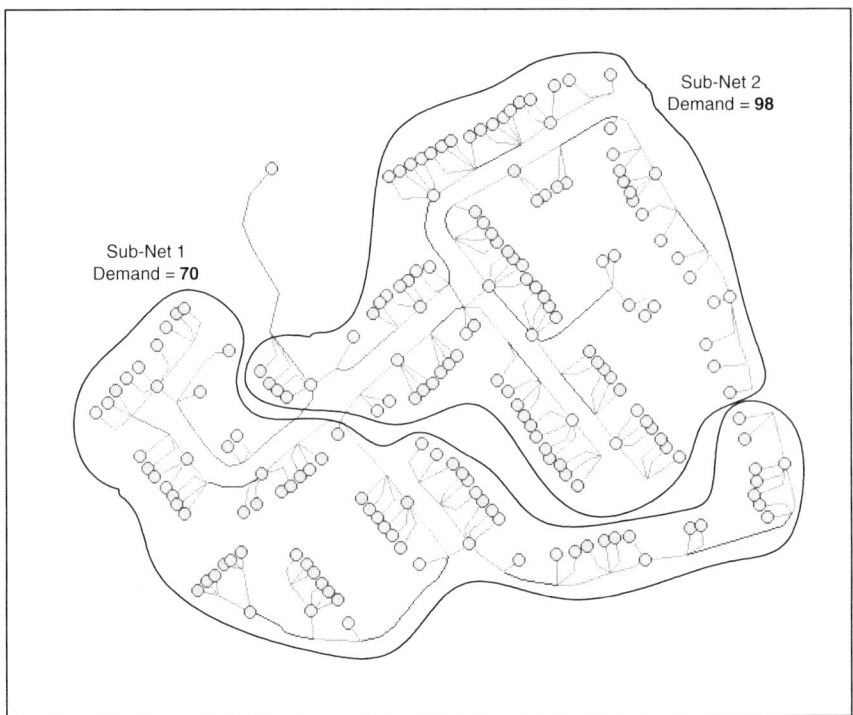

Figure 5.11 Optimised solution showing DP locations, with demand distribution bias factor disabled.

In this case, the total service demand for both networks is 168. As each sub-network can only accommodate 100 cable pairs, at least two sub-networks will be required. The criterion applied to form optimal sub-networks is based solely on the shortest distance between customers to DPs and DPs to a PCP. In this example, the demand distribution for each sub-network may be considered to be out of balance, with demand levels of 70 and 98. This design dictates that there is a relatively large spare capacity for the first network, but only a small amount for the second.

Figure 5.12 shows what could possibly be a more desirable DP layout, exhibiting a more even distribution of capacity. In this modified design, the service demand for sub-networks 1 and 2 becomes 83 and 85. A consequence of this design strategy is that the modified network will require longer cables and be marginally more costly to implement. In this mode of operation, the optimiser employs a MOGA to plan (i) a low cost network, which (ii) also exhibits an evenly distributed capacity between sub-networks.

5.7 General Discussion

As discussed, network planning problems are inherently difficult to solve. An added complication and an important factor to consider is the customer demand forecast level. It is sometimes difficult to apply forecasting methods to derive an accurate and reliable forecast level for planning purposes. Unfortunately, there is inevitably a degree of uncertainty associated with demand level forecasts.

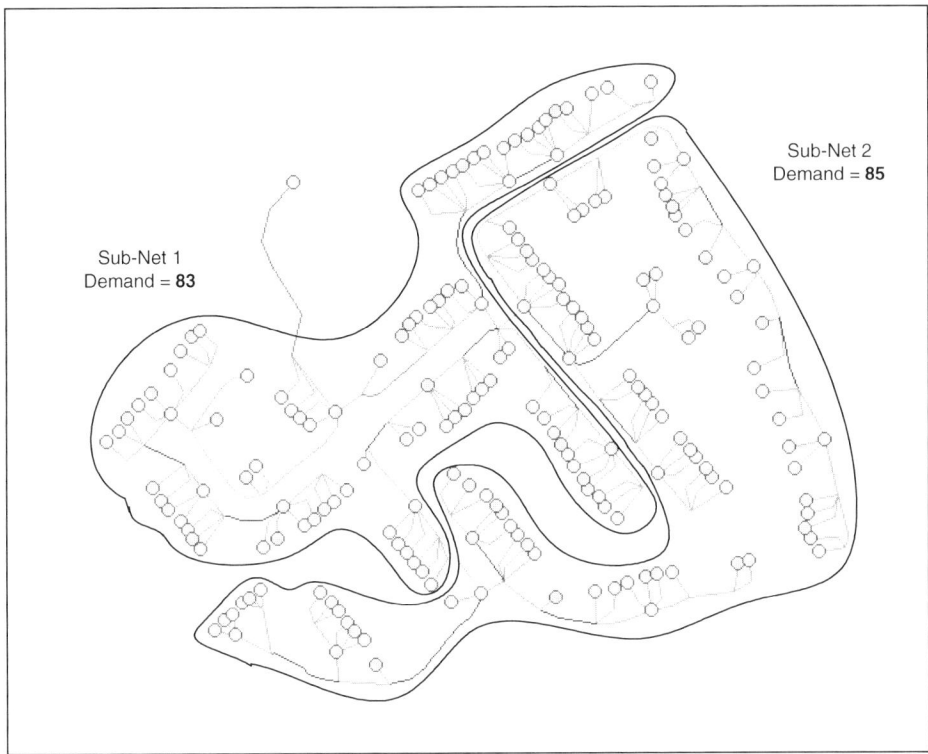

Figure 5.12 Optimised solution showing DP locations with demand distribution bias factor enabled.

The planner can control the level of a spare capacity in a sub-network by adjusting the demand distribution variable. The purpose of this feature is to aid the management of networks that are characterised by uncertain data. Smart computer based tools rapidly generate solutions, enabling the planner to select between alternative network configurations after interactively changing the design criteria.

To summarise, GenOSys has the following principal features:

- GenOSys is able to process networks defined as mesh or tree network structures. Graph theory (Floyd's Algorithm) is used to analyse network structures.
- A highly detailed cost model has been created which takes into account, (i) the co-location of equipment in a single cabinet, (ii) the cost of travelling and setup time. These factors have been shown to have a significant effect upon the structure of network solutions.

- Support for cable 'back feeding' is provided. In some cases, it is necessary that cables connected to customers follow a duct path directed away from the exchange.
- The optimiser uses a hybrid genetic algorithm and is intelligently seeded so that only valid solutions are added to the initial population.
- The system has been developed using Object-Oriented Methods (OMT) and a generic toolkit of software components for optimisations has been created.
- Rapid Application Development (RAD) techniques have been used throughout the development, using the programming language Delphi.
- The tool is easy to set up because users need only define the *demand distribution bias factor* to ensure an even distribution of spare capacity across the network.
- A cable aggregation function is built into the optimiser. An important feature is the ability to design optimal layout schemes for multicore cables.
- Very large and complex networks can be solved rapidly, with more than 500-nodes.
- Planners are now able to experiment and adapt network solutions using sensitivity analysis techniques.
- A MOGA allows the planner to customise network layouts to satisfy specific design criteria.

As this tool has been developed in collaboration with BT the dominant components of the CPP have been retained in the formulation.

5.8 Benefits for Management

GenOSys has been used throughout BT to plan greenfield sites. This tool has now been in use for about a year and reports from users relating to its performance have been very positive. It is evident that productivity levels have improved dramatically, between twenty and hundred fold. Concurrent working methods have also been successfully applied and work backlogs are gradually being removed. In the future, it is anticipated that fewer offices will be needed to process all new BT UK copper planning projects.

Furthermore, Asumu and Mellis (1998) report that, "automatic network optimisation offers an additional improvement in guaranteeing the cost optimal solution (typically 20% better than the average planner) but the main benefits of automatic planning are greater standardisation, consistency and quality, even with less skilled operators". The GenOSys tool is a component within a suite of smart tools under development in BT. They also stated that, "if the key requirements of modularity and data-sharing are adhered to, the results will be a cohesive set of network planning tools, offering massive planning productivity and improvements, the better integration of network planning and management, and the guarantee of optimally designed, high performance networks". Subsequently, they reported major savings in capital expenditure. Finally, 'the set of tools described here have been successfully proven in test-bed trials and the business benefits projected to accrue from their field deployment is about £10 million per annum, or about 10% of the total annual investment made in the copper greenfield network' (Asumu and Mellis, 1998).

For users, the tasks associated with network planning have changed and now mostly involve data preparation and creative interactive design. Computer based interactive design practices have largely eliminated inefficient manual methods. However, a great deal of work remains to be completed because current network records are usually held on paper drawings. The conversion and transfer of all paper records into the digital data domain is now under way.

This work has shown that EA methods can be usefully employed in a practical setting to design telecommunications networks, by demonstrating that they can rapidly produce solutions exhibiting near ideal low cost structures.

Smart planning tools allow designs to be produced more rapidly with automatically created component inventories and orders may be processed via electronic interfaces between suppliers and contractors. These advances make the application Japanese manufacturing methods in the telecommunications sector a real option. Just-in-time manufacturing methods are have the advantage of reducing response times, stock inventory levels, overheads and capital expenditure.

5.9 Strategic Issues

The underlying objective of UK Government policy is to provide customers with the widest range of telecommunications services at competitive prices. Current Government policy has concentrated on creating a framework for infrastructure development and investment. As the process of European liberalisation continues the level of competition is gradually increasing in the telecommunications sector. It is now essential that network operators have the necessary analytical, financial and customer oriented skills to meet the challenge of providing a wide range of reliable, low cost services.

In the telecommunications sector, sustainable competitive advantage can be achieved by applying new 'intelligent' technologies which empower managers, enabling them to analyse complex systems, determine appropriate management strategies and expedite projects and business cases. Implementation of the methods described in this section provides managers with new smart tools to manage complexity and change, thereby providing improved insight into, and control over, network operational issues.

Acknowledgements

This research has been conducted in collaboration with BT Labs. The authors would like to thank BT staff, Don Asumu, Stephen Brewis and John Mellis, for their technical input and support. In addition, the authors would also like to thank the following PhD researchers, Harald Paul and Christian Woeste, for their valuable work on the prototype PON optimisers.

6

Node-Pair Encoding Genetic Programming for Optical Mesh Network Topology Design

Mark C. Sinclair

6.1 Introduction

Telecommunications is a vital and growing area, important not only in its own right, but also for the service it provides to other areas of human endeavour. Moreover, there currently seems to be a demand for an ever-expanding set of telecommunication services of ever-increasing bandwidth. One particular technology that has the potential to provide the huge bandwidths necessary if such broadband services are to be widely adopted, is multi-wavelength all-optical transport networks (Mukherjee, 1997). However, the development of such networks presents scientists and engineers with a challenging range of difficult design and optimisation problems.

One such problem is mesh network topology design. In the general case, this starts with a set of node locations and a traffic matrix, and determines which of the node pairs are to be directly connected by a link. The design is guided by an objective function, often cost-based, which allows the 'fitness' of candidate networks to be evaluated. In the more specific problem of the topology design of multi-wavelength all-optical transport networks, the nodes would be optical cross-connects, the links optical fibres, and the traffic static. Suitable routing and dimensioning algorithms must be selected, with sufficient allowance for restoration paths, to ensure that the network would at least survive the failure of any single component (node or link).

Telecommunications Optimization: Heuristic and Adaptive Techniques, edited by D.W. Corne, M.J. Oates and G.D. Smith
© 2000 John Wiley & Sons, Ltd

In previous work, Sinclair (1995; 1997) has applied a simple bit-string Genetic Algorithm (GA), a hybrid GA and, with Aiyarak and Saket (1997) three different Genetic Programming (GP) approaches to this problem. In this chapter, a new GP approach, inspired by edge encoding (Luke and Spector, 1996), is described (a shorter version of this chapter was first published at GECCO'99 (Sinclair, 1999)). It was hoped that this would provide better results than the best previous GP approach, and perhaps even prove competitive with GAs. The eventual aim of the author's current research into GP encoding schemes for mesh network topologies is to provide a more scalable approach than the inherently non-scalable encoding provided by bit-string GAs.

The chapter is structured as follows. Section 6.2 presents some of the previous work on Evolutionary Computation (EC) for network topology design; section 6.3 provides a more detailed description of the problem tackled, including the network cost model; and section 6.4 summarises two previous solution attempts. In section 6.5 the node-pair encoding GP approach is described; then in section 6.6 some experimental results are outlined; and finally, section 6.7 records both conclusions and suggestions for further work.

6.2   EC for Network Topology Design

Over the years, at least 30 papers have been published on EC approaches to network topology design. Two of the earliest are by Michalewicz (1991) and Kumar *et al.* (1992). Michalewicz uses a two-dimensional binary adjacency matrix representation, and problem-specific versions of mutation and crossover, to evolve minimum spanning tree topologies for computer networks. Kumar *et al.* tackle three constrained computer network topology problems, aiming for maximum reliability, minimum network diameter or minimum average hop count. Their GA is bit-string, but uses problem-specific crossover, as well as a repair operator to correct for redundancy in their network representation.

For optical network topology design, in addition to the papers by Sinclair, mentioned above, there is the work of Paul *et al.* (1996) and Brittain *et al.* (1997). Both these groups of authors have addressed constrained minimum-cost Passive Optical Network (PON) topology design for local access. However, while problem-specific representations are used by both, as well as problem-specific genetic operators by Paul *et al.*, only Brittain *et al.* provide full details of their algorithm. This uses a two-part chromosome comprising a bit-string and a permutation, with each part manipulated with appropriate standard operators.

Other recent work of interest includes Dengiz *et al.* (1997; Chapter 1, this volume) on a hybrid GA for maximum all-terminal network reliability, Ko *et al.* (1997a), who use a three-stage GA for computer network topology design, routing and capacity assignment; and Pierre and Legault (1998), who use a bit-string GA for computer mesh network design.

6.3   Problem Description

Given the locations of the n nodes (optical cross connects) and the static traffic requirements, t_{ij}, between them, the problem is to determine which of the $n(n-1)/2$ possible bi-directional links (optical fibres) should be used to construct the network. The number of possible topologies is thus $2^{n(n-1)/2}$. To illustrate, Figure 6.1 shows a 15-node network design problem, for which an example mesh topology design is given in Figure 6.2.

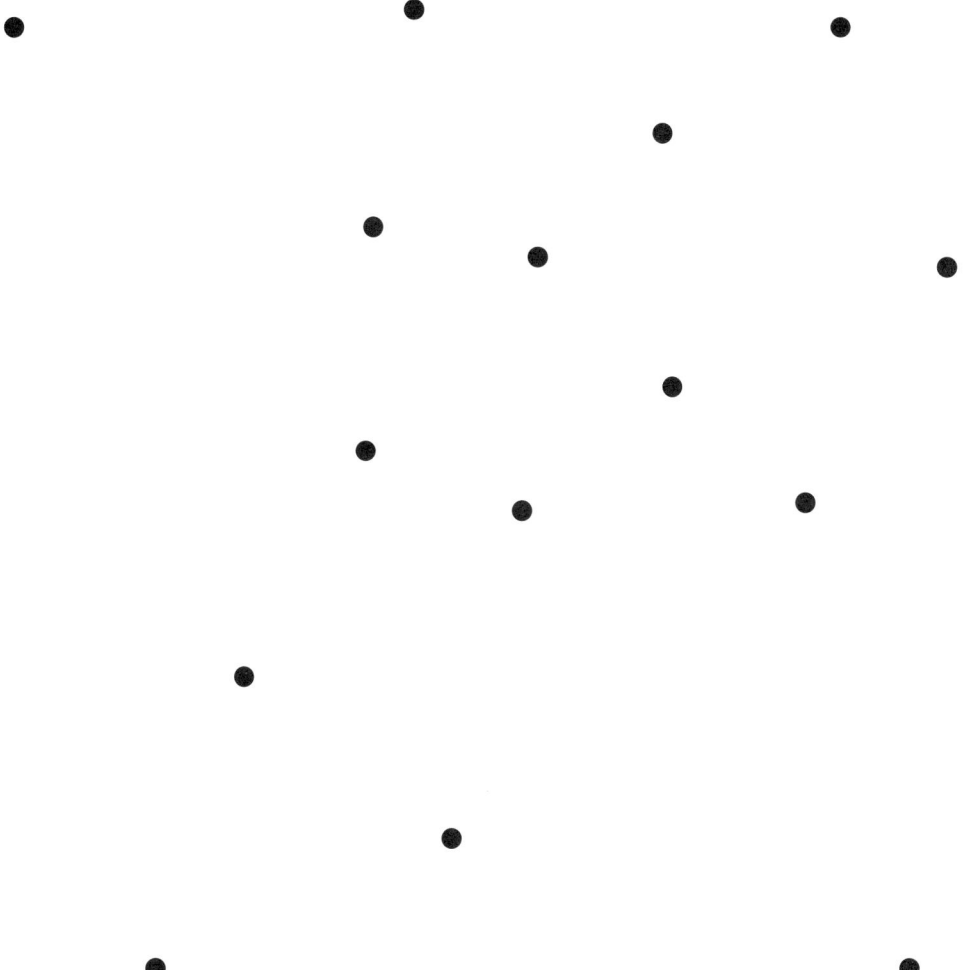

Figure 6.1 Example network design problem.

The cost model used to guide the design was developed by Sinclair (1995) for minimum-cost topology design of the European Optical Network (EON) as part of the COST 239 initiative (O'Mahony *et al.*, 1993). It assumes static two-shortest-node-disjoint-path routing (Chen, 1990) between node pairs, and that a reliability constraint is used. This is to ensure that there are two, usually fully-resourced, node-disjoint routes between node pairs, thus guaranteeing the network will survive the failure of any single component.

To determine the cost of a given topology, separate models for both links and nodes are required. The intention is to approximate the relative contribution to purchase, installation and maintenance costs of the different network elements, while ensuring the model is not too dependent on the details of the element designs, nor too complex for use in the 'inner loop' of a design procedure.

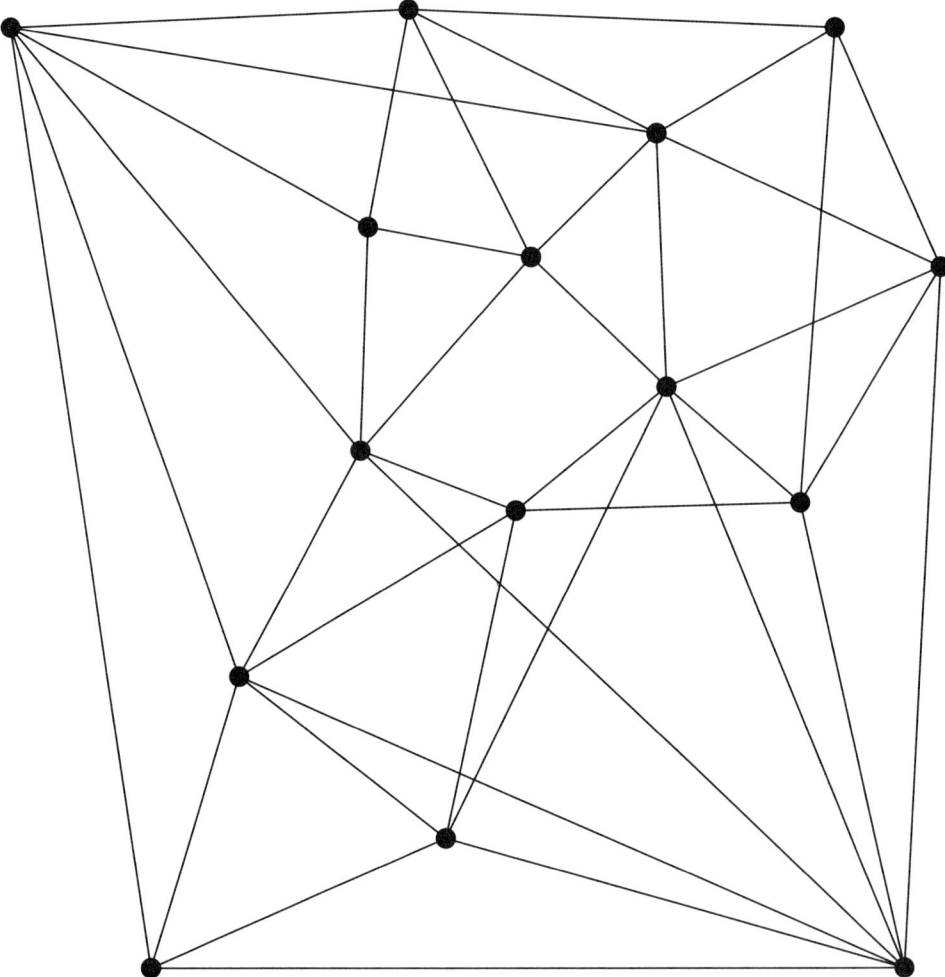

Figure 6.2 Example mesh topology.

First, the two-shortest-node-disjoint routes are determined for all node pairs. In this, the 'length' of each path is taken to be the sum of the contributing link weights, with the weight of the bi-directional link between nodes *i* and *j* given by:

$$W_{ij} = 0.5N_i + L_{ij} + 0.5N_j \qquad (6.1)$$

where N_i and N_j are the *node effective distances* of nodes *i* and *j*, respectively (see below) and L_{ij} is the length of link (*i,j*) in km. Then, the link carried traffic is determined for each link by summing the contributions from all the primary and restoration routes that make use of it. The restoration traffic is weighted by a parameter K_R, but restoration routes are usually taken to be fully-resourced (i.e. $K_R = 1.0$), and thus are assumed to carry the same traffic as the corresponding primary routes, i.e. the traffic they would be required to carry if their primary route failed. The capacity of link (i,j) is taken to be:

$$V_{ij} = \operatorname{ceil}_{K_G}(K_T T_{ij}) \qquad (6.2)$$

where T_{ij} is the carried traffic in Gbit/s on the link, and $\operatorname{ceil}_x()$ rounds its argument up to the nearest x – here, the assumed granularity of the transmission links, K_G (say 2.5 Gbit/s). The factor of K_T (say 1.4) is to allow for stochastic effects in the traffic. The cost of link (i,j) is then given by:

$$C_{ij} = V_{ij}^\alpha L_{ij} \qquad (6.3)$$

where α is a constant (here taken to be 1.0). Increasing capacity necessarily implies increased cost due, for example, to wider transmission bandwidth, narrower wavelength separation and/or increasing number and speed of transmitters and receivers. With $\alpha = 1$, a linear dependence of cost on capacity is assumed, but with $\alpha < 1$, the cost can be adjusted to rise more slowly with increases in the link capacity. The linear link length dependency approximates the increasing costs of, for example, duct installation, fibre blowing and/or the larger number of optical amplifiers with increasing distance.

Node effective distance was used as a way of representing the cost of nodes in an optical network in equivalent distance terms. It can be regarded as the effective distance added to a path as a result of traversing a node. By including it in link weights (equation 6.1) for the calculation of shortest paths, path weights reflect the cost of the nodes traversed (and half the costs of the end nodes). As a result, a longer geographical path may have a lower weight if it traverses fewer nodes, thereby reflecting the relatively high costs of optical switching. The node effective distance of node i was taken to be:

$$N_i = K_0 + n_i K_n \qquad (6.4)$$

where n_i is the degree of node i, i.e. the number of bi-directional links attached to it. The constants K_0 and K_n were taken to be, say 200 km and 100 km, respectively, as these were judged to be reasonable for the network diameters of 1400–3000 km used here. Node effective distance thus increases as the switch grows more complex. Node capacity is the sum of the capacities of all the attached links, i.e.

$$V_i = \sum_j V_{ij} \qquad (6.5)$$

where V_i is the capacity of node i in Gbit/s. The node cost was taken to be:

$$C_i = 0.5 N_i V_i \qquad (6.6)$$

The cost is thus derived as if the node were a star of links, each of half the node effective length, and each having the same capacity as the node itself. Further, if all the links attached to a node are of the same capacity, the node costs would grow approximately with the square of the node degree, corresponding, for example, to the growth in the number of crosspoints in a simple optical space switch. Overall, the relative costs of nodes and links can be adjusted by setting the values of K_0 and K_n appropriately.

The network cost is then taken to be the sum of the costs of all the individual links and nodes comprising the network. However, to ensure the reliability constraint is met, the actual metric used also includes a penalty function. This adds P_R (say 250,000) to the network cost for every node pair that has no alternative path, and P_N (say 500,000) for every node pair with no routes at all between them. This avoids the false minimum of the topology with no links at all, whose cost, under the link and node cost models employed, would be zero. The values selected for P_R and P_N must be sufficiently large to consistently ensure no penalties are being incurred by network topologies in the latter part of an evolutionary algorithm run. However, values substantially larger than this should be avoided, as they may otherwise completely dominate the costs resulting from the topology itself, and thereby limit evolutionary progress.

6.4 Previous Approaches

Two previous attempts at optical mesh topology design using the same cost model are outlined below. An additional attempt with a hybrid GA (Sinclair, 1997) has been excluded from consideration due to the highly problem-specific nature of its encoding and operators.

6.4.1 Genetic Algorithms

In 1995, Sinclair used a bit-string GA to tackle two variants of the EON: a small illustrative problem consisting of just the central nine nodes, as well as the full network of ~20 nodes. The encoding simply consisted of a bit for each of the $n(n-1)/2$ possible links. Clearly, this representation scales poorly as problem size increases, as the bit string grows $O(n^2)$, rather than only with the number of links actually required, m, i.e. $O(m) \approx O(n)$ provided node degree remains approximately constant with network size. The genetic operators were single-point crossover (probability 0.6) and mutation (probability 0.001); and fitness-proportionate selection (window-scaling, window size 5) was used. The GA was generational, although an elitist strategy was employed.

For the full EON, with a population of 100 and ten runs of less than 48,000 trials each, a network design was obtained with a cost (at 6.851×10^6) which was some 5.1% lower than an earlier hand-crafted design (O'Mahony et al., 1993). In addition, the GA network design was of superior reliability, at least in terms of the reliability constraint.

6.4.2 Connected-Nodes GP

More recently, Aiyarak et al. (1997) described three different approaches to applying GP to the problem. The most successful of these, Connected Nodes (CN), encodes the design as a program describing how the network should be connected. Only one terminal and one function are required. The terminal is the ephemeral random integer constant $\Re$, over the range of *n* node identification numbers (ids). The function is con, which takes two arguments, representing the ids of two nodes that are to be connected in the network topology represented by the program (note: in Aiyarak *et al.*, 1997), this function was called connect2). As each con function must also provide a return value, it simply returns its first argument. However, if its two id arguments are equal, it does nothing apart from returning their value. The program tree is evaluated depth-first: children **before** parents. For example, Figure 6.3 shows a small target network and Figure 6.4 a corresponding 'hand-crafted' connected-nodes GP tree. Executing the tree would connect those node pairs indicated above each of the con functions: (4,3), (1,4), (1,3), and so on, resulting in the desired network. Clearly, with this representation, minimum program size grows only with the number of links ($O(m) \approx O(n)$), as only a single con and at most two terminals are required for each node pair connected. In Aiyarak *et al.* (1997), the only genetic operator used was crossover, and tournament selection was employed.

Experimental results were obtained for both the nine central nodes, as well as the full EON, establishing the superiority of the CN approach over the two other GP encoding methods. In addition, the network design cost obtained with CN (6.939×10^6) was only some 1% above Sinclair's earlier GA result (1995). However, the computational burden of CN was far greater: the best design was obtained using two runs of 500,000 trials each on a population of 1000 individuals.

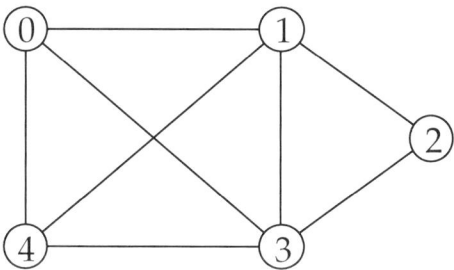

Figure 6.3 Target network.

6.5 Node-Pair Encoding GP

The two node-pair encoding (NP) GP approaches presented in this chapter are based on an earlier edge encoding for graphs described by Luke and Spector (1996). Their approach evolved a location-independent topology, with both the number of nodes and their interconnections specified by the GP program. Here, however, the number of nodes is fixed in advance, and the GP program is only required to specify the links.

As in the CN approach, the program again describes how the network should be connected. However, for NP, the program tree is evaluated top-down: children **after** parents. The functions operate on a node pair (represented by two node ids, a and b), then pass the possibly-modified node-pair to each of their children. Similarly, terminals operate on the node pair passed to them. Overall execution of the tree starts with (a,b) = (0,0).

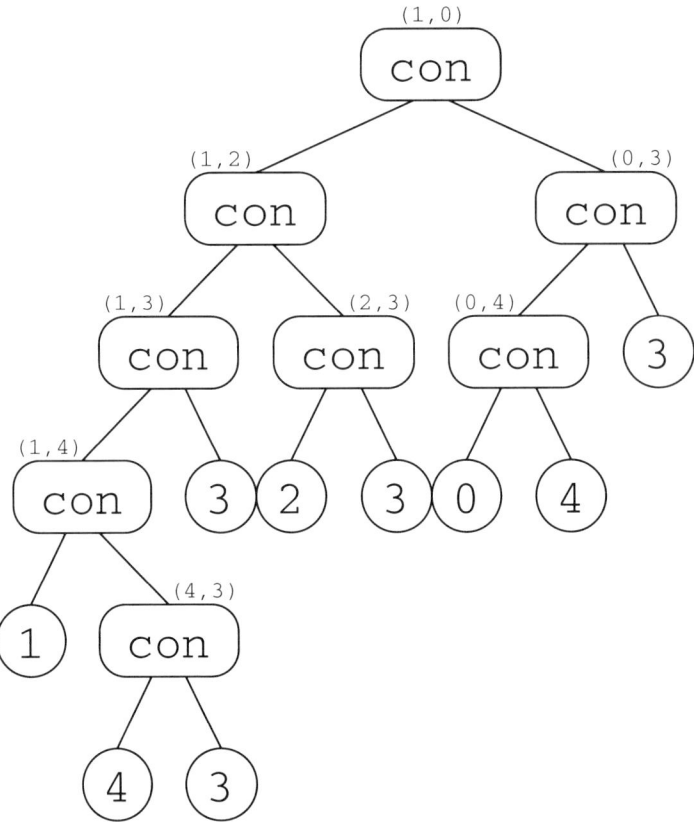

Figure 6.4 CN program for the target network.

This use of a current node pair is similar to the concept of the current edge in Luke and Spector (1996), but without any structure equivalent to their node stack. Two different variants of node-pair encoding are described here (NP1 and NP2); the difference is due to the arities of their functions/terminals. These are given, for both variants, in Table 6.1.

The function/terminal add adds a link between the current node pair (a,b), provided a ≠ b; whereas terminal cut removes the link between the current node pair, if there is one. To allow the current node pair to change, thus moving the focus of program execution to a different point in the network, five of the functions (da, ia, ia2, ia4, ia8) modify the value of variable a. The choice of 1, 2, 4 and 8 for the values by which a may be increased was motivated by minimum-description length considerations. By providing the reverse function (rev), similar modifications can be made, in effect, to variable b. The double (dbl) and triple (tpl) functions allow operations on node pairs that are numerically close together to be accomplished using a smaller depth of tree. For example, in Figure 6.5, the tpl in the lower left of the diagram, which refers to node pair (2,1), enables links (2,1), (3,1) and (4,1) to be added by its subtrees. Without this tpl, an equivalent subtree based on just adds, ia?s and nops would be two levels deeper. Finally, terminal nop does nothing to its node pair, thus allowing a program branch to terminate without either adding or removing a link.

Table 6.1 NP function/terminal sets.

Abbr.	Arity NP1	Arity NP2	Description
rev	1	1	Reverse: (a,b) = (b,a)
da	1	2	(decrement a) mod n
ia	1	2	(increment a) mod n
ia2	1	2	(increase a by 2) mod n
ia4	1	2	(increase a by 4) mod n
ia8	1	2	(increase a by 8) mod n
dbl	2	2	Double: pass current node pair to both children
tpl	3	3	Triple: pass current node pair to all three children
add	1	0	Add link (a,b)
cut	0	0	Cut link (a,b)
nop	0	0	Do nothing

In NP1, add is a function and has a single child, whereas in NP2 it is a terminal. Further, in NP1, the functions that modify variable a all have just one child, whereas in NP2, they have two. The effect of these differences is to encourage taller, narrower program trees in NP1, with further operations following the addition of a link, and shallower, broader trees in NP2, with link addition ending a program branch. This is illustrated by the 'hand-crafted' program trees, for the target network of Figure 6.3, given in Figures 6.5 and 6.6 using NP1 and NP2 respectively. In both diagrams, the current node pair is indicated above each of the add function/terminals. It should be noted that the tree in Figure 6.5 has a depth of 10 and uses 28 program nodes, whereas that in Figure 6.6 is only 7 levels deep and uses just 24 function/terminals.

6.6 Experimental Results

For the relative assessment of the different approaches to optical mesh network topology design, it was decided to use the full 20-node EON (Sinclair, 1995) and five additional network design problems. For the latter, the initial node locations and traffic requirements were generated using the approach described by Griffith *et al.* (1996), although further modified to ensure reasonable node separations. Each network has 15 nodes, covers a 1000km ×1000km area and carries an overall traffic of 1500 Gbit/s. Due to their smaller size, reduced penalties of $P_R = 50,000$ and $P_N = 100,000$ were used for the 15-node networks, rather than the larger values suggested in section 6.3, which were those applied for the EON.

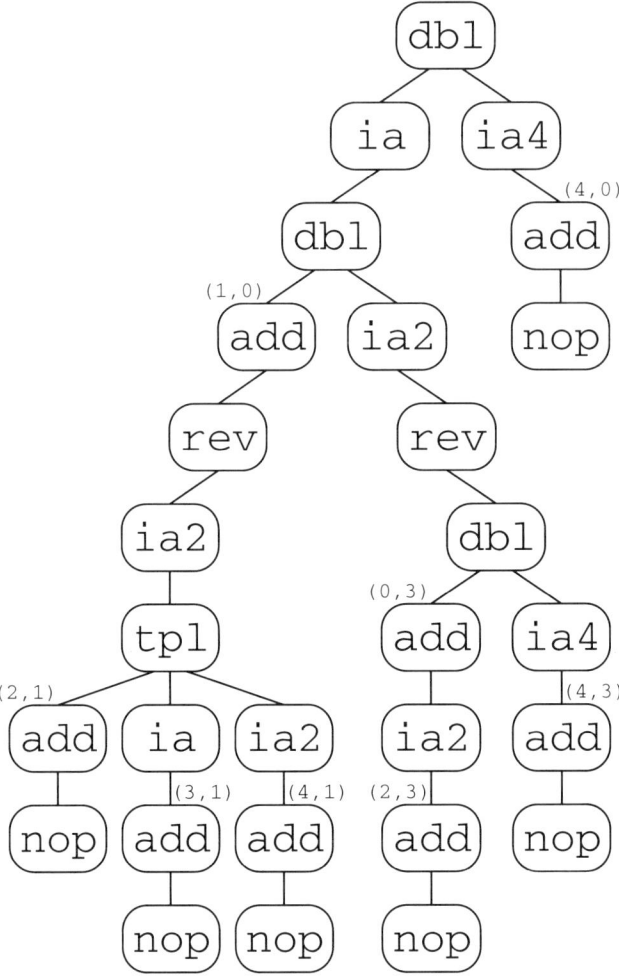

Figure 6.5 NP1 program for target network.

6.6.1 Genetic Algorithms

The bit-string GA developed by Sinclair (1995), and reviewed in section 6.4.1, is referred to here as GA1 (with a maximum of 15,000 trials) when used on the five 15-node problems, and as GA4 (with a maximum of 50,000 trials) on the EON (Table 6.2). In addition, in an attempt to improve on Sinclair's results, both a higher mutation rate and a larger population size were also tried for both the 15-node problems and the EON. From a few trials runs, reasonably successful algorithms with a higher mutation rate (GA2) and a larger population size (GA3) were discovered for the 15-node problems. Also, for the EON, algorithms with an increased mutation probability of 0.01, and increased population sizes of 200, 500 and 1000 (the latter three still with mutation probability 0.001) were tried, without good results, for runs of up to 1,000,000 trials. However, a reasonable algorithm was found with mutation probability 0.002 (GA5).

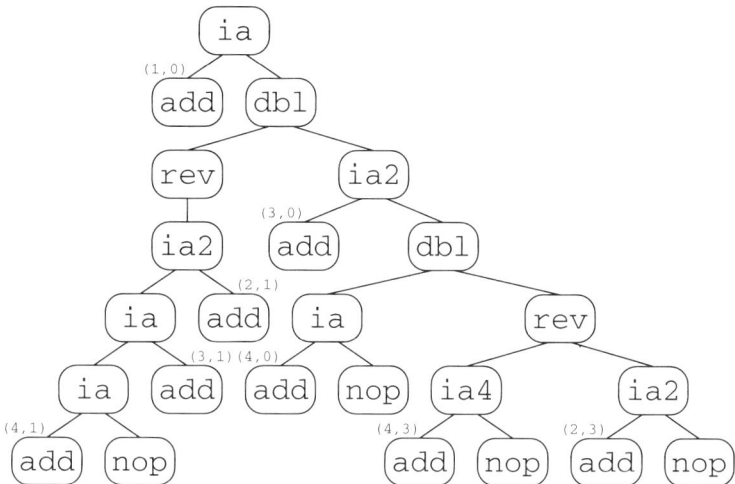

Figure 6.6 NP2 program for target network.

Table 6.2 Genetic algorithm parameters.

Algorithm	Population Size	Maximum Trials	Mutation Probability
GA1	100	15,000	0.001
GA2	100	25,000	0.01
GA3	700	210,000	0.001
GA4	100	50,000	0.001
GA5	100	100,000	0.002

To allow a statistical comparison of the genetic algorithms, GA1, GA2 and GA3 were applied to all five 15-node test problems, plus GA4 and GA5 to the EON. In every case, ten runs were made with different pseudo-random number seeds. GENESIS v5.0 (Grefenstette, 1990) was used for the implementation. A non-parametric median test (Sprent, 1992) was applied to establish if there were significant differences in the medians from the different GAs. The results for the EON are given in Table 6.3; both the best and median of each set of runs is recorded, as well as the significance levels for the median differences.

For the 15-node problems, GA3 is the best algorithm, providing not only all five of the best individual runs, but also the best median for four of the problems (with very highly significant differences). In addition, for the EON (Table 6.3), GA5 provided the better median (not a significant difference) although the best individual run still used GA4. However, it should be noted that both these improvements over Sinclair's GA1/GA4 (1995) were achieved at the cost of a larger number of trials for both network sizes (Table 6.2).

Table 6.3 Results for EON.

Algorithm	Best ($\times 10^6$)	Median ($\times 10^6$)	Significance Level (%)	
GA4	6.851	6.926	GA5<GA4	–
GA5	6.856	6.891	CN2<CN1	1.15
CN1	6.961	7.002	NP2<NP1	1.15
CN2	6.888	6.930	GA5<CN2	–
NP1	6.898	6.962	GA5<NP2	–
NP2	6.862	6.900	NP2<CN2	–

6.6.2   Connected-Nodes GP

For the connected-nodes GP approach, a tournament size of 4 was used, rather than the large value of 30 used in previous work (Aiyarak *et al.*, 1997). Also, as well as Aiyarak *et al.*'s original function set (CN1), a second, CN2, was introduced. CN2, in addition to the `con` function, also provides a `dis` function which takes two node id arguments, and removes any link between them. For the 15-node problems, the population size was 700 and the maximum number of trials 210,000; for the EON, 1000 and 500,000, respectively (after Aiyarak *et al.* (1997)). GP parameters are given above or followed Koza (1992).

Both CN1 and CN2 were applied to all five 15-node test problems and the EON. As with the GAs, ten runs were made with different seeds. The implementation used lil-gp v1.02 (Zongker and Punch, 1995). The results for the EON are again given in Table 6.3. On the 15-node problems, the CN2 connected nodes encoding provided the best individual runs in three of the five networks, and the best median in three. For the EON, CN2 also provided both the best individual run and the best median (highly significant difference). Thus including the new `dis` function, whose role in *removing* links may seem counter-intuitive, has resulted in a marked improvement in the results. This can perhaps be attributed to increased redundancy in the encoding, allowing greater freedom in tree composition.

6.6.3   Node-Pair Encoding GP

For the node-pair encoding GP approach, exactly the same parameters were used as for the corresponding CN runs. The results of applying both NP1 and NP2 encodings to the EON are recorded in Table 6.3. On the 15-node problems, the NP2 node-pair encoding provided the best individual runs in four of the five networks, and the best median in one (highly significant difference). For the EON, NP2 also provided both the best individual run and the best median (highly significant difference). These differences perhaps arise from the NP2 encoding's ability to use shallower trees, giving an advantage in tree composition, as the usual maximum tree depth of 17 was imposed on all runs (Koza, 1992).

6.6.4   Comparison

The results of the three leading approaches, GA3/GA5, CN2 and NP2, are summarised in Tables 6.4 and 6.5 for the 15-node problems, and in Table 6.3 for the EON. For the smaller networks, GA3 has shown itself to be the best approach, both in terms of individual runs and median differences. Nevertheless, one of the five best individual runs were performed using NP2 (Network 3), and for the other four networks the NP2 results were very close to those of GA3. For the larger EON, however, there is no overall best algorithm in terms of median differences, although both GA5 and NP2 performed well. The best individual run is still that from Sinclair (1995), using GA4, and is shown in Figure 6.7. However, execution time for all the different approaches was almost entirely determined by the number of trials, due to the time-consuming fitness assessment. Consequently, while all the runs on the 15-node networks required approximately the same time, the GP runs on the EON took some five times longer than the GA.

The best network cost for the initial part of typical runs of GA5, CN2 and NP2 on the EON are illustrated in Figure 6.8. The best individual in the initial random GA5 population has a reasonable cost, but those of the GP approaches, especially NP2 are very high. This is entirely due to the different initialisation approaches. For GA5, individual bits, corresponding to links, are set at random, with probability 0.5; this results in at least some reasonable topologies. For the GP approaches, the 'standard' ramped half-and-half initialisation with small trees (2–6 levels) (Koza, 1992) creates networks with very few connections. Consequently, many thousands of trials are needed by CN2/NP2 simply to evolve networks with costs as low as those initially present in the GA5 population. However, for a narrower range of network costs, the full runs can be seen in Figure 6.9. The GA ceases to make further progress at 70,000 trials or so, but by allowing a sufficiently large number of trials, the cost of this NP2 run (almost) matches that of GA5.

Table 6.4 Comparative results for problems 1–5.

Problem	GA3		CN2		NP2	
	Best ($\times 10^6$)	Median ($\times 10^6$)	Best ($\times 10^6$)	Median ($\times 10^6$)	Best ($\times 10^6$)	Median ($\times 10^6$)
1	4.935	4.940	4.944	4.959	4.937	4.950
2	4.737	4.744	4.743	4.750	4.743	4.752
3	4.404	4.407	4.404	4.416	4.404	4,414
4	4.387	4.589	4.589	4.597	4.590	4.595
5	4.416	4.418	4.417	4.428	4.417	4.436

Table 6.5 Median difference significance levels for problems 1–5.

Problem	Significance Level (%)		
	GA3<CN2	GA3<NP2	NP2<CN2
1	0.05	–	–
2	1.15	1.15	–
3	–	–	–
4	1.15	0.05	–
5	1.15	0.00	–

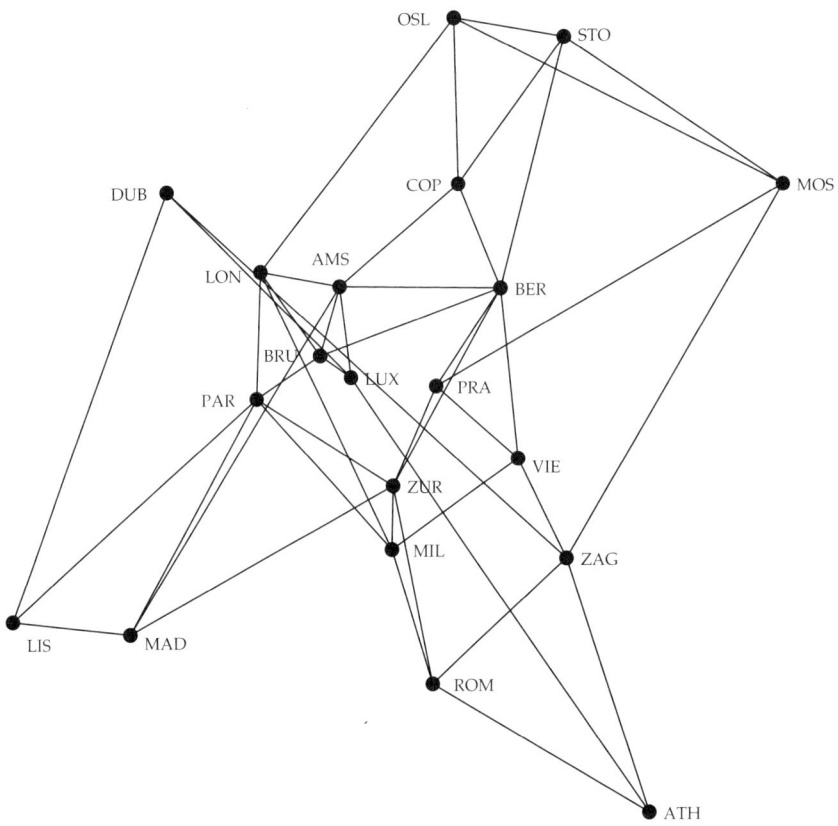

Figure 6.7 Best EON topology (using GA4).

6.7 Conclusions and Further Work

In this chapter two variants (NP1 and NP2) of node-pair encoding genetic programming (GP) for optical mesh network topology design have been described. In addition, a new variant (CN2) of the earlier connected-nodes encoding has been presented. Experimental results have shown that both NP2 and CN2 are comparable in design quality to the best bit-string genetic algorithms (GA) developed by the author, particularly for the larger network size examined, although in that case requiring much greater computational effort. While node-pair encoding has only been applied to optical network design here, it may also be useful for graph construction in other applications, such as the interconnection of artificial neural networks.

Future work with node-pair encoding could include investigating its sensitivity to the choice of cost model, starting node pair, initial population or GP parameters, such as tree depth. In addition, the author hopes to develop further encodings that are both computationally efficient and scale well with network size. In particular, it is anticipated that removing or reducing dependency on explicit node ids will result in more powerful encodings. Also, for regular or near-regular networks, ADFs (Koza, 1994) may well provide a mechanism to capture and succinctly express network regularities.

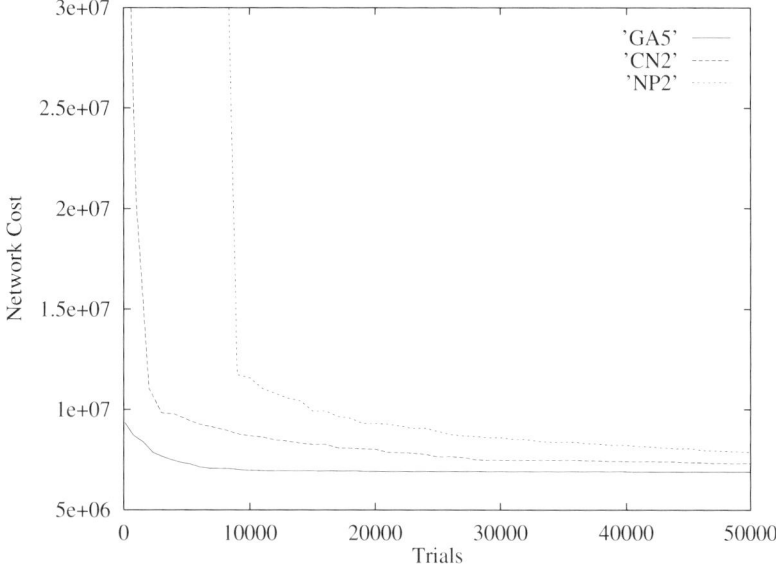

Figure 6.8 Best network cost for typical runs of GA5, CN2, NP2 on EON (initial part of run).

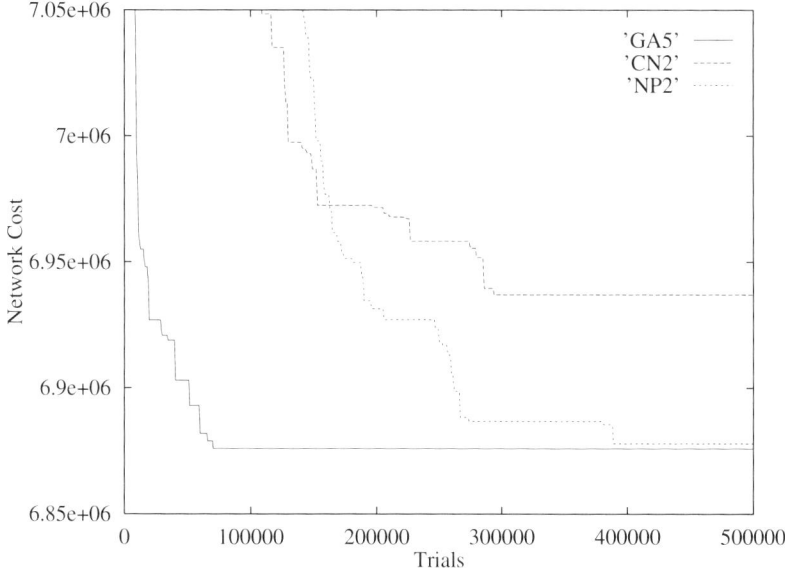

Figure 6.9 Best network cost for typical runs of GA5, CN2, NP2 on EON (full run).

Acknowledgements

The node-pair encodings used in this chapter are based on an earlier encoding developed by Christos Dimitrakakis as part of an MSc in Telecommunication and Information Systems, supervised by the author.

The author is grateful to Rob Smith (University of the West of England) and Brian Turton (Cardiff University) for their encouragement to further improve the bit-string GA results obtained using GA1/GA4.

7

Optimizing the Access Network

David Brittain and Jon Sims Williams

7.1 Introduction

The telecommunications access network is the section of the network that connects the local exchange to the customers. At present most of the access network is composed of low bandwidth copper cable. Electronic communications are becoming an essential feature of life both at home and at work. The increasing use of applications that require larger bandwidths (such as the internet and video on demand) are making the copper infrastructure inadequate. These demands could be met using optical fibre technologies.

At present, optical fibre is principally used in the trunk network to provide connections between exchanges, and to service customers with high bandwidth requirements. Point-to-point links are used. In the access network, customers generally have lower bandwidth requirements and also require a cost-effective service. For them, point-to-point links are not viable as both the link and the dedicated exchange-based equipment are expensive. A network based on point-to-multi-point links provides a high capacity service, together with shared equipment costs. Point-to-multi-point networks can be created using optical fibre and passive splitting devices. They are known as Passive Optical Networks (PONs).

With this new architecture and a commercial environment that is increasingly competitive, there is a need for improved methods of network planning to provide cost-effective and reliable networks. Currently, much of the access network planning that takes place is performed by hand which may mean networks are not as cost-effective as they could be. There have been many attempts to produce systems that optimise the topology of copper networks, although they often make use of much simplified network models. The task of designing optical networks is significantly different to that for copper networks and little work has been published in this area. Most access networks are installed gradually, over time, and so a dynamic approach to planning is also required.

This chapter describes a methodology for optimising the installation cost of optical access networks. The plans produced specify where plant should be installed, the component sizes and when the plant should be installed. In fact, the system is not limited to optical networks and could easily be extended to plan most types of access network. Two optimisation methods are presented: genetic algorithms and simulated annealing; the first of these is found to have the best performance on the problems considered.

The chapter starts with a presentation of the problem of network planning by describing the network technologies involved and the decision variables. This is followed by a brief review of Genetic Algorithms (GA) and simulated annealing (SA), and a description of how they can be applied to designing a network; some results and comparisons are then presented. Finally, the issue of demand uncertainty in the planning process is discussed.

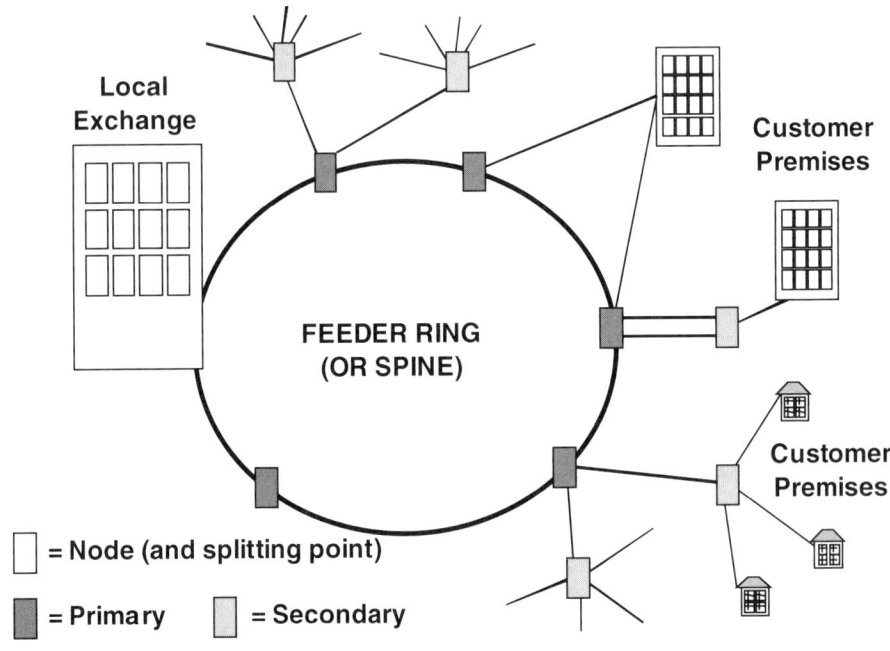

Figure 7.1 A schematic of a generic optical access network. If the nodes contain splicing units then it is a point-to-point network and if they contain splitting units it is a Passive Optical Network (PON).

7.1.1   Access Network Architectures

The simplest possible optical fibre architecture for the access network is a point-to-point (PTP) network in which each customer has a dedicated fibre pair from the exchange. These networks still have the same topology as traditional tree-and-branch topologies (see Figure 7.1). There are cost advantages in aggregating a number of fibres into a single cable, both because of the material cost of the cable and because of the cost of installation. Thus a larger single cable is split into pairs or groups of fibres at remote sites. This problem will be called the *point-to-point planning problem*.

The disadvantage of the point-to-point network architecture is that the cost is currently prohibitively high for *domestic* telecommunications services. A large amount of expensive exchange equipment must be duplicated for every customer (including the laser source and receiver). This has lead to the development of Passive Optical Networks (PONs) (Homung et al., 1992). These allow exchange-based equipment and a large fraction of the cabling cost to be shared between customers. Figure 7.1 shows a schematic of a PON. The topology of the network is based on two remote sites between the exchange and customer, the primary and secondary nodes, at which the optical signal is split. The splitters are passive devices and split the signal on the input across the output fibres. Commonly, splitters are available with a limited number of split ratios: 1:2, 1:4, 1:8, 1:16 and 1:32. The attenuation of the splitter is proportional to the split; this leads to a constraint on the combined split-level of the primary and secondary nodes. The constraint used in the work described in this paper is that the product of the split-levels must be equal to thirty-two. This is because there is a maximum *and* minimum attenuation constraint. As an example, if the primary node split-level is chosen to be 1:4 then the secondary node split-level must be 1×8. This constraint makes planning the network significantly harder.

7.1.2   Access Network Planning

This section summarises the problems that the access network planning process presents. It is assumed that the initial information available to a planner is:

- the location of the exchange,
- the location of *potential* customers,
- a forecast of the demand from these customers in terms of number of lines and year.

In addition, for the purpose of this work, it is assumed that there is information concerning:

- available duct infrastructure,
- and potential node locations,

Also available to the planner is a selection of plant and cables with which to implement the network, and their associated costs.

The aim of the planner is satisfy both the network's customers and the network operator, by producing a reliable cost-effective network. To achieve these aims the planner must decide on the following:

- Primary and secondary node locations.
- Concentrator capacities or PON splitter sizes.
- Cable sizes and routes.
- Assignment of customers to secondary nodes.
- Assignment of secondary nodes to primary nodes.

The typical cost functions and constraints are summarised below:

First installed cost is used to define the cost of installing the network and does not consider the network's maintenance cost. It can be applied to both networks installed on day one (static) or networks that are rolled-out over a number of years (dynamic).

Whole life cost accounts for the cost of installation, operation, maintenance, personnel training and disposal. Aspects such as maintenance are based on the network reliability.

Reliability has two aspects:

- the reliability as perceived by a customer of the network
- the overall network reliability as experienced by the operator.

However, it is usually treated as a constraint with a minimum acceptable value.

Attenuation is an engineering constraint; equipment in the network will be designed to work within a range of signal strengths. If the signal strength is outside these bounds then network performance will be unpredictable.

Access networks are rarely installed all at once; they are gradually installed to meet demand over a number of years. This delay in investing reduces overall project costs. It is common to consider the dynamic nature of the problem for the installation of high capacity trunk networks (see Minoux (1987) for an overview). There is little work that considers this in access network optimization, exceptions being Jack *et al.* (1992) and Shulman and Vachini (1993), who describe a system developed at GTE for optimising copper access networks. Common to this work is the use of Net Present Value (NPV) as the objective function. If a series of investments $I(1), I(2).....I(T)$ are made over time then the NPV is (Minoux, 1987):

$$C_{NPV} = \frac{I(1)}{1+\tau} + \frac{I(2)}{(1+\tau)^2} + ... + \frac{I(T)}{(1+\tau)^T}$$

where τ is the actualisation rate = [0.05...0.15] which is decided based on economic conditions (the actualisation rate is different from, but related to the interest rate; it is often decided based on company or government policy). NPV is used to represent the fact that capital can be invested and so increase in value over time. If capital is used to purchase hardware and it is not needed until later, then interest is being lost. Instead of investing in plant now, the money could be invested at a time when the plant is actually needed. A related factor that can be considered is price erosion, which is particularly important when considering investing in new technologies (such as optical fibre). The price of new products often decreases rapidly as they are deployed more widely. Also, technological advances in the product's manufacturing process may lead to price reductions.

7.1.3 Techniques for Planning Networks

Much of the published work in the field describes the concentrator location problem. This problem typically involves the location of a number of concentrators at remote sites and the allocation of customers to these sites. A concentrator is a component within a copper or optical network that multiplexes the signals from a number of cables onto one and conversely de-multiplexes a single signal into many.

An early survey of research in this area is given by Boorstyn and Frank (1977). The methods that they describe are based on decomposing the problem into five sub-problems:

1. the number of concentrators to install,
2. the location of the concentrators,
3. connection from exchange to concentrators,
4. assignment of terminals to concentrators,
5. and connection from terminals to concentrators.

Items 3 and 5 are simple to determine; connections are made through the shortest route in the graph. A solution to sub-problems 1, 2 and 4 is presented, which is based on the clustering algorithm of McGregor and Shen (1977). This algorithm decomposes the problem into three stages:

- *clustering* physically close terminals, and representing them by a new node at the centre of mass of the cluster (a COM node),
- *partitioning* of COM nodes into sets which are supplied by a common concentrator,
- and *local search* optimisation of concentrator location.

Much work in this area concentrates on using integer programming relaxations of the problem; a review of problems and solution techniques is presented in Gavish (1991). Balakrishnan *et al.* (1991) survey models for capacity expansion problems in networks. These problems differ from concentrator location problems because when capacity is exhausted at a node, extra plant can be added to meet the demand. This means that cost functions for concentrators and cables are non-linear, usually step-wise increasing.

Literature describing methods of optimising passive optical networks is sparse. This network architecture is a recent development, which perhaps explains this. The majority of published work is produced from collaboration between the University of Sunderland and British Telecommunications Plc. Paul and Tindle (1996), Paul *et al.* (1996) and Poon *et al.* (1997) each describe genetic algorithm based systems. The first makes use of custom genetic operators and the second uses a graph-based hybrid genetic approach. Woeste *et al.* (1996) describe a tabu search approach and Fisher *et al.* (1996) a simulated annealing system. The work described focuses on static network problems in which the network is optimised for cost. However, none of the papers give details of the algorithms used or detailed performance data due to commercial confidentiality.

None of the published work described so far have considered the dynamic nature of the network planning process. In particular, the interest here is with capacity expansion problems, where plant is gradually installed into the network to meet demand (see Luss (1982) for a general discussion of the field of capacity expansion). Jack *et al.* (1992) and Shulman and Vachini (1993) both describe the algorithms that form the basis of an optimisation system NETCAP™ developed for GTE. The papers describe a technique for producing capacity expansion plans for concentrator networks over multiple time periods. This is formulated as a 0-1 integer program, and optimised by splitting the task into two sub-problems. The first sub-problem (SP1) is a static optimisation based on the final year demand. The second sub-problem (SP2) determines the time period in which the concentrators and cables should be placed and the size of these components. As, to solve (SP1), it is necessary to know the cost of the plant used, and this information is specified by (SP2), (SP1) is parameterised by a factor that is used to represent cable costs. The overall problem is solved by iterating between (SP1) and (SP2) as this factor is varied.

7.1.4 Uncertainty in Planning Problems

Traditionally, network planning in the access network has not involved much uncertainty. State owned telecommunications providers were in a monopoly position and all customers had to connect through them. Also, these organisations tended to provide a line when it was convenient for them and not when required by the customer. This situation still exists in many parts of the world where the waiting time for a telephone line may be many years.

However, many countries are now deregulating the telecommunications supply market and introducing competition. Competition in the UK consists of cable companies offering telecommunications services in competition with BT, the incumbent supplier. Therefore, a service provider can no longer guarantee that a customer will use their network.

Further uncertainty exists in the demand forecast; a prediction is made in this forecast about when a customer will require a service. However, in reality the customer may require a service earlier or later than this date.

Powell *et al.* (1995) identify five types of uncertainty relevant in planning a network:

- uncertainty in the demand forecast
- uncertainty in the availability of components/manpower
- external factors, e.g. weather
- randomness in factors affecting the management and operation of the network, e.g. economic factors
- errors in data that is provided to the model

Ideally, a network operator would like to build a network that performs well in the face of this changing demand. One measure of this would be the average cost of the network across different instances of demand. Another measure would be obtained by considering the standard deviation of the cost across a set of demand scenarios. This second metric could be considered as a measure of the quality or robustness of a network plan (Phadke, 1989).

7.2 The Genetic Algorithm Approach

An introduction to genetic algorithms is provided in Chapter 1. The aim of this section is to introduce the methods that are used in this chapter for applying them to the problem of access network planning.

A common misconception of genetic algorithms is that they are good 'out of the box' optimizers. That is, a GA may be applied directly as an optimiser with little consideration for the problem at hand. With difficult problems, there is no evidence that this is the case and, in general, the representation and associated genetic operators must be carefully chosen for suitability to the problem. The key to producing a good genetic algorithm is to produce a representation that is good for the problem, and operators that efficiently manipulate it. An ideal representation for a genetic algorithm has a number of features:

- There should be no redundancy – each point in the problem search space should map to a unique genetic representation. Redundancy in the representation increases the size of the genetic search space and means that many different genomes representing a single point in the problem space will have the same fitness.
- It should avoid representing parts of the problem space that break constraints or that represent infeasible solutions.

If a representation can be found that can meet these criteria then the challenge becomes that of designing appropriate operators. The next section describes Forma theory, a method that is designed to help in this process.

7.2.1 Forma Theory

One of the difficulties of applying a GA to a *particular* problem can be that suitable operators simply do not exist. Forma theory was developed by Radcliffe (1994) for designing problem specific representations and genetic algorithm operators for manipulating these generated representations. It allows the development of operators that work with and respect the representation for arbitrary problems. It is based around *equivalence relations* which are used to induce equivalence classes or *formae* (plural of forma); these formae are used to represent sets of individuals. An example of an equivalence relation is 'same colour hair', which can be used to classify people into groups. Each group has the property that all members have the same hair colour, and these groups can then be represented using formae such as ξ_{blonde}, ξ_{black} and ξ_{ginger}.

Once a representation has been developed for a problem one of a number of crossover operators can be chosen. The choice is based on certain properties of the representation. Important properties of a crossover operator are respect, assortment and transmission. *Respect* requires that formae common to both parents are passed on to all children. If an operator exhibits *assortment* then this means that it can generate all legal combinations of parental formae. Finally, *transmission* means that any generated children are composed only of formae present in the parents. Radcliffe (1994) introduces a number of operators that each exhibit some of these properties.

7.2.2 Local Search

A common approach to improving the performance of genetic algorithms is to combine them with other search methods or heuristics. Nearly all the GAs that have produced good results on difficult problems have used this approach. The basic operation of these algorithms is that a local search operator is applied to each individual in the population at each generation after crossover and mutation. A local search algorithm is one that, given the current point in the search space, examines nearby points to which to move the search. The algorithm is often *greedy* in that it will only accept moves that lead to an improvement in the cost of a solution. The search is guaranteed to find a local minimum, but if there is more than one of these it may not find the global minimum.

The name memetic algorithm is used to describe these algorithms, and comes from Dawkins' idea (1976) that as humans we propagate knowledge culturally, so our success goes beyond genetics. He called the information that is passed on from generation to generation the *meme*. It was adopted by Moscato and Norman (1992) for GAs that use local search, as they considered that in one sense the algorithm was learning, and then passing this learnt information on to future generations. A modern introduction to memetic algorithms and their applications can be seen in Moscato (1999).

7.2.3   Overall Approach

The approach taken to solving the access network planning problem is to treat it as a multi-layer location-allocation problem. A location-allocation problem is a general term for any problem where there are a set of customers that must be allocated to supplier sites, where the location of these sites must also be determined. Network planning can be considered as a multi-layer version of this problem because customers must be allocated to secondary nodes and secondary nodes must be allocated to primary nodes. At the same time these primary and secondary nodes must be located. The problem is formulated so that the primary and secondary locations must be selected from a finite set of possible locations.

7.2.4   Genetic Representation

A simple representation for the problem is one where for each customer there is a set of alleles which represent all the possible sites that can supply the customer, the actual allele chosen represents the site which supplies the customer. This representation was used by Routen (1994) for a concentrator-location problem. It is good as there is no redundancy – one genome maps to a unique instance of a location-allocation.

This representation allows manipulation of the genome using standard crossover operators such as n-point crossover and uniform crossover. Uniform crossover is the most appropriate, as there is no information contained in the ordering or position of the genes within the genome.

A natural representation for a location-allocation problem is one in which sets (or clusters) of customers are formed along with an associated location. The set of customers is allocated to this associated location. The objective of the optimisation is then to form good clusters of customers and to find good locations as centres for these clusters. The problem can be decomposed so that first a cluster of customers is found and then a location is selected on which to centre them.

Using the terminology of forma theory, the equivalence relation used is therefore

$$\psi_{ab} = \text{'customer } a \text{ shares a set with customer } b\text{'}$$

So, if there are three customers a, b and c, the following equivalence classes (or formae) can be induced:

$$\Xi_{\psi_{ab}} = \{\xi_{ab}, \xi_{\overline{ab}}\}$$
$$\Xi_{\psi_{ac}} = \{\xi_{ac}, \xi_{\overline{ac}}\}$$
$$\Xi_{\psi_{bc}} = \{\xi_{bc}, \xi_{\overline{bc}}\}$$

where, e.g. ξ_{ab} means that a shares a set with b, and $\xi_{\overline{ab}}$ is the negation of ξ_{ab}.

A simple method has been chosen for representing the site that is associated with each set of customers. The association is made through the customers; each customer has an associated gene that represents their preferred supplying site. When a set is created, the first customer to be assigned to the set has its associated location used to supply the whole set of customers. A similar scheme is used for determining which primary node supplies each set of secondary nodes. If the target network is a PON then the split-level is represented in the same way – each customer specifies a preferred split. Both the representation of the primary and secondary sites and of the splitter sizes, are strings of genes. As such, these strings can be manipulated using standard operators such as uniform crossover.

7.2.5 Crossover and Mutation

The set-based representation is such that traditional crossover and mutation operators will not perform well. The representation is not orthogonal, as for example $\xi_{ab} \cap \xi_{bc} \cap \xi_{\overline{ac}} = \varnothing$. The non-orthogonality displays itself as dependencies between forma; this means that a traditional operator would generate many invalid individuals. As a consequence of this non-orthogonality, and the fact the formae are separable (as assortment and respect are compatible, see the previous section) an operator based on Random Transmitting Recombination (RTR – see Radcliffe (1994)) was developed. The operator is a version of uniform crossover adapted for problems where the alleles are non-orthogonal.

The operator functions in a number of stages:

Step 1: Gene values that are common to both parents are transmitted to the child.

Step 2: For each remaining uninitialised gene in the child. Randomly select a parent and take the allele at the same loci, set the child's gene to this allele. Update all dependant values in the array.

Step 3: Repeat Step 2 until all values are specified.

The aim of a mutation operator is to help the algorithm *explore* new areas of the search space by generating new genetic material. Given the representation described in the previous section it is necessary to devise a compatible mutation operator. The implementation chosen is based on three components that could provide useful changes to the membership of the sets. The three components are:

Split: This chooses a set randomly and splits the contained individuals into two new sets.

Move: Two sets are chosen at random and a member of the first set is picked at random and moved to the second set.

Merge: Two sets are chosen at random and the members of both sets are combined to form a single new set.

When mutation is applied to an individual, each of these components is applied with a certain probability. Thus, a mutation operation may actually consist of a combination of one or more of the above operators.

The probability with which each component is chosen is determined by the genetic algorithm itself, that is the algorithm self-adapts its mutation rate. Smith and Fogarty (1996) use such a system to adapt the mutation rate for each gene independently, that is, for each gene an extra gene is added which specifies the mutation rate for its companion. The results of their work showed that, on complex fitness landscapes, the self-adapting algorithm out-performed standard algorithms. Experiments on the access network planning problem show similar results, with a self-adapting mutation operator out-performing a simpler implementation (Brittain, 1999).

7.2.6 Local Search

Two forms of local search are used within the algorithm. One tries to improve the allocation of customers to secondary nodes, and secondary nodes to primary nodes. And the other aims to improve the position of secondary and primary nodes with respect to the nodes that are allocated to them.

The implementation of the first of these is simple; given the current position in the search space, the algorithm proceeds as follows:

Step 1: Choose two customers a and b from different sets

Step 2: If swapping a and b leads to an improvement in cost, add a triple of (a, b, cost_reduction) to a list of possible exchanges.

Step 3: Repeat Steps 1 and 2 for all pairs of customers. Sort the list of possible exchanges based on cost reduction, with largest cost reduction first. Move through the list implementing the exchanges, once a customer has been moved ignore all later exchanges involving this customer.

The second type of search attempts to improve the position of a secondary (and primary) node sites. The algorithm works as follows:

Step 1: The initial values for the secondary nodes locations are provided by the genetic algorithm. Then for each secondary node:

Step 2: Take each adjacent node in the graph. Calculate the gradient of the cost function with respect to the distance between it and the current node.

Step 3: Choose the node that gives the steepest descent in the cost function as the new node position. Repeat Steps 2 and 3 until no improvement can be made.

Note: The calculation of the cost function is computationally cheap because it is decomposable. It depends only on the position of the customers assigned to the secondary and the position of the primary node. The algorithm that is used begins by finding the *cost per metre* of the input cable for the node, and the *cost per metre* of all the output cables. These are constant with respect to the position of the secondary node. Calculation of the cost function therefore only requires the distance from the secondary to the primary and the customers to be re-calculated at each iteration of the algorithm. These distances are multiplied by the relevant cable cost-per-metre and summed.

7.2.7 Simulated Annealing

A simulated annealing algorithm must have a move-generator that defines the local search neighbourhood. The moves that are described here are similar to the search neighbourhood used with the local search in the genetic algorithm. They are

- change the position of a primary/secondary node,
- assign a customer/secondary node to a different secondary/primary node,
- and in the case of PONs change the splitter size.

Ernst and Krishnamoorthy (1996) use these first two moves for solving a related problem – the p-hub median problem.

It is also necessary to have a way of creating new sub-trees in the network; this is achieved by the operator for moving a section. Firstly, the SA is initialised so that each customer is supplied by their own primary and secondary node. For example, if there are four customers, then there will be four primary nodes, each supplying a single secondary node which in turn supplies a customer. This means that no new sections need to be created, as the maximum possible number are created at the start; Figure 7.2 illustrates this.

7.2.8 Dynamic Installation Strategy

Given a network definition provided by a genome, the time of installation of the components must be decided. This is achieved using the cost function. Network cost is calculated by working from the customers upwards through the network tree. Starting from the customer, a cable is connected to the secondary node in the first year that the customer requires a service. The secondary node that the customer connects to is added when the first customer connects to it, and so on up through the network. This is illustrated in Figure 7.3.

Cable sizing is based on the assumption that all future demand is known. So, when a cable is installed, its size is chosen to satisfy all of the forecast demand. Although, if the demand exceeds the maximum size available, then the largest cable is installed and additional cables are added in future when required. For example, imagine the demand for fibres at a node is six in the first year, two in year two, four in year four, and the maximum cable size is eight. Then, a cable of size eight would be installed in the first year and an extra cable to supply the remaining demand (size 4) would be installed in year four.

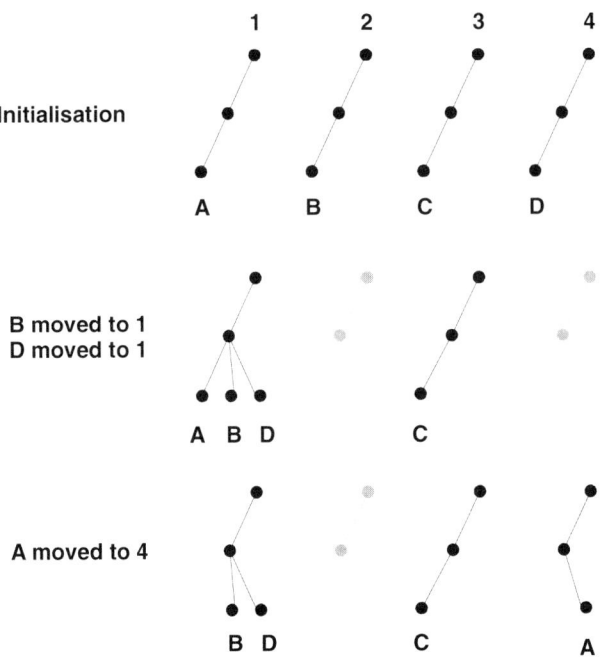

Figure 7.2 Illustration of a series of simulated annealing moves. Those section greyed out are not included in the cost, as they have no customer connected. Initially, each customer is supplied by their own primary and secondary node. This means that a move specifically for creating new sections is not needed. As shown in the final step, customer A is assigned to another section with no customers attached, this effectively creates a new section.

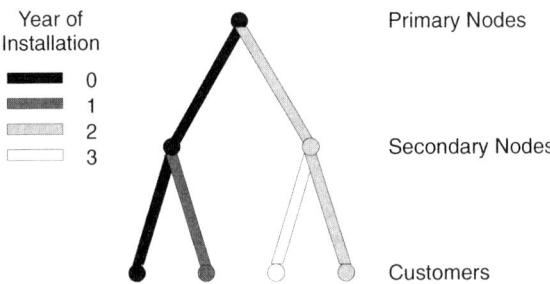

Figure 7.3 Illustration of how a network is installed over time. Colour coding shows when the cables and nodes are installed.

The same approach is adopted for sizing nodes, and installing splicing and splitting units within these nodes. So given the above example, for a splicing node (where splicing units are of size four), two splicing units would be installed in year one, and one would be installed in year four.

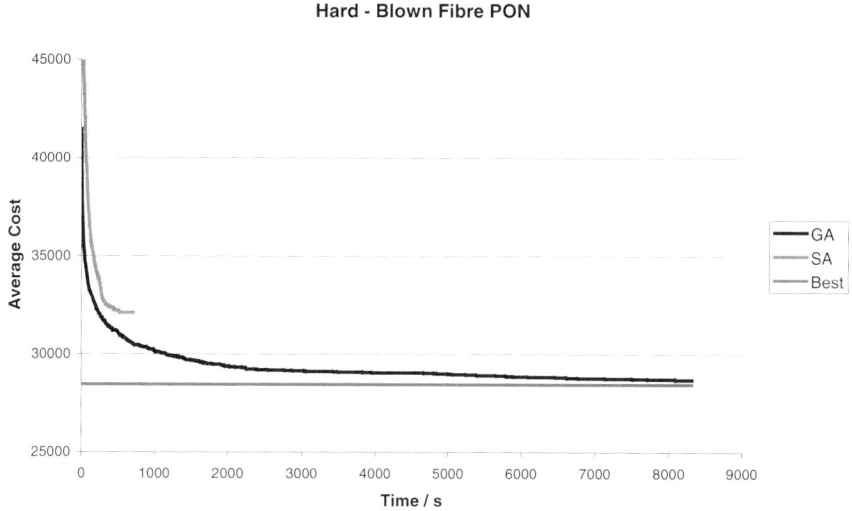

Figure 7.4 The performance of the GA vs SA on a passive optical network planning problem.

7.3 Results

This section summarises the results that were obtained for some example planning problems. The algorithms have been tested against a wide range of problems from those with twenty to thirty customers, to the example below (called *hard*) with seventy customers, sixty possible secondary node sites and eight possible primary sites.

In the results described, the algorithms (GA with local search and SA) are used to produce two plans: one for a point-to-point network architecture; and the other for a passive optical network. The genetic algorithm uses the novel set-based representation, as it has been shown that this performs more robustly across a range of problems than the simpler representation described at the start of the previous section (Brittain, 1999).

Figure 7.4 shows the results of a comparison of the GA with SA averaged across ten runs, with a new random seed for each run. It is clear that the GA's performance is much superior to that of simulated annealing. Analysing the results of the simulated annealing algorithm, it was clear that the results where poor because it only found large clusters of customers. The best solution to the problem consists of smaller clusters of customers. The results shown are similar across all the planning problems, except that the performance of SA is closer to that of GA when there are fewer customers.

Figures 7.5 and 7.6 show the best solutions that were found for the *hard* planning problem for PTP and PON respectively by the GA. It is interesting to examine the difference between the solutions to the PON planning problem and the PTP problem. It is clear that the PON solution contains a small number of large clusters of customers compared to the PTP solution that contains a large number of small clusters. This is easy to explain as the passive splitting devices are expensive compared to splicing units (they may be up to fifty times more expensive). Therefore, for PON planning the optimisation attempts to minimise the number of splitting devices installed. From the diagram it is clear that nearly every splitting device is connected to its maximum number of customers. In PTP networks, the splicing units cost less and cable is comparatively expensive, so the optimisation attempts to reduce the total amount of cable installed into the network. This difference means that finding the optimum for the PTP problem is much harder as, for example, the number of possible clusters of size four, is much greater then the number of clusters of size eight.

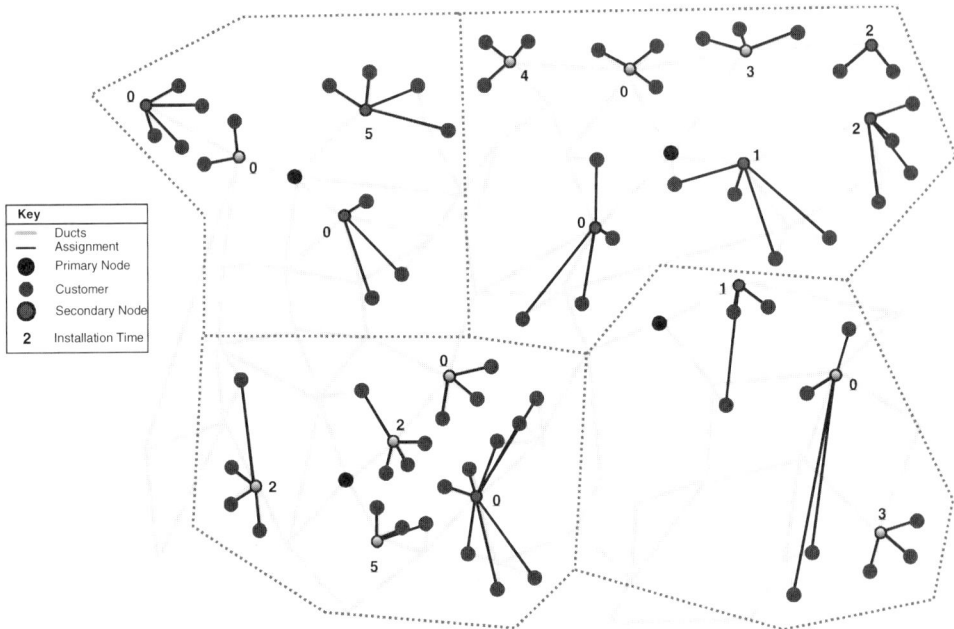

Figure 7.5 The cheapest network found by the genetic algorithm for point-to-point dynamic planning on the *hard* test problem.

7.4 Uncertainty in Network Planning

All of the work described so far in this chapter has assumed that the demand forecast is accurate. As was argued in an earlier section, this is unlikely to be true; customers may connect to another telecommunications operator, or they may connect to the network earlier or later than forecast. Therefore, in this section a method of optimising the access network in the presence of uncertainty is presented.

Two models are used, one for *whether* and the other for *when* a customer connects to the network. These are used to generate *demand scenarios*, which represent possible instances of customer demand. It is assumed that *whether* and *when* a customer connects are independent. This is a reasonable assumption given that *whether* is considered to be a decision by the customer to use one service supplier or another, whereas *when* is an independent decision over when they have a need for this service.

Optimizing the Access Network

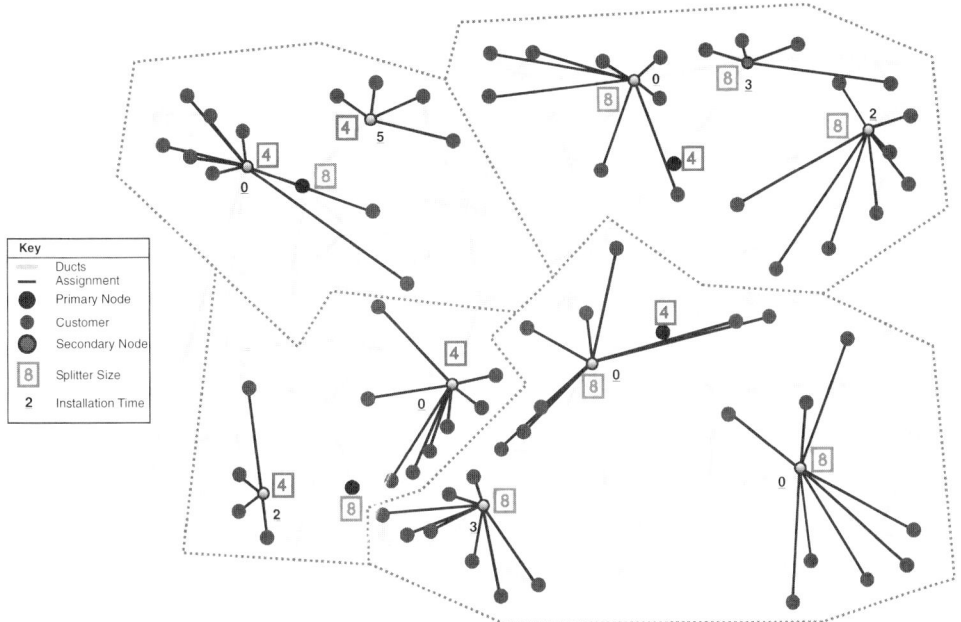

Figure 7.6 Diagram showing the cheapest network found by the genetic algorithm for Passive Optical Network dynamic planning on the *hard* test problem.

To model the probability of *whether* a customer connects, a simple model is used. Given a figure for an expected service penetration (the percentage of customers expected to connect to the service) a biased coin is tossed were the probability of a head is equal to the penetration. So if expected penetration is 65%, then the coin is biased so that there is a probability of 0.65 that there will be head. This model can be used to model which customers connect by equating connection with a coin flip which results in a head.

To model *when* a customer connects a different model must be used. A probability density function is needed that models the probability that a customer will connect in a particular year. Any model is likely to have a peak representing the most probable year that they will connect and a decreasing probability either side of this peak. A triangular Probability Density Function (PDF) was chosen. The apex of the triangle of the PDF is coincident with the most likely year of connection. The length of the base of the triangle represents the degree of uncertainty in when the customer connects. The wider the base of the triangle, the more uncertainty there is over the year. An example of a PDF is given in Figure 7.7, where the customer is most likely to connect in year five and there is a decreasing chance that they will connect earlier or later than this date.

Next we clarify what a solution to the planning problem under uncertain demand represents and how cost is calculated. As before, a genome represents the information needed to define a network connecting the customers to an exchange. In the uncertain case, all customers that might connect to the network are represented. So, the genome represents a network that could connect all customers to the exchange if they all required a service.

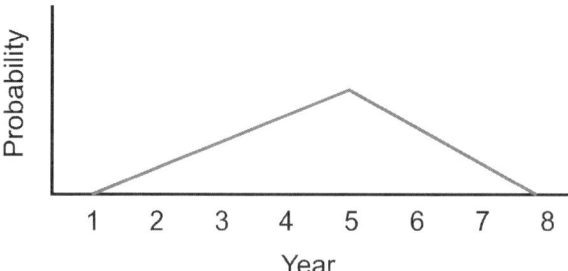

Figure 7.7 Illustration of the probability density function used to represent when a customer will connect to the network.

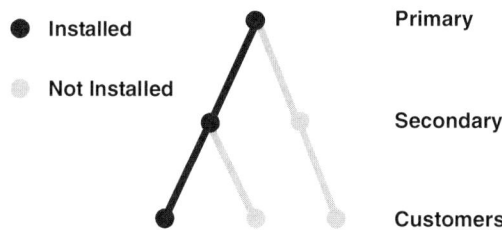

Figure 7.8 Illustration of how the cost is calculated when not all customers connect to the network.

Figure 7.8 illustrates how the cost of a network is calculated. The black circles representing customers show that there is a demand. Plant that is installed into the network using the *build* method is shown in black and the grey parts of the diagram illustrate the plant that would be installed if the other customers had demand.

Another factor that must be considered is how much plant should be installed in anticipation of future demand. The cost model used in the dynamic optimisation assumed that the demand forecast for the future is totally accurate; this means that when installing a cable, the size that is chosen is enough to meet the future, as well as the current demand. For the uncertain case the level of future demand is unknown.

The approach taken in the cost model used here is to install cable to meet expected future demand plus 25%. This figure was chosen based on a rule of thumb used by planners when designing networks by hand. It should be noted that, as with the dynamic cost model, if the demand exceeds the maximum size of the plant available, and if this maximum size meets demand for the current year, then that is all that would be installed. The installation of further capacity would be deferred until a later year.

For example, if a secondary node is being installed that has ten customers assigned to it, two of these require a service in the current year, and the penetration is 50%. Then the expected number of future customers will be four. To this figure, 25% is added to give five, so plant is installed into the node to meet the demand of five future customers plus two current customers. However, if the maximum cable size were four, then a cable of size four would be installed and extra capacity would be added at a later year if required.

7.4.1 Scenario Generation

Previous sections have presented a genetic algorithm combined with local search methods that can be effectively used for planning the access network. Bäck *et al.* (1997) note that:

> "Natural evolution works under dynamically changing environmental conditions, with nonstationary optima and ever changing optimization criteria, and the individuals themselves are also changing the structure of the adaptive landscape during adaptation."

and suggest that GAs which are commonly used in a static environment should be applied more often to problems where the fitness function is dynamically changing as the evolution progresses. Fitzpatrick and Greffenstette (1988) show that GAs perform robustly in the presence of noise in the fitness function. By changing the demand scenario against which a network's cost is evaluated, a noisy cost function is obtained. This suggests that the GA described in the previous sections can be applied almost completely unchanged to stochastic problems.

In the experiments that follow, two sets of demand scenario are used. One set is used in the evaluation of the cost function by the GA and is used to train it; this will be called the *training set*. The other set is used to test the performance of the solutions generated by the GA; this will be called the *test set*, and is described later.

For every evaluation of an individual, the GA takes a new scenario from the *training set* and its cost is evaluated against it. This cost is used to produce the individual's fitness score. At the end of the optimisation, after a fixed number of generations, all of the individuals in the population are evaluated against the *test set*. Two individuals are taken from this final population, the one with the lowest average cost and the one with lowest SD. The choice of the first is clear; the second is also considered as it has a robust performance across all the demand scenarios.

The other important consideration is how local search is performed. The aim of the optimisation is to find a solution that performs well across a range of demand scenarios. Therefore, local search cannot be performed on a network for a given instance of demand, as this is likely to improve the solution for that particular instance alone. The approach used is to instantiate the access network model with the expected value for year of demand and connect all customers to the network. A local search is then performed based on this level of demand.

7.4.2 Results

The algorithm was tested against a problem with 30 customers, 50 possible secondary nodes sites and a single primary node. The installation of a point-to-point network was optimised. Penetration was taken to be 50% and the triangular PDF for each customer was centred on the year of demand used for the dynamic optimisation problems described earlier in the paper. A population size of one hundred was used and the GA evolved for two hundred generations. Five experiments were run, each with a different seed for the generation of the demand scenarios.

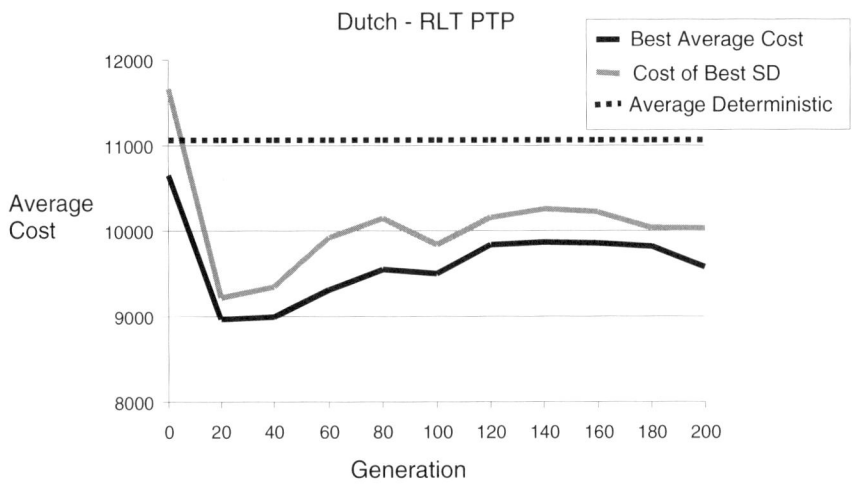

Figure 7.9 Average performance of the genetic algorithm when optimising against changing demand scenarios. Also shown are the average cost of the deterministic solutions and the average cost of the solution that has the best standard deviation in each test.

The performance of the algorithm with respect to cost is illustrated in Figure 7.9. The graph shows the average performance across the five experiments of the best individual in each population. In every run of the algorithm the lowest average was found at either twenty or forty generations. The average cost of the best solution then reliably gets worse. On average, the final solution found by the algorithm has a 13.5% lower cost than the average cost of the deterministic solution. The best *deterministic* solution is the best solution found during a dynamic optimisation. This best solution was then tested against the demand scenarios used in the stochastic optimisation.

The best solutions found during the course of a run of the algorithm are on average 19% lower cost. Also shown in the graph is the average cost of the individual with the best SD at each test point. It can be seen that the cost of this individual closely tracks the cost of the lowest cost individual.

These results show that GAs perform well in the presence of uncertain data. The algorithm presented here is capable of producing networks that have, on average, a significantly lower cost than the best network from a deterministic optimisation.

7.5   Discussion and Conclusion

This chapter has introduced a set-based genetic representation that can be used for dynamic access network planning problems. A Simulated Annealing (SA) algorithm is also described for the same problem, and its performance is compared to the GA. It is found that the GA consistently outperforms SA algorithm. The crossover operator that was used by the GA was developed using Forma theory, a representation independent method for designing operators. This has shown itself to be an effective method for designing genetic operators.

The main advantages of a GA-based approach are its flexibility and robustness across a range of problems. The cost model used by the GA is based on an object-oriented model of access networks. This models both point-to-point networks and passive optical networks, and each of these can be composed of arbitrary cable and node types. The GA can be used to optimise all of these types of network regardless of their composition. In fact, the model could easily be extended so that other network technologies – such as concentrator-based networks – could be optimised. Another feature of the GA approach is that the model used is an integrated one, where the cost function used by the optimiser accurately reflects the cost of installing the network. This contrasts with the previous work in the field that was described earlier in the chapter, where the problem is broken down into approximate sub-problems which are then solved.

In the final sections, it is argued that the consideration of demand uncertainty in access network planning is essential in the current economic environment where competition for customers is intense. The section describes how the GA can be applied, virtually unchanged, to a robust network design problem where there is uncertainty in customer demand. The results show that this approach can lead to networks that, on average, cost less and have a lower standard deviation than a deterministic solution, over a range of demand scenarios.

The introduction of computer based methods for network planning has pay-offs that are not directly related to the fact that the solutions found may be better or cheaper than manually generated solutions. The cost saving in labour and materials in the machine generated solution may be only a few percent of the total cost, but because the solution has been machine checked for consistency with the rules of installation, they will need less corrective work. Additionally the solution is in machine readable form and so can easily be passed down to the next stages of stores requests, installation, etc. and upwards to the database of installed networks.

Acknowledgements

We would like to thank Pirelli Cables Limited for their sponsorship of this research. Also thanks are due to Peter Hale for his help in formulating the access network planning problem.

Some of the software for this work used the GAlib genetic algorithm package, written by Matthew Wall at the Massachusetts Institute of Technology.

Part Two

Routing and Protocols

The point of having telecommunications networks in the first place is, of course, to carry voice and data traffic between users. Almost always, a large number of calls are happening simultaneously, and the result on busy networks is that users experience delays, poor reception, or degraded quality of service of some form or other. Routing and protocols are central to the issue of minimizing such problems. Routing is essentially the job of deciding how a packet of data or voice should get from a to b. A good routing strategy will adapt its decisions according to the current network traffic enviroment. Good routing also involves good restoration – that is, how to re-establish a connection (and very quickly) if one of the links in the connection suddenly fails, or becomes unacceptably slow. By 'protocols' we refer to the basic data-link protocols which govern how communication is handled in terms of error-checking, acknowledgement packets, timing issues, and so forth. The sad fact is that most protocols are out of date, and new ones must be developed to handle modern traffic requirements efficiently.

In this part, three chapters look at routing and two look at protocols. The routing chapters use either fuzzy logic, genetic algorithms, or both, to address dynamic routing problems, in which quick next-hop decisions will be made based on adaptive routing tables. Two are online methods, although one is an 'offline' method, where the genetic algorithm is shown to provide the best solutions, but its use is as a *benchmark* for online strategies to aspire to.

The protocols chapters address two issues: first, an adaptable protocol scheme is devised, in which the protocol used between two nodes will be altered dynamically according to the traffic requirements on the link.sand prevailing environmental conditions. A neural network is used to decide which new protocol to adopt. The second looks at offline protocol design, using a genetic algorithm to automate the complex task of protocol validation.

8

Routing Control in Packet Switched Networks using Soft Computing Techniques

Brian Carse, Terence C. Fogarty and Alistair Munro

8.1 Introduction

This chapter describes the joint application of two soft computing methods – evolutionary algorithms and fuzzy reasoning – to the problem of adaptive distributed routing control in packet-switched communication networks. In this problem, a collection of geographically distributed routing nodes are required to adaptively route data packets so as to minimise mean network packet delay. Nodes reach routing decisions locally using state measurements which are delayed and necessarily only available at discrete sampling intervals. Interactions between different nodes' routing decisions are strong and highly non-linear. Extant routing methods in packet-switched networks (Khanna and Zinky, 1989) mostly employ, in one form or another, some direct form of least-cost or 'shortest-path' algorithm (Dijkstra, 1959) operating at each routing node. Such methods pay little attention to the dynamic interactions of routing decisions made at different nodes and do not directly address the temporal effects of delayed measurements and persistence of the effects of routing choices.

This contribution proposes a very different approach to the routing problem. The routing policy of routing nodes is determined by a fuzzy rule base, which takes as inputs various network state measurements and provides route selection probabilities as outputs.

A Genetic Algorithm (GA) is used to optimise these fuzzy rule bases. At each generation of the GA, identical copies of each candidate fuzzy rule base are deployed to routing nodes. Fitness evaluation for each individual fuzzy rule base is based on the network-wide performance of the distributed assembly of routing controllers.

The layout of the chapter is as follows. The following section (section 8.2) concentrates on some preliminaries and offers a background on the adaptive distributed routing problem, and on fuzzy logic and fuzzy control. Section 8.2 also gives a brief overview of some other approaches which use the genetic algorithm to optimise fuzzy rule bases. Section 8.3 describes two versions of a fuzzy classifier system (a non-temporal version and a temporal version) used as the basis for experiments; this section then provides experimental results in applying the fuzzy classifier systems to routing control. Performance of the evolved fuzzy routing controllers is compared to that of routing using other methods. Finally, section 8.4 concludes and offers suggestions for further work.

8.2 Preliminaries

8.2.1 Adaptive Distributed Routing in Packet Switched Networks

Communication networks employ two major methods of switching: circuit-switching and packet-switching (refer to Schwartz (1987), Stallings (1994) and Tanenbaum (1996) for descriptions of these switching methods). In the former, a dedicated amount of network bandwidth is allocated to a source-destination pair during a circuit set-up phase and end-to-end delays are usually small and fixed. These characteristics have lead to the widespread adoption of circuit-switching for telephony and real-time video. However, circuit-switching has the drawback of making inefficient use of network resources when information sources generate 'bursty' or sporadic traffic. Packet-switching attempts to overcome this problem by employing a distributed form of statistical or dynamic multiplexing. Each network user offers packets to the network and these packets are routed through the network by Packet-Switching Exchanges (PSEs) on a store-and-forward basis. Link bandwidth is no longer pre-allocated at connection set-up time, but instead each PSE maintains a queue of packets to be delivered over a particular outgoing link. Two main ways of implementing packet-switched networks have emerged: virtual-circuit and datagram. In virtual-circuit packet-switched networks, a connection set-up phase establishes a fixed path through the network between a source-destination pair (although it does not allocate network bandwidth). For the duration of a connection, all packets follow the same path through the network. In datagram networks, no connection set-up phase is involved and subsequent packets between a source-destination pair may take different routes through the network. While packet-switching makes better use of network resources for bursty traffic sources, end-to-end delays are variable and depend on the level of traffic offered to the network. This characteristic has meant that such networks have, until recently, been ruled out for conveyance of real-time information sources such as telephony and real-time video. The currently developing Broadband Integrated Services Digital Network (B-ISDN) is intended to convey telephony, video and computer–computer information over the same network. It is almost certain that B-ISDN networks will employ Asynchronous Transfer Mode (ATM), implying that B-ISDN will be a packet-based network (using fixed size packets called 'cells').

Routing policies in computer networks may be static, dynamic or adaptive; and centralised or distributed. Static routing uses fixed routing policies which do not change with time or network conditions. Dynamic routing alters routing policies in time (e.g. according to the time of day). Adaptive routing allows routing decisions to take into account the changing nature of network traffic distributions. With centralised routing, a single centralised node (Routing Control Centre) gathers network status information relating to topology and traffic distribution, calculates routing policies for individual nodes based on this information, and then informs nodes in the network of these policies. Distributed routing, however, has network nodes (PSEs) reaching their own routing decisions based upon the information available to them. Adaptive distributed routing has the advantage in that calculation of routes is spread over many nodes; there is no convergence of routing information to and from an individual routing control centre (causing congestion on links in its vicinity); and routing decisions can be made to adapt to changes in the network status.

Virtually all packet-switched networks base their routing decisions using some form of least-cost criterion. This criterion may be, for example, to minimise the number of hops, or to minimise packet delay. Two elegant algorithms in widespread use in both centralised and distributed form, are those of Dijkstra (1959) and Ford and Fulkerson (1962), both of which translate to shortest-path routing algorithms in the communication network context. We now briefly discuss the development of the USA ARPANET packet-switched network, since the problems encountered, and solutions to these problems, exemplify the difficulties of adaptive, distributed routing techniques. The ARPANET is also significant since it formed the initial basis from which the current world wide 'Internet' has evolved.

The original ARPANET routing mechanism used an adaptive, distributed approach using estimated delay as the performance criterion and a version of the Ford and Fulkerson algorithm (sometimes called the Bellman-Ford algorithm). Each switching node exchanged current estimates of minimum delays to each destination, with its neighbours every 128 ms. Once this information was received, a node calculated the likely least-delay next node for each destination and used these for routing. This original approach suffered from many problems, in particular, the distributed perception of the shortest route could change while a packet was *en route*, causing looping of packets. The second generation ARPANET routing mechanism (McQuillan *et al.*, 1980), also adaptive and distributed, measured delays to each neighbour directly by time-stamping packets. Every 10 seconds, the measured link delays were averaged and then flooded (i.e. transmitted to every other node) through the network. Each node was then in possession of a (time-delayed) map of delays in the complete network. Nodes re-computed routing tables using Dijkstra's shortest-path algorithm. This strategy was initially found to be more responsive and stable than the old one. However, as network load grew, new problems arose, and instabilities in routing decisions were observed, whereby routes currently measured as heavily used were simultaneously avoided by all nodes and routes measured as lightly used were simultaneously selected, thus causing unwanted oscillations in routing decisions and inefficient network usage. One conclusion reached from these observations was that every node was attempting to obtain the best route for all destinations and that these efforts conflicted. As a result, the ARPANET routing method was further changed and in a later form (Khanna and Zinky, 1989) measures were introduced to damp oscillations through the use of digital filtering to smooth estimates of link utilisation, and the linearisation of projected delay as a function of link utilisation.

8.2.2 Fuzzy Logic and Fuzzy Control

Fuzzy logic is based on the concept of fuzzy sets (Zadeh, 1965). A fuzzy set is a generalisation of a classical set in the sense that set membership is extended from the discrete set {0,1} to the closed real interval [0,1]. A fuzzy set A of some universe of discourse X can be defined in terms of a membership function μ_A mapping the universe of discourse to the real interval [0,1]:

$$\mu_A : X \to [0,1]$$

Fuzzy set theory redefines classical set operations. For example, the most common forms of fuzzy union, intersection and complement are

$$\mu_{A \cup B}(x) = \max(\mu_A(x), \mu_B(x))$$
$$\mu_{A \cap B}(x) = \min(\mu_A(x), \mu_B(x))$$
$$\mu_{\overline{A}}(x) = 1 - \mu_A(x)$$

Given a universe of discourse X, it is possible to identify fuzzy sets with linguistic variables. For example if X is identified with linear displacement, then fuzzy sets over X might be {Negative-Large, Negative-Small, Zero, Positive-Small, Positive-Large}, or {NL, NS, Z, PS, PL}. Given universes of discourse for a set of system input and output variables, rules can be written in terms of fuzzy sets. For example, a rule for linear position control might be:

if (x is NL) and (v is Z) then (f is PL)

where the rule inputs are position (x), velocity (v) and the rule output is force (f). Once a fuzzy rule base and associated fuzzy set membership functions have been defined, the mapping of actual (crisp) input values to output values is achieved by fuzzification, fuzzy inference and defuzzification. One widely used fuzzification and inference method is the max-min or Mamdani method (Mamdani, 1976). Fuzzification evaluates every crisp input parameter with respect to the fuzzy sets in rule antecedents. For the above example rule, these are evaluated and combined by:

$$s = \min(\mu_{NL}(x), \mu_Z(v))$$

The output of the fuzzy rule is the fuzzy set defined by the function:

$$\mu_R(f) = \min(\mu_{PL}(f), s)$$

This method produces a fuzzy set for each rule R. Aggregation of these resulting fuzzy sets using fuzzy union (max) produces a single fuzzy set ($\mu_{R1} \cup \mu_{R2}...$). A single crisp output is obtained by applying a defuzzification operator to this aggregate fuzzy set.

Fuzzy logic has been applied in control systems for a wide variety of applications; an excellent overview, which includes a historical perspective together with many references to work on fuzzy control, may be found in (Kruse *et al.*, 1996). The main choices which need to be made by a fuzzy controller designer include (Lee, 1990):

1. Fuzzification and defuzzification strategies and operators.
2. Knowledge Base
   - universe of discourse
   - fuzzy partitions of input and output spaces
   - fuzzy set membership functions
3. Rule Base
   - choice of input and output variables
   - source and derivation of fuzzy control rules
   - types of fuzzy control rules
4. Decision Making Logic
   - definition of fuzzy implication
   - interpretation of connectives *and* and *or*
   - definition of compositional operator and inference mechanism.

Commonly, fuzzification/defuzzification methods, implication and inference methods are fixed at the outset of the design and the main design element then is ascertaining a suitable 'knowledge base' (fuzzy sets) and rule base. A number of different methods are available for devising appropriate fuzzy sets and rules. These include:

1. Extracting a human expert's experience and knowledge (if available) through knowledge elicitation techniques.
2. Observing and modelling a human operator's actions (possibly using automatic supervised and/or unsupervised learning operating on the observed data sets).
3. Understanding the physics of the process to be controlled and creating a model of the process from which fuzzy sets and rules for the controller may be designed.
4. Automatic generation of fuzzy sets and rules employing a directed search strategy in combination with some form of performance measurement.

Which approach to employ clearly depends upon whether a human expert exists and how easy it is to model the process to be controlled. Most real-world fuzzy controllers have been derived using one or more of the first three methods. However, in cases where no human expert knowledge nor input/output data sets are available, and additionally it is not possible to derive an accurate model of the process, these methods cannot be used and it becomes necessary to employ some sort of exploration strategy together with performance measurement to learn fuzzy sets and rules. One possible method of doing this is to employ reinforcement learning, which uses an adaptive critic for evaluating controller performance (e.g. Sutton's Adaptive Heuristic Critic (Sutton, 1984) and Watkins' Q-learning (Watkins, 1989)). Recently, researchers have investigated the use of evolutionary algorithms such as the genetic algorithm as a basis of learning fuzzy sets and rules. The next section summarises this work.

8.2.3 Previous Work using the GA to Optimise Fuzzy Rule Bases

Optimisation of fuzzy rule based systems using a GA is clearly a constrained optimisation problem; the main source of constraint being linguistic interpretability (the genotype must represent a valid rule base and fuzzy set membership functions should make sense). In an attempt to summarise previously published research in this area, this work is next categorised in decreasing order of constraints imposed on the (genetic) learning system. The categories used are: (1) learning fuzzy set membership functions only; (2) learning fuzzy rules only; (3) learning both fuzzy rules and fuzzy set membership functions in stages; and (4) learning both fuzzy rules and fuzzy set membership functions simultaneously. For a more detailed account of the brief summary given here, please see Carse *et al.* (1996).

Learning Fuzzy Membership Functions with Fixed Fuzzy Rules

Karr (1991) applied GAs to fuzzy controller design by adaptation of membership functions of a fixed rule set. This work demonstrated the success of the approach in generating both non-adaptive and adaptive fuzzy controllers for the cart-pole problem.

Learning Fuzzy Rules with Fixed Fuzzy Membership Functions

Thrift (1991) described the design of a fuzzy controller for centring a cart on a one-dimensional track by evolving fuzzy relations using fixed membership functions. Thrift's system was able to evolve a fuzzy control strategy which compares well with the optimal 'bang-bang' control rule. Pham and Karaboga (1991) described a system which learns fuzzy rules and output membership functions simultaneously using fixed input membership functions. Optimisation of the controller was carried out in two stages. In the first stage, different populations of controllers were independently evolved (using different initial random seeds) to produce 'preliminary' designs. The second stage combined the best individuals from the first stage into a single population to which the GA is applied to evolve a 'detailed' design.

Learning Fuzzy Rules and Membership Functions in Stages

Kinzel *et al.* (1994) described an evolutionary approach to designing fuzzy controllers, and applied the technique to the cart-pole problem. They argued that learning fuzzy rules and membership functions simultaneously is difficult due to complex interactions and propose an alternative three stage task solving process. An initial rule base and membership functions were selected heuristically, rather than randomly and the initial population was seeded with mutations of this initial rule base. The GA was then applied to rules (keeping membership functions fixed). The final stage was to apply the GA for fine tuning of membership functions within good evolved rule bases.

Learning Fuzzy Rules and Membership Functions Simultaneously

Lee and Takagi (1993) employed the genetic algorithm to optimise simultaneously a variable size fuzzy rule base and fuzzy set membership functions of a Takagi–Sugeno controller. The system was applied with success to the cart-pole problem. Cooper and Vidal (1994) used a variable length genome to represent a fuzzy rule- et and accompanying membership functions. They argued that domain-based representations which imply complete coverage of the input space cannot be expected to scale well to high-dimensional problems, and that using variable size rule sets in conjunction with rule creation and deletion operators allows the GA to evolve rule sets which do not include superfluous rules. Liska and Melsheimer (1994) used a GA for simultaneously discovering fuzzy rules and membership functions, with a final stage of fine-tuning membership functions using conjugate gradient descent. They applied the system to learning a dynamic model of plant using known input-output data. After the GA approached convergence, conjugate gradient descent was employed to further improve good solutions by fine-tuning membership function parameters. Linkens and Nyongesa (1995) described off-line evolutionary learning of fuzzy rules and associated membership functions for a multivariable fuzzy controller for medical anaesthesia.

8.3 Evolving Fuzzy Routing Controllers with the GA

8.3.1 Fuzzy Classifier System Details

The fuzzy classifier system employed here is version of P-FCS1 (a Pittsburgh Fuzzy Classifier System #1), described and evaluated in Carse et al. (1996). P-FCS1 is a synthesis of the classifier system (Holland, 1976; Booker et al., 1989) and fuzzy sets (Zadeh, 1965) which employs the genetic algorithm in the 'Pittsburgh'-style (Smith, 1980) in which individuals in the population operated on by the genetic algorithm are complete rule sets. P-FCS1 employs variable length rule sets, uses a real-numbered representation of fuzzy membership function centres and widths, and applies modified recombination operators which are particularly suited to fuzzy as opposed to discrete classifier systems.

In PFCS-1, each rule R_k for an n-input, m-output system, is represented as:

$$R_K : (x_{c1k}, x_{w1k}); \ldots (x_{cnk}, x_{wnk}) \Rightarrow (y_{c1k}, y_{w1k}); \ldots (y_{cmk}, y_{wmk})$$

a similar representation to that used in Parodi and Bonelli (1993). The bracketed terms represent the centres and widths of fuzzy set membership functions over the range of input and output variables. The genome representing a complete rule set is a variable length concatenated string of such fuzzy rules.

The two-point version of the crossover operator used in P-FCS1 involves the generation of two crosspoints C_{1i} and C_{2i} as follows:

$$C_{1i} = \min_i + (\max_i - \min_i) \cdot (R_{1c})$$
$$C_{2i} = C_{1i} + (\max_i - \min_i) \cdot (R_{2c})^{1/n}$$

where R_{1c} and R_{2c} are selected randomly in the range [0,1] with uniform probability density. The range [$\min_i$, $\max_i$] is the universe of discourse of the ith input variable. After crossover, Child 1 contains rules from Parent 1 such that

$$\forall i, ((x_{cik} > C_{1i}) \wedge (x_{cik} < C_{2i})) \vee ((x_{cik} + \max_i - \min_i) < C_{2i})$$

together with rules from Parent 2 which do not satisfy this condition. Child 2 contains the remaining rules from both rule sets. This crossover operator is designed to enhance the probability that good fuzzy 'building blocks' (i.e. high-fitness assemblies of fuzzy rules with overlapping input membership functions) survive and proliferate in future generations.

Mutation in P-FCS1 applies real number 'creep' to fuzzy set membership function centres and widths. In addition, a cover operator is employed to create a new classifier if inputs are encountered which no existing rules match.

In the first set of experiments described below, P-FCS1 is applied to routing control in a simple simulated three node network. In the second set of experiments, an extended version of P-FCS1 called FCDACS (a Fuzzy Clocked Delayed Action Classifier System) is applied to routing control in a more complex network. In FCDACS, the fuzzy classifier syntax is

$$R_K : (x_{c1k}, x_{w1k}); \ldots (x_{cnk}, x_{wnk}) \Rightarrow (y_{c1k}, y_{w1k}); \ldots (y_{cmk}, y_{wmk}); (t_{ck}, t_{wk})$$

where (t_{ck}, t_{wk}) is a tag which encodes the centre and width of a time membership function which is used to modulate the contribution of an activated classifier over time. This allows the evolution of temporal fuzzy classifiers with linguistic interpretations such as:

```
IF (Route1 has lower delay than Route2)
      THEN increase proportion of traffic over Route1
      OVER the medium future
```

A full description of P-FCS1 and FCDACS can be found in Carse (1997).

8.3.2   Routing Control in a Small Scale Network using P-FCS1

In these experiments an assembly of fuzzy controllers are required to perform adaptive, distributed routing control in a simulated fully-connected 3-node datagram packet switched network, as illustrated in Figure 8.1). All links in the network are bidirectional full duplex. Packets requiring transmission over a particular link are queued using a first-come first-served discipline. Packets arrive from outside at network source node $i \in \{A,B,C\}$, to be delivered to destination node $j \in \{A,B,C\}, j \neq i$, at an average rate of λ_{ij} packets/second.

A fuzzy controller at each node decides whether to route each packet directly to its destination or via an intermediate node. Controller decisions are based on packet delay measurements over different paths. The goal is to minimise average global packet delay (i.e. the mean delay between packet arrival at the source node and packet delivery to the destination node for all packets which arrive during the period of simulation, irrespective of source and destination). The learning system is therefore required to determine a routing policy, copies of which are deployed at each switching node and operate in parallel, which minimises global packet delay.

Each routing controller is implemented as a variable size fuzzy classifier system with four inputs and two outputs. At each node the controller inputs are:

DelayLeftDirect: The measured packet delay from the source node for packets destined for the node to the left of the source node and which are routed directly.

DelayLeftIndirect: The measured packet delay from the source node for packets destined for the node to the left of the source node and which are routed indirectly.

DelayRightDirect: The measured packet delay from the source node for packets destined for the node to the right of the source node and which are routed directly.

DelayRightIndirect: The measured packet delay from the source node for packets destined for the node to the right of the source node and which are routed indirectly.

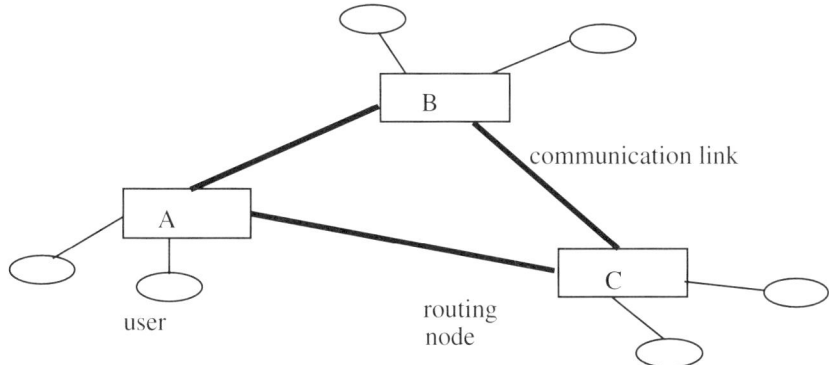

Figure 8.1 Three-node packet switched network used in simulation. Routing nodes A, B and C are connected by bidirectional communication links.

Packet delays are measured at the destination node (each packet is time-stamped on arrival in the system) and averaged over the last $N_{Measure}$ packets for each route taken for each source node. In the simulation, we assume this information is transmitted without delay to source nodes once the averages have been taken, and transmission of control information does not consume network bandwidth. In a real network such information would be sent as control packets which would incur a finite delay and utilise network bandwidth (unless a separate signalling network is in place). $N_{Measure}$ is a parameter varied in the experiments described later, and determines the granularity of measurements. Also, in a real network, a trade-off would have to be made in choosing the value of $N_{Measure}$. If too small a value is chosen, the network becomes swamped with control packets which compete with user data packets for use of the shared bandwidth. If too large a value is chosen, measurements become out of date and meaningless.

At each node, the controller outputs are:

PLeftDirect: The probability that a packet arriving at the source node which is destined for the node to the left of the source node is routed directly.

PRightDirect: The probability that a packet arriving at the source node which is destined for the node to the right of the source node is routed directly.

By dynamically adjusting local *PLeftDirect* and *PRightDirect* control outputs based on network delay measurements, the distributed assembly of controllers should attempt to spread the network load to minimise global mean packet delay in response to changing traffic conditions in the network.

Each network simulation is run for a simulation time of 500 seconds. The data rates of all network links are set to 10,000 bits per second. Mean packet arrival rates used in the simulation, and their variation in time, are given by:

$$\lambda_{AC}, \lambda_{BA}, \lambda_{BC}, \lambda_{CA} = 3 \text{ packets/sec}, 0 < t < 500$$
$$\lambda_{AB} = 3 \text{ packets/sec} \quad 0 < t < 125, t > 250, \text{ and } 15 \text{ packets/sec } 125 < t < 250$$
$$\lambda_{CB} = 3 \text{ packets/sec} \quad 0 < t < 250, t > 375, \text{ and } 15 \text{ packets/sec } 250 < t < 375$$

These patterns were chosen to test the dynamic capabilities of the routing controller in moving from relatively light network load, when direct routing is optimal, to heavy load when controllers must balance the offered load between direct and indirect network paths. In the simulation, packets arriving at an intermediate node are always forwarded to their destination to avoid a 'ping-pong' effect. The evaluation function for each rule set returns the inverse of the mean measured packet delay for all packets delivered during the simulation. Experiments were done with both deterministic and probabilistic (Poisson) packet arrival processes with packet sizes exponentially distributed with mean 1000 bits.

To evaluate the controllers evolved by the fuzzy classifier system, we compared their performance with a shortest-path routing algorithm, which routes all packets along the route whose measured delay is least between a particular source/destination pair. A range of measurement intervals, $N_{Measure}$ from two packets to 100 packets were used. Ten independent runs of P-FCS1 were conducted with different initial random seeds. In addition, different initial random seeds were also used for each of the network simulations used in evaluating a particular individual. The latter introduces noise in the evaluation function and we were interested in whether the system could learn in the face of this potential difficulty.

Each of the 10 evolved fuzzy controllers using P-FCS1 were evaluated in 20 subsequent simulations and the results are presented in Table 8.1 where they are compared with the shortest path routing algorithm. This table shows mean packet delay over the complete simulation interval. The results shown in Table 8.1 indicate that, when the measurement interval is small, the shortest-path algorithm outperforms the learned fuzzy controllers, although not by that large a margin. As the measurement interval increases, the learned fuzzy controllers begin to outperform the shortest path algorithm significantly. As mentioned earlier, an important characteristic of a routing algorithm is that routing control information should not consume excessive network bandwidth. A value of $N_{Measure}$ greater than (at least) 20 is realistic for a real network, and the results using a GA-derived fuzzy controller appear better than the simple shortest-path algorithm in this region of rate of feedback. Inspection of the measured delays on direct and indirect paths demonstrated that the shortest-path controller shows pronounced instabilities as $N_{Measure}$ is increased, while the evolved fuzzy controllers appear to shift traffic in a much more stable manner.

8.3.3 Routing Control in a More Complex Network using FCDACS

Although experiments with the simple 3-node network, described in the previous section, offer some enlightenment regarding the behaviour of learned fuzzy controllers, we need to scale up to more complex networks. To further test the viability of the approach of evolutionary fuzzy control applied to adaptive distributed routing, experiments were carried out on the simulated 19-node network shown in Figure 8.2. This network is taken from earlier work described in Thaker and Cain (1986).

Table 8.1 Mean packet delay (in seconds) of fuzzy routing controllers learned by P-FCS1 compared with mean packet delay using Shortest-Path (SP) routing algorithm (Standard deviations shown in brackets).

Measurement Interval ($N_{Measure}$)	SP-Routing (Deterministic arrivals)	SP-Routing (Probabilistic arrivals)	Fuzzy Control (Deterministic arrivals)	Fuzzy Control (Probabilistic arrivals)
2	0.58 (0.09)	1.14 (0.32)	0.73 (0.12)	1.06 (0.42)
5	1.11 (0.22)	1.61 (0.29)	1.16 (0.16)	2.78 (0.41)
10	1.96 (0.23)	2.46 (0.42)	1.31 (0.32)	2.98 (0.48)
20	3.52 (0.39)	4.27 (0.75)	1.29 (0.28)	3.21 (0.45)
50	6.14 (0.42)	6.48 (1.29)	1.38 (0.60)	3.66 (0.51)
100	8.53 (3.00)	10.85 (1.49)	1.50 (0.38)	4.10 (0.65)

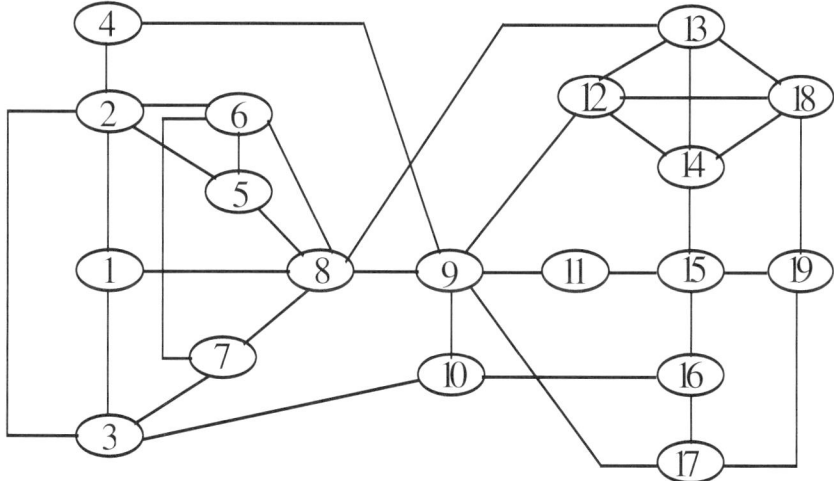

Figure 8.2 19-node network used in simulations.

Clearly, the input/output variables selected for the simple 3-node network controller in the previous section do not scale to this size of network. In the 3-node network, measured delays to all possible destinations over all possible routes were taken into account. To limit the 'state-space' explosion which would occur for larger networks, we propose a hybrid routing controller for operation in the larger scale network. This hybrid scheme scales well to very large networks.

The approach we propose is to employ evolutionary fuzzy control in conjunction with the shortest-path heuristic. However, rather than selecting a single shortest path for packet transmissions (with its attendant stability problems), two shortest-hop paths are selected for each session source-destination pair and the degree of usage of each is varied by the fuzzy controller based on measured delays over those paths. We call this 'fuzzy bifurcated routing'. Shortest hop paths are ascertained before learning using Dijkstra's least cost algorithm, summarised here:

```
M = {s}
FOR each n in N - {s}
    C(n) = l(s,n)
WHILE (N ≠ M)
    M = M∩{w} such that C(w) is minimum for all w in (N-M)
    FOR each n in (N-M)
        C(n) = min(C(n),C(w)+l(w,n))
```

where N is the set of nodes in the network, s is the source node, M is the set of nodes so far incorporated by the algorithm, $l(i,j)$ is the link cost from node i to node j, and $C(n)$ is the cost of the path from the source node to node n.

Inputs to the fuzzy controller are the measured delay and the time derivative of this delay over both shortest-hop paths. The controller output is the fraction of traffic to be routed over the first path (the remainder of traffic being routed over the second path). In an attempt to model a real network in simulation, link delay measurements are sampled every T seconds, and the measurements offered to each routing controller are the previous samples.

To evaluate the performance of evolved temporal fuzzy controllers, the simulator was also run using two alternative routing strategies. The first (which we call Static Shortest-Hop routing) routes packets along a fixed shortest-hop path for each traffic session. Routes are established in advance using Dijkstra's algorithm. The second (which we call Adaptive Shortest Path routing) is similar to the traditional shortest-path routing scheme used in the ARPANET. Each routing node runs Dijkstra's algorithm every T seconds using link delay measurements from the previous sampling interval. Thus each router decides on an outgoing link over which to route traffic to each destination for the next time interval.

Each link is assigned a data rate of 10,000 bits/second and exponentially distributed packet sizes with mean 1000 bits were employed in all experiments. Offered traffic (measured in packets/second) and the link delay measurement update interval, T, are parameters varied in the experiments. T is varied from 0.1 seconds to 10.0 seconds. Offered traffic is based on the traffic profiles employed in Thaker and Cain (1986). Offered traffic for each session is determined by an 'Offered Load Traffic Multiplier' (λ_{Offd}); the amount of traffic offered to the network is directly proportional to this multiplier.

Fuzzy controllers are evolved for different link delay measurement update intervals (T) in the range 0.1 to 10.0 seconds with offered load traffic multipliers in the range 1.0 to 7.0. The fitness evaluation is the inverse of the mean packet delay over the simulation interval of 100 seconds.

Table 8.2 shows mean packet delay versus offered traffic for static shortest-hop, adaptive shortest path and fuzzy bifurcated routing when the link delay measurement update interval is 0.1 seconds. This corresponds to the situation when routers are in possession of almost perfect link state information (i.e. very small delay in measurements). It can be seen that the static shortest-hop algorithm performs poorly. Since the routes selected using this algorithm are fixed, routers cannot direct packets over other routes once the fixed routes become saturated. The adaptive shortest path and fuzzy bifurcated routers yield similar performance to each other, with the adaptive-shortest path method demonstrating slightly better performance.

Table 8.2 Mean packet delay(s) versus offered traffic (λ_{Offd}) for $T = 0.1$ s.

Routing Method	λ_{Offd}						
	1.0	2.0	3.0	4.0	5.0	6.0	7.0
Static Shortest Hop	0.378	0.729	4.687	7.880	11.70	-	-
Adaptive Shortest Path	0.368	0.469	0.669	0.923	1.550	5.441	9.005
Fuzzy Bifurcated	0.377	0.485	0.657	0.936	2.260	7.441	11.71

However, a value of $T = 0.1$ seconds is not realistic for a network with this topology and traffic since the amount of control information in transit is likely to be excessive. Table 8.3 shows the mean packet delay for different routing algorithms for $T = 1.0$ second, a more realistic link delay measurement update interval. In this case, it is seen that the performance of the adaptive shortest path routing method degrades significantly at high offered loads, to the extent that it is comparable to that of the static shortest-hop router. When the dynamics of routing decisions were inspected, it was seen that the adaptive shortest path router showed marked routing instabilities. The degradation in performance of the learned fuzzy bifurcated router, while present, is much less serious.

Table 8.3 Mean packet delay(s) versus offered traffic (λ_{Offd}) for $T = 1.0$ s.

Routing Method	λ_{Offd}						
	1.0	2.0	3.0	4.0	5.0	6.0	7.0
Static Shortest Hop	0.378	0.729	4.687	7.880	11.70	-	-
Adaptive Shortest Path	0.368	0.755	2.830	6.940	8.495	11.90	-
Fuzzy Bifurcated	0.382	0.556	0.880	1.472	3.090	7.685	11.95

To further investigate the effect of link delay measurement interval (T) on performance, we ran the three types of controller in simulations with various values of T with a traffic multiplier of 5.0. The results of this experiment are shown in Table 8.4. The static shortest-hop router gives constant mean packet delays since it does not employ updates (it is not adaptive). The adaptive shortest path router outperforms the fuzzy bifurcated router for $T < 0.2$ seconds, but quickly degrades for larger values of T. The fuzzy bifurcated router's performance degrades much more gracefully as the update interval is increased.

As a further measure of performance, the ratios of number of packets delivered successfully to their destinations to the total number of packets offered to the network during the simulation interval were observed. These results are shown in Table 8.5, for a value of T equal to 1.0 second for various traffic loads.

Table 8.4 Mean packet delay(s) for $\lambda_{Offd} = 5.0$ versus measurement update interval (s).

Routing Method	T						
	0.1	0.2	0.5	1.0	2.0	5.0	10.0
Static Shortest Hop	11.70	11.70	11.70	11.70	11.70	11.70	11.70
Adaptive Shortest Path	1.552	1.913	4.170	8.492	15.80	15.90	13.85
Fuzzy Bifurcated	2.260	2.445	3.094	3.097	4.320	6.365	8.561

Table 8.5 Ratio of delivered packets to offered packets, $T = 1.0$ s.

Routing Method	λ_{Offd}						
	1.0	2.0	3.0	4.0	5.0	6.0	7.0
Static Shortest Hop	0.980	0.980	0.895	0.792	0.705	-	-
Adaptive Shortest Path	0.990	0.970	0.950	0.905	0.892	0.701	-
Fuzzy Bifurcated	0.990	0.990	0.986	0.982	0.965	0.931	0.735

8.4 Conclusions

An approach to adaptive distributed routing using fuzzy control, using a genetic algorithm to evolve the controllers, has been introduced. Experimental results obtained using network simulations have demonstrated the viability of fuzzy bifurcated routing using evolutionary learning. Application of the method to physical communication networks requires further investigation, and it would be valuable to experiment with using the technique on a real testbed network. The method might also extend to circuit switched network routing. The effect of link or node failures has not been discussed in this contribution. Clearly, using a fuzzy controller to select between two shortest *hop* paths will experience problems if a link or node on one or both of these paths fails. Extending the approach to using the fuzzy controller to select between two or more minimum *delay* paths would alleviate this problem, although care would have to be taken to avoid potential packet looping.

9

The Genetic Adaptive Routing Algorithm

Masaharu Munetomo

9.1 Introduction

A routing algorithm constructs routing tables to forward communication packets based on network status information. Rapid inflation of the Internet increases demand for scalable and adaptive network routing algorithms. Conventional protocols such as the Routing Information Protocol (RIP) (Hedrick, 1988) and the Open Shortest-Path First protocol (OSPF) (Comer, 1995) are not adaptive algorithms; they because they only rely on hop count metrics to calculate shortest paths. In large networks, it is difficult to realize an adaptive algorithm based on conventional approaches. This is because they employ broadcasts to collect information on routing tables or network link status, which causes excessive overheads in adaptive algorithms.

In this chapter, we describee an adaptive routing algorithm called the Genetic Adaptive Routing Algorithm (GARA). The algorithm maintains a limited number of alternative routes that are frequently used. Instead of broadcasting a whole routing table or link status, it only observes communication latency for the alternative routes. It also tries to balance link loads by distributing packets among the alternative routes in its routing table. The alternative routes are generated by *path genetic operators*.

9.2 Routing algorithms in the Internet

From the early history of the Internet, vector distance routing algorithms based on Bellman-Ford's algorithm (Bellman, 1957; Ford and Fulkerson, 1962) have usually been employed.

The Routing Information Protocol (RIP) (Hedrick, 1988) based on the vector-distance algorithm with hop count metric is commonly used even now in small local area networks. The algorithm yields many communication overheads in larger networks because they periodically send broadcast messages that contain a whole routing table. More precisely, the number of messages to broadcast routing tables is proportional to n^2 (n is the number of nodes in a network) and the size of a routing table is proportional to n, which means that the total communication overhead for the routing information exchange becomes $O(n^3)$. The vector distance algorithm also suffers from its slow convergence because it is necessary for the routing tables to reach all the nodes in the network to obtain the correct distance.

To reduce communication overhead, routing algorithms based on link status information such as the SPF (Shortest Path First protocol) send broadcast messages that only contain information on link status. Based on a topological database generated from the collected link status, the algorithm calculates the shortest paths employing Dijkstra's shortest path algorithm (Dijkstra, 1959) in each node. The SPF broadcasts messages that only contain the nodes' link status instead of the entire routing table. Therefore, the size of a message that contains link status information becomes $O(1)$ when the degree of nodes is constant such as in a mesh network, and the size becomes $O(\log n)$ in a hypercube network. However, the number of messages is also proportional to n^2, which leads to extensive overheads in larger size network. Therefore, the OSPF (Open Shortest Path First protocol) (Moy, 1989), a widely used network routing protocol for Interior Gateway Protocol (IGP) (Comer, 1995) based on the SPF, relies on hop count metric and detects topological changes such as link failure.

It seems easy to collect information on the communication latency of links and calculate routes with minimum delay; however, it is almost impossible in large networks. This is because we need to collect the communication latency of all the links frequently by broadcast messages, which leads to extremely heavy communication overheads. In addition, delayed information for the latency may create far from optimal routes. In a huge network such as the Internet, it is essentially important for routing algorithms to be scalable. To achieve scalability in adaptive network routing algorithms, it is necessary to observe communication latency with as few communication overheads as possible.

9.3   GAs in Network Resource Management

Before introducing our routing algorithm by GA, this section gives a brief review of related work concerning the application of GAs to network resource management problems. Since many of the problems are classified into combinatorial optimization problems, they can directly be solved by GAs. For example, an application of GA to network routing in telecommunication networks is proposed in the *Handbook of Genetic Algorithms* (Davis, 1991) by Cox *et al.* (1991). The authors solved a constrained optimization problem that allocates link bandwidth to each connection. For another approach, Munakata and Hashier (1993) solved the maximum network flow problem. The objective of this optimization problem is to maximize the network flow based on the global information of the network. These two algorithms cannot be applied directly to routing in packet-switching networks such as the Internet because they are based on circuit-switching networks, in which the bandwidth of the network is allocated to circuits – connections between nodes (Tanenbaum, 1988). On the other hand, Carse *et al.* (1995; and Chapter 8) applied a Fuzzy Classifier System (FCS) to a distributed routing algorithm on packet-switching networks in which each packet is routed independently by a routing table. This routing algorithm, however, only decides whether a packet should take a direct route or an indirect route by using a FCS. Therefore, the algorithm cannot be directly applied to actual routing algorithms.

9.4 Overview of the GARA

The Genetic Adaptive Routing Algorithm (GARA) is an adaptive routing algorithm that employs genetic operators to create alternative routes in a routing table. It is based on source routing algorithms, which determine the entire route of a packet in its source node. Each node has a routing table containing a set of alternative routes, each of which is created by genetic operators. Each packet selects one of the alternative routes to the destination. The algorithm observes the communication latency of routes frequently used.

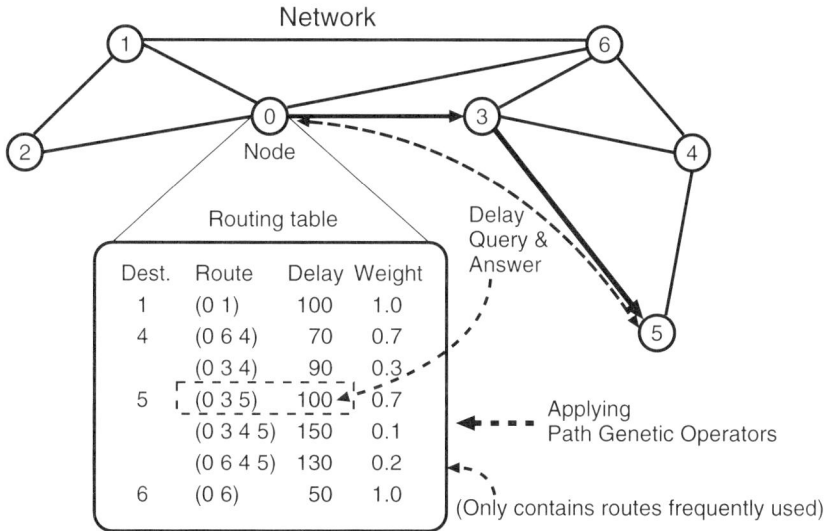

Figure 9.1 Overview of the GARA.

Figure 9.1 shows an overview of the GARA. Each node has a routing table that contains routes to destinations. For each destination, a set of alternative routes is assigned. Each route has its weight value (as its fitness) that specifies the probability for the route to be selected from a set of alternative routes. The weight is calculated from the communication latency, observed by sending delay query packets. At the beginning, the routing table is empty. When a packet must be sent to a destination, a *default route* to the destination is generated by using Dijkstra's shortest path algorithm based on hop count metric. After a specified number of packets are sent along a route, a packet is sent to observe the communication latency of the route. To generate alternative routes, *path genetic operators* are invoked at a specified probability after every evaluation of weight values. To prevent overflow of a routing table, selection operators are applied to reduce its size. A selection operator deletes a route that has the lowest weight among the routes to a destination. Moreover, another selection operator may be applied; this deletes all the routes to a destination to which the number of packets sent is the smallest among all the destinations in a routing table.

The GARA can greatly reduce communication overheads for dynamic observation of network status by limiting the observation to alternative routes frequently used. It can also reduce the possibility of congestion by distributing packets probabilistically among the alternative routes.

9.5 Path Genetic Operators

For the GARA, path genetic operators such as *path mutation* and *path crossover* are designed to generate alternative routes based on the topological information of the network. The path mutation operator causes a perturbation of a route in order to create an alternative route. The path crossover exchanges sub-routes between a pair of routes.

A route (path) is encoded into a list of node IDs from its source node to its destination. For example, a route from node 0 to node 9 is encoded into a list of nodes along the route: (0 12 5 8 2 9). The route is constrained to the network topology; that is, each step of a route must pass through a physical link of the network.

9.5.1 Path Mutation

The path mutation generates an alternative route by means of a perturbation. Figure 9.2 shows how to perform this mutation operator. To perform a mutation, first, a node is selected randomly from the original route, which is called a mutation node. Secondly, another node is randomly selected from nodes directly connected to the mutation node. Thirdly, an alternative route is generated by connecting the source node to the selected node and the selected node to the destination employing Dijkstra's shortest path algorithm.

More precisely, the path mutation proceeds according to the following sequence where r is the original route and r' is its mutated one.

1. A mutation node n_m is selected randomly from all nodes along the route r, except its source and destination nodes.
2. Another node n_m' is selected randomly from neighbors of the mutation node, i.e. $n_m' \in \varepsilon(n_m)$.
3. It generates a shortest path r_1 from the source node to n_m', and another shortest path r_2 from n_m' to the destination.
4. If there is no duplication of nodes between r_1 and r_2, they are connected to have a mutated route $r = r_1 + r_2$. Otherwise, they are discarded and no mutation is performed.

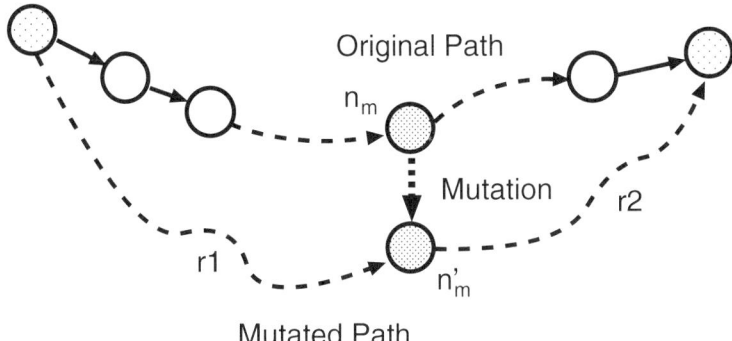

Figure 9.2 Path mutation applied to a route.

Suppose that the mutation operator is applied to a route $r = (0\ 3\ 5\ 6\ 7\ 10\ 12\ 15)$. First, we select, for example, a node 7 as a mutation node. Secondly, another node 8, for example, is randomly selected from the neighbors of the mutation node. Thirdly, by Dijkstra's shortest path algorithm, we connect the source node 0 and the selected node 8 to generate a path $r_1 = (0\ 2\ 4\ 8)$. We also connect node 8 and the destination node 15 to generate another path $r_2 = (8\ 10\ 12\ 15)$. We finish the mutation by connecting the routes r_1 and r_2 to eventually have $r' = (0\ 2\ 4\ \boxed{8}\ 10\ 12\ 15)$. We do not generate a route r' if any duplication exists between r_1 and r_2. This is because we need to avoid creating any loop in a route that passes through the same nodes twice or more.

9.5.2 Path crossover

The path crossover exchanges sub-routes between a pair of routes. To perform the crossover, the pair should have the same source node and destination node. A crossing site of the path crossover operator is selected from nodes contained in both routes. The crossover exchanges sub-routes after the selected crossing site. Figure 9.3 shows an overview of the operator.

The crossover operator applied to a pair of routes r_1 and r_2 proceeds as the following sequence.

1. A set of nodes N_c included in both r_1 and r_2 (excluding source and destination nodes) are selected as potential crossing sites. If N_c is an empty set, we do not perform the crossover.
2. A node n_c is selected randomly from N_c as a crossing site.
3. The crossover is performed by exchanging all nodes after the crossing site n_c between r_1 and r_2.

Suppose that a path crossover is applied to a pair of routes r_1 and r_2 from node 0 to node 20 as in the following:

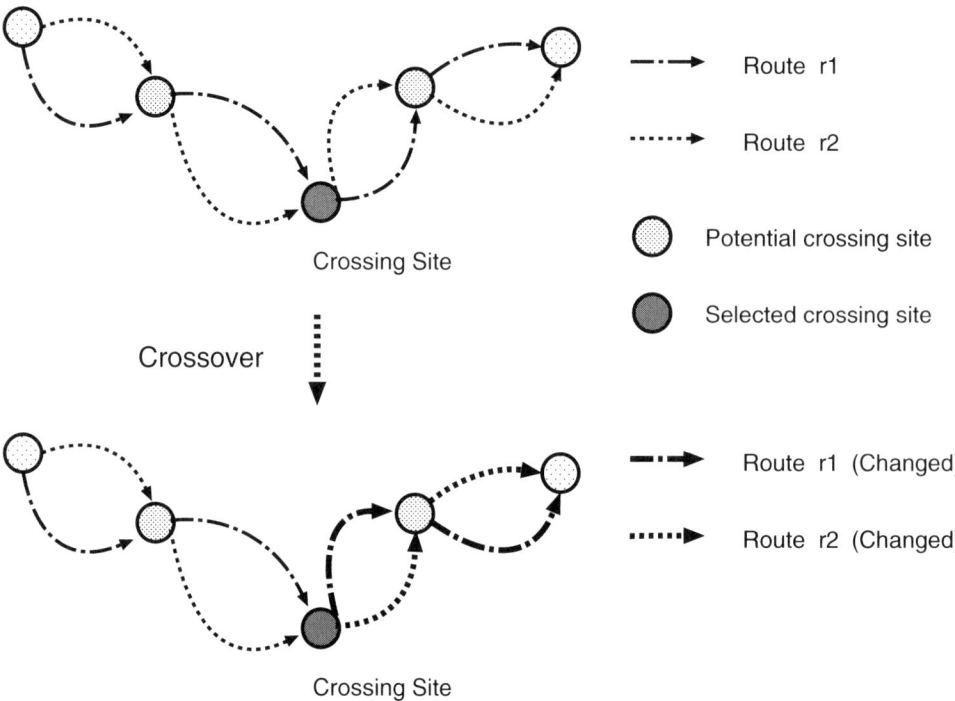

Figure 9.3 Path crossover applied to a pair of routes.

$$r_1 = (0\ 2\ 3\ \boxed{7}\ 9\ \boxed{11}\ 12\ \boxed{15}\ 17\ 18\ 20),$$
$$r_2 = (0\ 4\ 5\ \boxed{7}\ 10\ \boxed{11}\ 13\ \boxed{15}\ 16\ 20).$$

Their potential crossing sites are nodes 7, 11 and 15. When we select, for example, the node 11 as a crossing site, the offspring are generated by exchanging the sub-routes after the crossing site:

$$r_1 = (0\ 2\ 3\ 7\ 9\ \boxed{11}\ \mathbf{13\ 15\ 16\ 20}),$$
$$r_2 = (0\ 4\ 5\ 7\ 10\ \boxed{11}\ \mathbf{12\ 15\ 17\ 18\ 20}).$$

When we do not have a common node in a pair of routes, we cannot select a crossing site and we do not perform a crossover. In our earlier work such as Amano *et al.* (1993) and Murai *et al.* (1996), in order to perform a crossover even when the pair has no common nodes, we connect crossing sites by calculating the shortest path between them. However, a pair of routes without any similarity is not worth crossing-over, because the operator may generate routes which are too far from their parents for such routes, which may lead to random regenerations of routes.

9.6 Maintaining Routing Tables

Table 9.1 shows a routing table used in the GARA. The routing table consists of five entries named *destination*, *route*, *frequency*, *delay* and *weight*. For each destination, we have a set of alternative routes. The *frequency* of a route specifies the number of packets sent to the destination along the route. The *delay* entry stores the communication latency of packets sent along the route. The *weight* of a route is calculated by its communication latency, which specifies the probability for the route to be selected when a packet is sent. From a GA's point of view, the weight corresponds to a fitness value.

Table 9.1 A routing table.

Destination	Route	Frequency	Delay	Weight
2	(1 3 2)*	7232	50	0.7
	(1 3 4 2)	2254	60	0.2
	(1 3 4 5 2)	1039	70	0.1
6	(1 8 6)*	20983	100	0.4
	(1 10 11 6)	34981	105	0.6
8	(1 8)*	30452	40	0.9
	(1 7 8)	3083	40	0.1

In the table, routes marked with asterisks are the default routes which are the shortest paths based on hop-count metric. Initially, the routing table is empty. When it becomes necessary to send a packet to a destination, and if there is no route to the destination, a default route is generated by Dijkstra's algorithm and is inserted to the routing table. After sending a specified number of packets along a route, we send a packet that observes the communication latency of the route. The fitness value is calculated after receiving its answer. After the observation, path genetic operators are applied to the routes at a specified probability to generate alternative routes.

9.7 Fitness Evaluation

A weight value (fitness value) of a route specifies the probability for the route to be selected from alternative routes. It is calculated from the communication latency along the routes. To observe delay along the route, a delay query message is issued at a specified interval. Using the delay obtained, we calculate weight values w_i using the following equations:

$$w_i = \frac{1/\eta_i}{\sum_{j \in S} 1/\eta_j}, \quad \text{where} \quad \eta_i = \frac{d_i}{\sum_{j \in S} d_j} \tag{9.1}$$

where d_i is the delay for route i and S is a set of routes to the same destination. In the above equations, we first normalize the delay values d_i by dividing by their total sum to yield η_i. Second, we calculate the reciprocal number of η_i and normalize them to have a weight value. This is because we need to have a larger weight of w_i for a smaller delay of d_i. Consequently, a route with smaller delay is frequently employed in sending packets. However, note that the selection of routes is a randomized one; routes with longer delay also have a chance to be selected.

9.8 Execution Flow

In the GARA, each node executes the same algorithm independently. Figure 9.4 shows a pseudo PASCAL code of the algorithm. Each packet has entries such as type, route and next, where type specifies type of a packet, route entry is the route of the packet, and next indicates next hop in its route. Types of packets are DataPacket (which contains data), DelayRequest (for requesting delay of a link), and DelayAnswer (for answering a DelayRequest packet). DelayRequst and DelayAnswer packets have DelayEntry which records the communication latency of the packets.

Every time a packet is created at a node, the node determines a route for the packet based on its routing table. For a packet arriving from another node, if its type is DataPacket, the node simply forwards the packet according to its route. In the initial state, a routing table is empty. If the routing table does not contain a route for the destination of a packet created, a default route is generated by employing Dijkstra's shortest path algorithm and is inserted to the table.

Each time after a specified number of packets are sent along a route, a DelayRequest packet is sent to observe the communication latency along the route. If the packet arrives at the destination, a DelayAnswer packet is sent back. After receiving the answer, the communication latency of the route is obtained by calculating the average amount of time sending a DelayRequest packet and receiving a DelayAnswer packet.

After obtaining the delay of a route, weight values of routes are calculated according to equation 9.1. After every evaluation of weights, genetic operators are invoked at a specified probability to create alternative routes in the routing table.

If the size of the routing table exceeds a limit, we perform a selection to reduce its size. We have two types of selection operators: *local selection* is invoked if the number of strings exceeds a limit, and it deletes a route with the smallest weight among routes with the same destination; and *global selection* is invoked when the number of destinations in the routing table exceeds a limit, and it deletes all the routes to a destination of which frequencies of sending packets is the smallest among all destinations in the routing table.

9.9 Empirical Results

To evaluate routing algorithms, it is important to perform experiments for networks large enough. In the following experiments, also discussed in Munetomo *et al.* (1998a), a sample network with 20 nodes is employed (Figure 9.5). The bandwidth of a link is represented by its thickness. The thicker link is for the link with 4.5 Mbps, and the thinner one has bandwidth of 1.5 Mbps.

The Genetic Adaptive Routing Algorithm

```
begin
Initialize routing table;
while not terminated do
  begin
    wait for a packet to be received or input from user;
    if packet.type = DataPacket then
      begin
        if the packet is sent from other node then
          begin
            if packet.destination = this node ID then receive packet;
            else send the packet to the node of packet.next;
          end
        else (* if the packet is input from user at the node *)
          begin
            if routing table for the destination is empty then
              begin
                create a default route by using Dijkstra's algorithm;
                add the default route to the routing table;
                reset frequency entry of the default route;
                packet.route := default route;
                if the number of destination > limit then
                    delete routes of a destination least frequently used;
              end
            else
              begin
                select a route from the routing table by roulette wheel selection;
                increment frequency entry of the selected route;
                packet.route := the selected route;
                if route.frequency mod EvaluationInterval = 0 then
                  begin
                    packet.type := DelayRequest;
                    packet.route := route;
                    send the packet according to the route;
                  end
              end
          end
      end
    if packet.type = DelayRequest then
     begin
      packet.DelayEntry := CurrentTime - packet.CreateTime;
      packet.CreateTime := CurrentTime;
      packet.type := DelayAnswer;
      send packet to its source;
     end;
    if packet.type = DelayAnswer then
     begin
      packet.DelayEntry := packet.DelayEntry + CurrentTime - packet.CreateTime;
      packet.delay_entry := packet.DelayEntry/2;
      change delay of the route based on packet.DelayEntry;
      if random < Pmutation then
        begin
          apply a mutation to a route in the routing table;
          add the string created to the routing table;
          if table_size > limit then delete a string of the lowest weight;
        end
      if random < Pcrossover then
        begin
          apply a crossover to a pair of route in the routing table;
          add the string created to the routing table;
          if table_size > limit then delete a string of the lowest weight;
        end
     end;
  end
end.
```

Figure 9.4 Pseudo PASCAL code for the GARA.

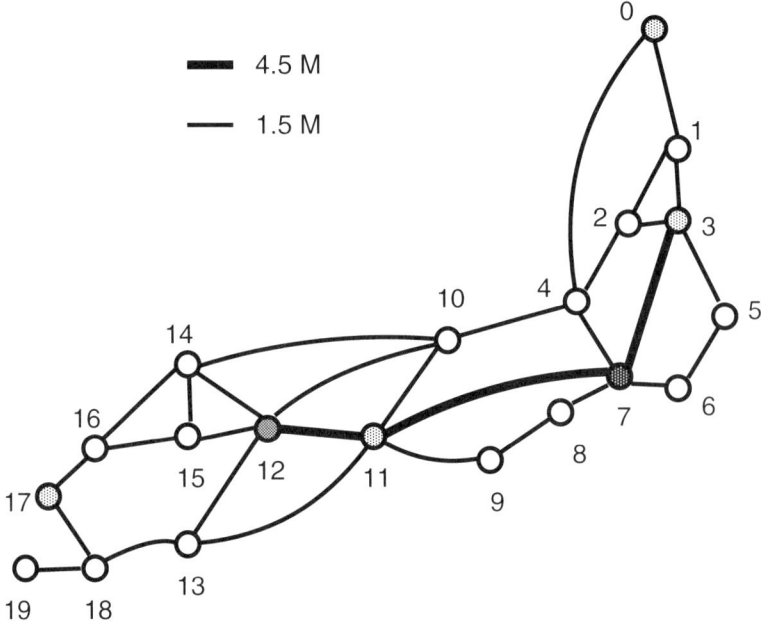

Figure 9.5 A sample network for the experiment.

We compare the GARA with conventional routing algorithms under the following conditions: at each node in the network, data packets are randomly generated at a specified interval that is exponentially distributed. The destination of a packet is randomly selected from nodes that are represented by circles with gray inside (see Figure 9.5). Delay query packets are sent every after 10 data packets are sent. The probability to apply a mutation after an evaluation is 0.1 and that to apply a crossover is 0.05. The maximum population size is 100. We continue the simulation for 3000(s) to obtain the following results.

In Figure 9.6, we compare the mean arrival time of data packets for the RIP, for the SPF, for an adaptive version of the SPF, and for the GARA. The adaptive SPF observes communication latency of links to calculate the shortest paths (we set the delay observation interval at 30(s) and 60(s) for the experiments). The mean arrival time indicates the mean value of the time it took to arrive at destinations of packets from their creation on the source nodes. We can change the frequency of generating data packets by changing the mean generation interval of data packets which is exponentially distributed.

This figure shows that the mean arrival time of packets is smallest when we employ the GARA algorithm for routing. The SPF is slightly better than the RIP. Adaptive SPFs suffer from communication overhead caused by their frequent observation of the link status of all links, especially when the network is lightly loaded (longer mean generation interval). For a 2000 (ms) interval of creating packets, which leads to heavily loaded links in the network, the GARA achieves about 20% of the mean arrival time compared with those of the RIP and the SPF. This means that packets sent by the GARA arrived at their destinations five times faster than those sent by the other algorithms. In this experiment, the adaptive SPF with a 30 (s) observation interval becomes the best in a heavily loaded situation; however, in larger networks, the observation of delay of all links in the networks must cause a serious communication overhead.

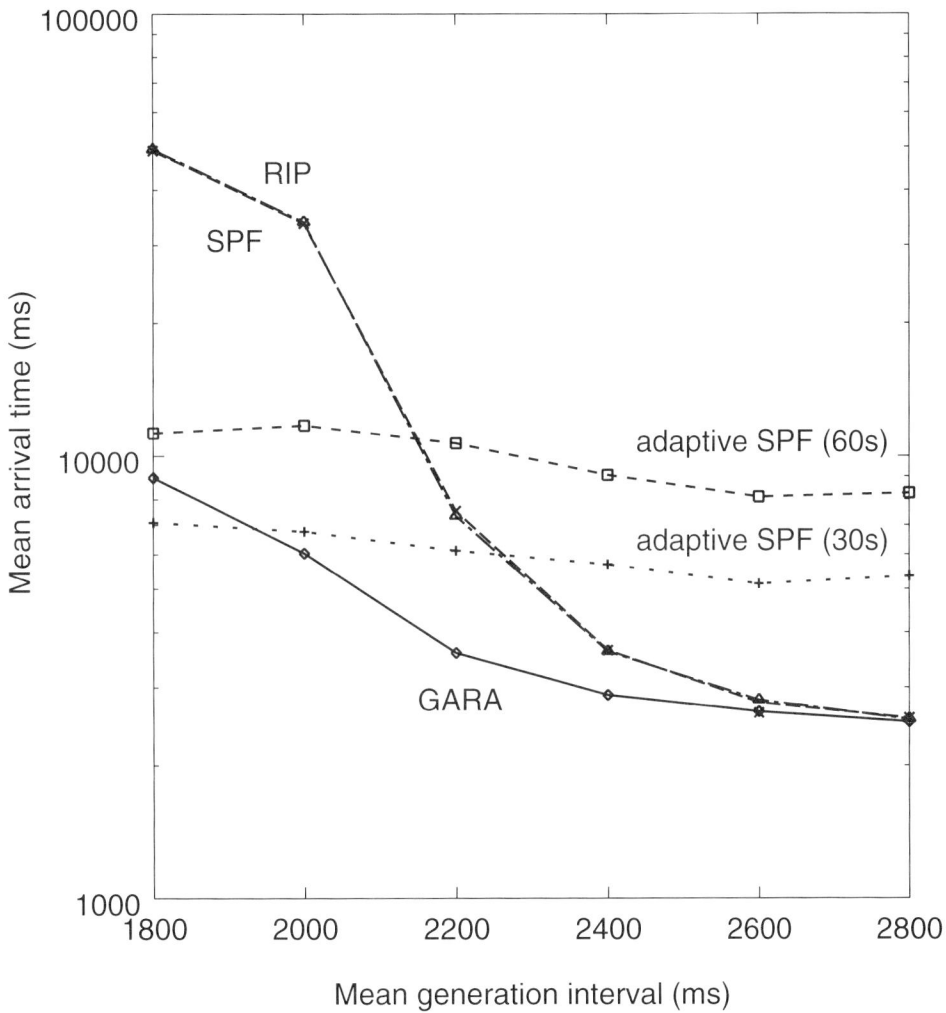

Figure 9.6 Mean arrival time of packets.

Figure 9.7 shows the number of packets sent by the routing algorithms in the network. For frequent observations, the adaptive SPF needs to send a number of packets. The RIP and the SPF send almost the same number of packets (the size of each packet is different). To reduce the number of packets in the adaptive SPF, it is necessary to increase the observation interval. In this experiment, the adaptive SPF with a 60s interval could reduce the number to nearly the same as the RIP and the SPF. However, less frequent observation causes inaccurate observed latency of links, which may not generate the shortest paths. This degrades total arrival time as shown in Figure 9.6. On the other hand, the GARA achieves a smaller arrival time with much lower communication overheads. The number of packets sent becomes less than 20% of those sent for the RIP and the SPF.

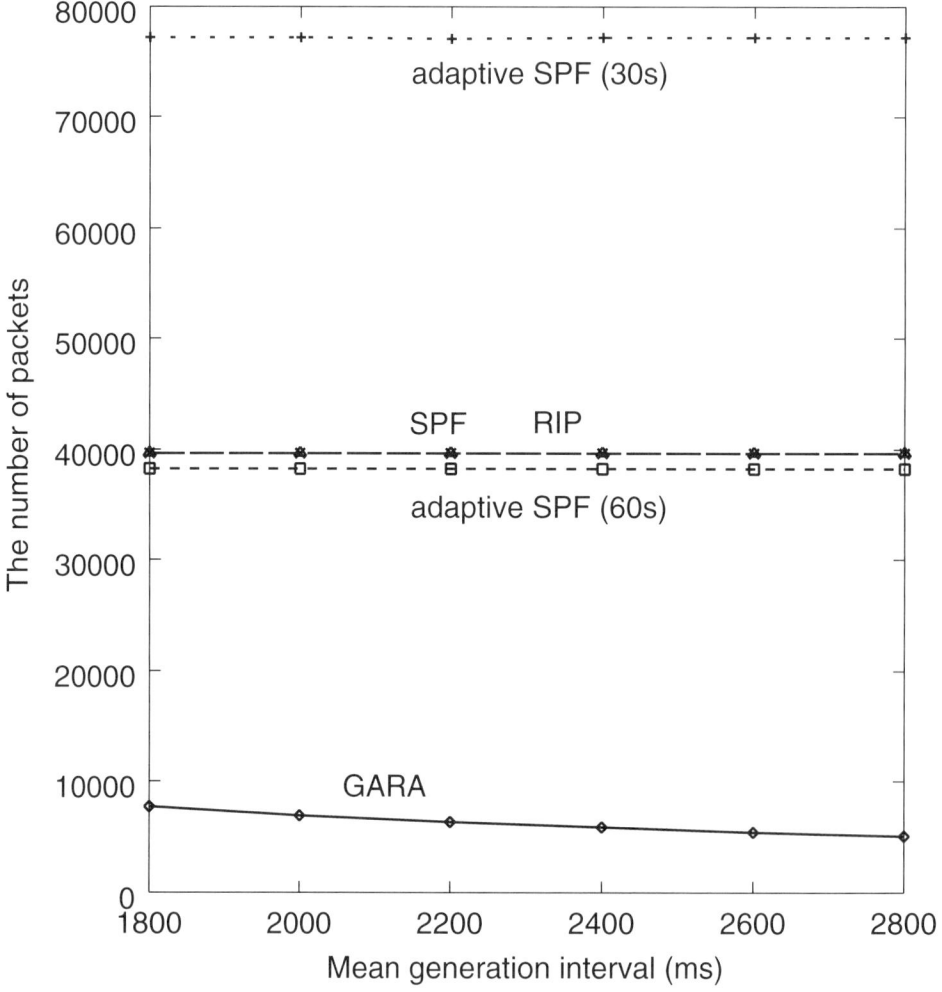

Figure 9.7 The number of packets sent by the routing algorithms.

To see the communication overheads of the network, Figures 9.8–9.11 display load status of links for the RIP, the SPF, the adaptive SPF and the GARA. The width of each link represents the log-scaled mean queue length for the link.

The Genetic Adaptive Routing Algorithm

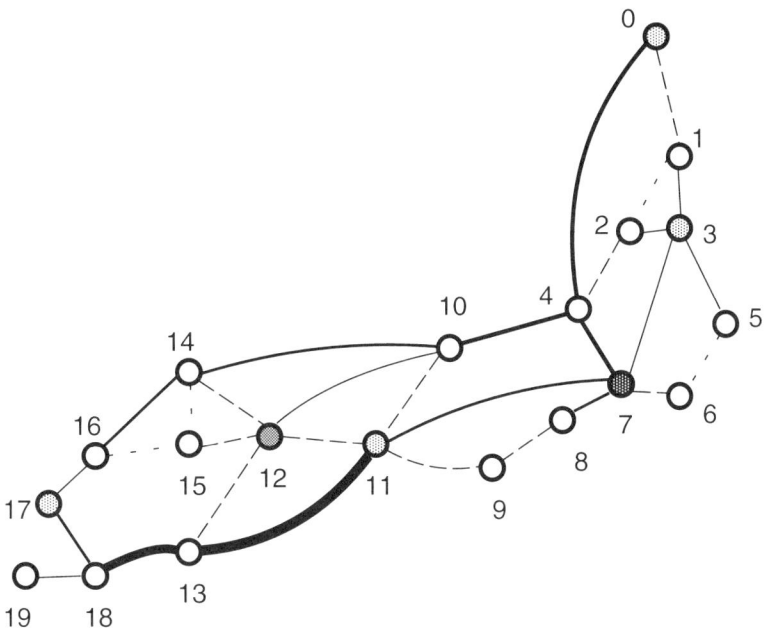

Figure 9.8 Load status of links (RIP).

In the routing information protocol, a path such as 11 ⇒ 13 ⇒ 18 becomes extremely heavily loaded. On the other hand, a path such as 11 ⇒ 12 ⇒ 15 ⇒ 16 ⇒ 17 ⇒ 18, which is an alternative to 11 ⇒ 13 ⇒ 18, is not frequently used at all; therefore, links for this more roundabout path are lightly loaded. By employing the shortest-path first protocol, we can slightly reduce the load on the links. For the adaptive shortest path first protocol, the majority of the links become heavily loaded because of the frequently broadcast messages to observe the communication latency of all links. On the other hand, the GARA can greatly reduce the load of links, especially on heavily loaded ones. This is not only because the GARA algorithm calculates paths which minimize communication latency with less communication overhead, but also because it distributes packets among alternative paths in the routing table which avoids heavily loaded links.

Table 9.2 shows a routing table generated in the node 0 after a simulation run under a 2200 (ms) arrival interval of creating packets. With reference to Table 9.2, we can observe the following points. For a destination node 12, the best route in terms of delay is clearly (0 4 7 11 12). An alternative route (0 4 10 12) is the shortest path from node 0 to node 12 in terms of the hop-count metric, but it is not the path with minimum delay because the links from node 7 to 11 and from node 11 to 12 have more bandwidth than the other links. This result also shows that a load balancing among alternative routes is realized by distributing packets probabilistically based on their weight values.

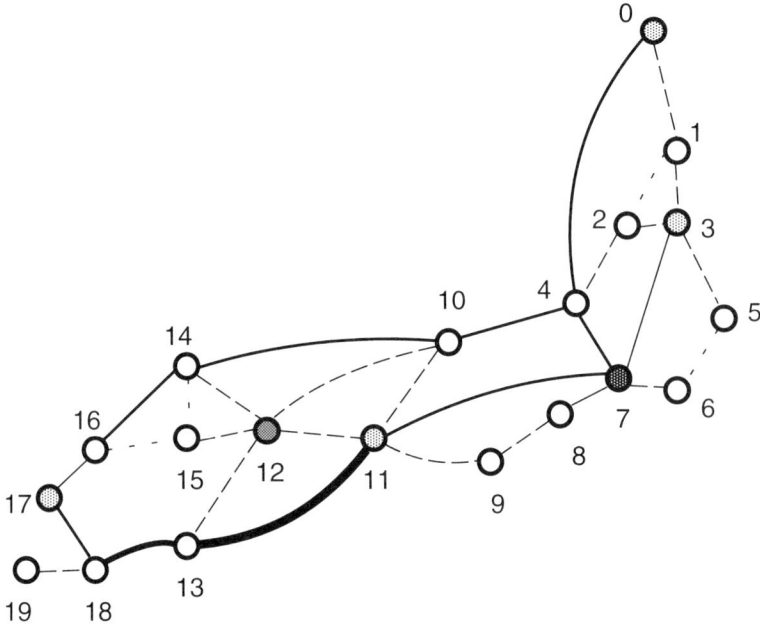

Figure 9.9 Load status of links (SPF).

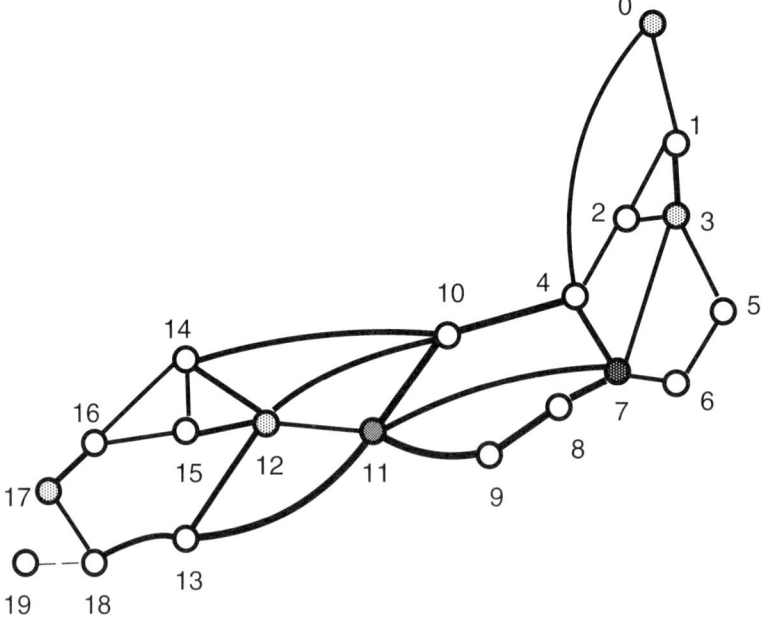

Figure 9.10 Load status of links (adaptive SPF, 30 s interval).

The Genetic Adaptive Routing Algorithm

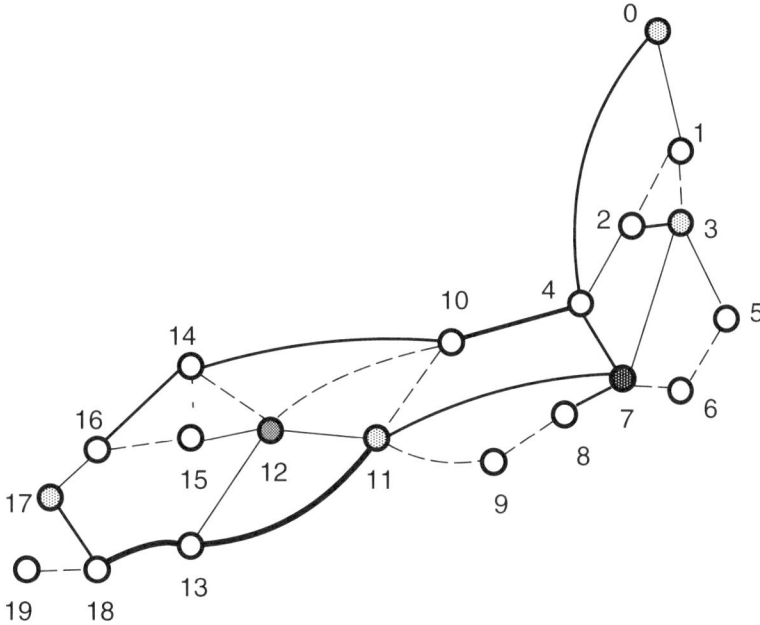

Figure 9.11 Load status of links (GARA).

Table 9.2 The routing table in node 0 generated after the simulation.

Destination	Route	Delay	Weight
3	(0 1 3)	554	0.802636
	(0 4 2 3)	2253	0.197364
7	(0 4 7)	5052	1.000000
11	(0 4 7 11)	4423	0.533488
	(0 1 3 7 11)	5058	0.466512
12	(0 4 7 11 12)	2941	0.564116
	(0 4 10 12)	6210	0.267160
	(0 1 2 4 10 12)	9833	0.168724
17	(0 4 10 14 16 17)	2 859	1.000000

9.10 Conclusions

The GARA realizes an effective adaptive routing with less communication overhead by the following mechanisms:

- It observes the communication latency of routes frequently used.
- It generates an appropriate set of alternative routes employing genetic operators.
- It distributes packets among the alternative routes, which realizes a load balancing mechanism among them.

Concerning communication overhead of the GARA, only $O(kn)$ messages are necessary to observe the load status of routes, where k is the number of data packets sent from a node and n is the number of nodes. On the other hand, the RIP needs $O(n^3)$ communication overhead and the SPF needs at least $O(n^2)$ (it depends on the network topology) for the observation. This means that the GARA is a scalable routing algorithm which works much better in a larger size network than the conventional routing algorithms.

A possible problem in applying the GARA to the Interior Gateway Protocols (IGPs) will be in its source routing approach. The Internet Protocol (IP) (Black, 1995) has a source routing option, but is not usually employed in the IGPs. Instead, it specifies the next hop for a packet to be forwarded. Obtaining the next hop from the route in the routing table generated by the GARA seems not so difficult, but it may cause some problems such as loops or invalid routes that never reach their destinations. Despite the above problem, the GARA is easily applied to Exterior Gateway Protocols (EGPs) such as the BGP4 (Border Gateway Protocols) which employ source routing mechanisms.

10

Optimization of Restoration and Routing Strategies

Brian C.H Turton

10.1 Introduction

All point-to-point communication networks have a means of directing traffic from a source to a destination via intermediate nodes. This routing function must be performed efficiently as it affects all aspects of network communications including jitter, latency and total bandwidth requirement. Other issues related to routing, such as policing and traffic control are not dealt with in this chapter.

There are two distinct approaches to 'routing' traffic, namely connection-oriented and connectionless. Connectionless networks use information contained within the data packet itself. Typically the destination address is all that is required, however explicit routing information may also be contained within the packet. Connection oriented networks establish the route first and subsequently use it for all traffic associated with the connection until it is torn down. In both cases, a method is required for the efficient identification of either a route for the packet or a route for the connection being established.

10.1.1 The Traditional Approach

Traditionally, routing has been static with permanent routes established in the switches or dynamic with routes established according to the traffic flow at the time. In static routing manual updates to the routing tables is still a common occurrence for connection oriented networks, and is frequently based on the experience of the network manager. Different forms of shortest path routing method (Steenstrup, 1995) are usually employed to assist in designing these routes. It should be noted that the 'length' of a link may be associated with time delay, blocking probabilities or financial cost, not just a hop count. In particular flow deviation (Kershenbaum, 1993) can be very effective if a single route between node pairs is required. There are no strong time constraints on this activity.

An alternative approach is to establish routes dynamically, allowing the network to respond to changes in traffic, either when a route is established or a packet is received. Different forms of link-state and distance vector routing (Tanenbaum, 1996) are used which are designed to cope with the network's distributed nature. These are combined with a form of hierarchical addressing in order to make routing tables both manageable and efficient.

Occasionally node or link failures occur, which requires rerouting a large number of connections, this activity is known as 'restoration'. Unlike normal routing, a large number of routes must be re-established in real-time. Although the normal techniques for dynamic routing can be applied, it is likely to be far more efficient to look for the best set of routes for the system as a whole rather than individually route according to the normal methods. However the network must also ensure that the sudden activity generated by the failure does not overwhelm the switches and that restoration occurs on an appropriate time scale (<2 seconds is a generous time scale). Consequently, a new type of distributed algorithm have been developed to disperse the work of restoring the links across the network. An alternative to this is to have fixed restoration routes (protection switching) that were established when the routes were first defined (static routing).

10.1.2 *Possibilities for Evolutionary Algorithms*

Dynamic routing at the time the connection is established, or when the packet is sent, is unlikely to benefit from the use of Evolutionary Algorithms (EAs). There is only one packet to forward or connection to establish, and consequently the existing techniques find the optimum 'greedy' choice. EAs are useful when a large number of routes must be established and a global optimum is required in order to efficiently utilise the available bandwidth and ensure a suitable quality of service. This is particularly evident in ATM where link utilisation levels may be over 85% and consequently any bandwidth savings have a major effect on performance. This problem has led to research in:

- Static routing
- Restoration algorithms
- Routing Strategies and Benchmarks
- Aids to routing algorithms.

10.2 Problems Associated with Routing

10.2.1 *Complexity*

The simplest form of routing assessment to consider is to establish a traffic matrix using the number of connections that can be established as the goal. This produces a number of difficulties, for example, time delay is ignored even though it is a critical parameter for real-time interactive voice and video traffic. In addition a simple connection count will not take account of prioritisation. One of the major concerns associated with costing networks is that the users' view of the worth of a connection does not correlate well with bandwidth. Traffic is not constant bandwidth, so connections may be assigned an effective bandwidth that is believed to be sufficient to ensure the connection can be maintained. In practice, the smallest effective bandwidth cannot be calculated in isolation from the rest of the traffic. Two Variable Bit Rate (VBR) connections can share significant bandwidth if there is little correlation between the two on peak usage. However two Constant Bit Rate (CBR) connections or a VBR plus CBR cannot. At present, traffic on networks is not well enough characterised to allow detailed calculations and even if it can be accurately simulated the time-scale would be prohibitive for evolutionary techniques due to the large number of evaluations required.

In practice the following metrics are used, ignoring some of the complexities of real traffic flow.

- Average number of hops (or average number of hops per call)
- Maximum or average residual link capacity
- Maximum or average network delay
- Number of unserved circuit demands or packet discards a closely linked alternative is the average utilisation or maximum utilisation levels
- Total network throughput

As the M/M/1 queuing model is often assumed the average network delay (T) is:

$$T = \frac{1}{\gamma} \cdot \sum_{i=1}^{l} \left(\frac{f_i}{C_i - f_i} \right)$$

where γ is the total arrival rate for the network (messages/sec), l is the number of links in the network, f_i is the flow rate for link i (bits/sec), and C_i is the capacity of link i (bits/sec). In practice the total network delay is faster to calculate than the mean and differs only by γ.

Any more detailed calculations are likely to be extremely network-dependant. Establishing benchmarks for realistic traffic generation and accurate simulations is still a problem for researchers in this area. The following papers contain or refer to some data on topolgies and traffic (Dutta and Kim, 1996; Mann, 1995; Turton and Bentall, 1998; Gavish and Neuman, 1989; Mann and Smith, 1996). In practice the simple M/M/1 queuing model (Mazda, 1996) is used rather than detailed simulation due to time constraints.

10.2.2 Timing and Efficiency

All evolutionary techniques require significant computational time. Consequently, the approaches used have either had to deal with small numbers of routes or pre-calculate solutions. A reasonable commercial backbone network may well have over fifty nodes and a degree of four. The typical number of routes broken by a single link failure depends on the technology used. In ATM over a hundred paths could be disrupted. Each path is likely to have hundreds of alternate routes if capacity constraints are ignored. If these values are used as reasonable guides to the size of the problem the inherent time scale difficulties are exposed. Despite these problems there are some positive points. Computer power is rising exponentially and networks are not random meshes, operators want to reduce costs and often base their networks on rings with additional chords. Consequently, the number of loop-free alternate routes is much smaller than a worst case scenario indicates. Nevertheless, EAs often have a run time of hours when sub-second times would be ideal.

10.3 An Evolutionary Approach to Routing and Restoration

A number of issues have been tackled by researchers relating to routing and restoration. This section will outline some of the approaches taken, followed by more detailed examples. Despite the real-time nature of restoration algorithms, it is possible to use evolutionary algorithms to set up a strategy that is continually updated and provides a set of pre-planned routes to a restoration algorithm. If the system can be informed of traffic requirements before they need to be routed, the same algorithms can be used for routing. Chng *et al.* (1994) describe a multi-layer restoration strategy which uses this technique and demonstrates its effectiveness. Consequently an evolutionary algorithm is only disadvantaged by the fact the routes are always based on slightly dated information. The nature of the network and traffic will therefore determine the usefulness of this approach.

10.3.1 Evolving Paths

A number of authors encode an index to a table of alternate routes (Mann and Smith, 1996; Shimamoto *et al.*, 1993; Sinclair, 1998; Al-Qoahtani *et al.*, 1998; Tanderdtid *et al.*, 1997). Typically the *k*-shortest paths are used within the table, but as pointed out by Mann and Smith (1996), this may not be the ideal technique. The advantage of this method is that the alternate routes available can be selected appropriately and then uniform, single point or two-point crossover can be used to recombine the chromosomes. Mutation is achieved by simply swapping to another randomly selected path. Results from several authors indicate that this approach compares well with traditional techniques and performs at roughly the same level when compared with simulated annealing. The key disadvantage is that only a limited number of alternate routes are stored and the user, not the algorithm, chooses the set of routes. By making this choice, before running the GA, some viable and effective options may not be considered. Alternatively, if all routes are available in the tables, possibly with limited lengths, a single population cannot contain indices to all routes. The algorithm will find it difficult to search the set of viable routes. Pre-seeding the populations where useful solutions are available has been shown to improve performance by ensuring some good sets of paths are included.

Seo and Choi (1998) developed an evolutionary algorithm designed to find a good set of alternate paths. Paths between a particular source and destination are stored as an ordered list of nodes. Where common nodes exist sections of path between the common nodes can be interchanged. A repair operator is used to remove loops caused by the recombination. Mutation selects two nodes at random and generates a new partial path between the two nodes. The fitness function is based on the number of common nodes, the common links ratio and area surrounded by the set of alternate paths. Apart from the computational cost the authors reported good results.

Cox, Davis and Qui (1991) used a request permutation technique to route the call requests. For their problem, a series of traffic demands are sent that must be fulfilled at some future time. A set of paths has been predetermined for each source destination pair and call type. They suggeet a k-shortest paths ($k \sim 4$) approach is used with constraints for the particular call type. In addition a set of path assignment probabilities based on the traffic patterns is stored. Call requests are initially assigned paths according to the static set of path selection probabilities. If the initial path selected is feasible then it is added to the list of pending requests ready to be connected at the assigned time. If the path is not feasible then an evolutionary algorithm is called that uses crossover and mutation to permute the order in which the call requests will be considered. The chromosome decoder will take each request in turn and check if the request can be assigned to any of the k-shortest paths previously calculated for the appropriate source destination pair. If a suitable path is found, its bandwidth is subtracted from the capacity available and the next request is considered. Once all the paths that can be assigned have been placed, a fitness value is calculated based on the cost of the capacity used for the set of paths and the cost of failing to route the outstanding requests. Uniform order-based crossover is used with scramble sublist mutation and an exponential fitness technique (Cox et al., 1991). Once the algorithm has terminated, the solution undergoes a simple pair wise swapping of each adjacent pair of requests in turn. If any swap improves the schedule it is incorporated into the schedule. At this point the new call can be accepted if all requests can be accommodated, or rejected. For small networks greedy heuristics, integer programming and 'random' search were effective. The evolutionary approach proved best technique for larger networks (~50 nodes or more).

An alternative approach was taken by Bentall et al. (1997), using a two-dimensional encoding with an unusual permutation based crossover technique. They use loopless paths, limited in maximum length, within the chromosome. The procedure is quite different from those discussed so far and so a detailed explanation is given at the end of the chapter.

10.3.2 Evolving Routing Tables

An alternative to defining paths for each traffic requirement or source/destination pair is to evolve the routing tables themselves. Sinclair (1993) uses a set of integers for each valid node-pair. The number of integers corresponds to the degree (d) of the originating node. The integers are ordered and the first two integers must be in the range 0 to $d-1$. All subsequent integers must be in the range 0 to $d-2$. Each node has an ordered list of connected nodes. The first integer x refers to the ($x+1$)th node as the first choice node to transmit to. Subsequent integers refer to the xth node where a zero means no more choices are available. When trying to establish a call, if none of the choices result in a successful transmission the call will be referred back to the previous node to ascertain if it can successfully try another route. In the event of backtracking to the source node, and having no choices available, the call is lost. Despite using step change and offset inverse penalty functions to discourage loops within the network, results obtained by the author were not as good as previous non-evolutionary work.

Munetomo *et al.* (1998; Chapter 9, this volume) have produced an algorithm which uses the distributed nature of a network to assist in forming routes. Each node is responsible for evolving routes from itself to other nodes. Each chromosome consists of an ordered set of nodes starting at the node holding the population and finishing at any of the other nodes in the network. Recombination is only allowed for chromosomes that share the same destination. The potential crossing sites are those nodes that are held in common by the two parents. The segment between the two common nodes is swapped. Crossover cannot take place if only the source and destination nodes are held in common. Mutation is performed by randomly selecting a node in the network and finding the shortest path from the source to the node and from the node to the destination. If these paths are not disjoint, the mutation is rejected. Network nodes can send information between their respective populations. The best chromosomes are 'migrated' from a node to an adjacent node. However, the transferred chromosomes will, initially, have the wrong start node listed in the chromosome. If the receiving node is referred to in the 'migrated' chromosome then it is simple to convert by simply deleting the nodes between the original source node and the new node in the chromosome. If the node is not contained within the chromosome it can be added to the front of the chromosome as the receiving node must be connected to the transmitting node (migration occurs between two adjacent nodes).

Fitness is assessed by periodically sending messages down the paths and measuring the delay. The delay is then converted to a fitness value. Routing is done by weighting each route according to the fitness of the chromosome. The packets are then statistically transmitted based on the relative weights of the routes. To limit the number of routes, one with a low weight may be deleted from the population. The authors claim that the technique is novel and robust. No comparative results are presented.

An alternative approach to encoding the routes in a routing table is to design routing rules that are reasonably fault-tolerant (Kirkwood *et al.*, 1997; Shami *et al.*, 1997). Genetic programming can be used to devise a set of rules that can be used to route traffic. In Shami *et al.* (1997) the problem specific expression is defined as follows, where *X*, *Y* and *Z* refer to nodes that the traffic should be sent to. An example rule might be (IF-CUR-GO W X Y Z); this means: if the current node is *W*, send the traffic to *X*, as long as *W* and *X* are directly connected, and node *X* has not been visited twice before. If *W* and *X* are not directly connected, or *X* has been visited twice before, then send the traffice to *Y*. If the current node is *not W*, then send the traffic to *Z*.

A multiple population approach was used so that each epoch five individuals were sent to replace the worst five individuals in an adjacent population. Three sub-populations were used in the form of a ring and a hierarchy. Initial results have proved to be disappointing. However, the authors are planning to investigate evolving software agents in order to make the solution more scaleable.

10.3.3 Ring Loading

This area could be considered to be capacity planning or routing. In a ring there are only two possible routes. Once these have been determined the capacity of the ring is forced. As all links have the same capacity the routing algorithm must distribute the load as evenly as possible. More formally (Mann and Smith, 1997):

$$\min\left\{\max_{e \in E}\left\{\sum_{v,w \in V}\{t(v,w)x_{t(v,w)} + t(v,w)y_{t(v,w)}\}\right\}\right\}$$

subject to:

$$x_{t(v,w)} + y_{t(v,w)} = 1$$
$$x_{t(v,w)}, y_{t(v,w)} \in \{0,1\} \forall t(v,w) \in T$$

where:

$$x_{t(v,w)} = \begin{cases} 1 & \text{clockwise} \\ 0 & \text{anti-clockwise} \end{cases}$$

$$y_{t(v,w)} = \begin{cases} 1 & \text{anti-clockwise} \\ 0 & \text{clockwise} \end{cases}$$

in which e is one of m members of the set of edges E. Each edge has a cost $c(e)$. Nodes v and w are members of the set of n nodes N, $t_{(v,w)}$ is the traffic between the two nodes v and w from the set of traffic requirements T, and is routed along the path $P(v,w)$. This problem is difficult because the traffic may not be split. The chromosome is represented by a binary value (clockwise or anti-clockwise) in a gene per node pair. Fitness is evaluated as a weighted sum of the path cost and the maximum load, which allows unnecessarily long paths to be penalised. Uniform, single or multi-point crossover can be used along with single bit mutation. Mann and Smith (1997) have found that simulated annealing and genetic algorithms perform well on this type of problem. The genetic algorithm did not require fine tuning, whereas simulated annealing did, but could potentially produce slightly better solutions by fine tuning. Karunanithi and Carpenter (1997) have also produced good results using the same basic technique, but have compared it to a commercial linear and mixed integer linear programming package. Again, the best results did not differ significantly but the genetic algorithm approach was computationally expensive.

10.3.4 Unusual Applications

Concurrent Point-to-Multi-point Routing
Multicasting is expected to become an important area of telecommunications as large bandwidth applications between groups of people, such as video conferencing, become more common. Unlike the problems addressed earlier, a permutation based approach was used by Zhu et al. (1998) to identify the best set of trees to route a number of multicast

requests. In essence, the chromosome represents the order in which each request shall be considered. Each request in turn will be checked against the remaining capacity in the network. Any links that do not have sufficient capacity for the request are removed from consideration. A separate routine is then called to find an optimal or near-optimal Steiner tree. In practice a non-GA approach to this stage may be advisable due to time constraints, however a number of GA methods exist for this sub-problem (Leung et al., 1998). To focus the evolution on the appropriate genes, only the first k out of n requests are considered in forming the fitness function. A GA is run for increasing values of k until the GA is no longer able to find a solution that satisfies the indicated number of requests. For a particular k the object is to minimise the cost (capacity x cost for each link summed over all links). PMX crossover was used for permutation with roulette wheel selection.

Adaptive Feedback to Distribute Loading

Congestion does not necessarily respond well to non-adaptive feedback mechanisms, because of the time delays inherent in a network. Olsen (1997) suggests that adaptive feedback mechanisms should change their behaviour in the light of previous events, not merely react to changes in link costs. Dijkstra's Shortest Path First (DSPF) is the algorithm used by Olsen to update routes depending on the link costs. The genetic algorithm determines which routes are best suited for updating. A binary string is used to indicate if a particular route should be updatable by DSPF. The GA uses single point crossover, single bit mutation and roulette wheel selection. Every chromosome in the population is used once, and the routing performance is analysed during that period. The inverse of the total link costs achieved during that period is used to set the fitness of the chromosome. The next generation can then be created. In addition to the aforementioned features, the GA could also adapt its crossover rate and enforce a minimum update rate irrespective of the values within the chromosome. The results were very impressive, 4-to-12 fold reductions in average packet delays and a 44% to 140% increase in average throughput. Some implementation details regarding how the information would be transmitted are also included in Olsen's study.

10.3.5 Routing Combined with Design

An unusual way of routing is to use it to drive the design process, Huang et al. (1997) looked into designing a 3-connected telecommunication network by finding three disjoint routes and using them to help design the network. The objective is to minimise total link costs subject to guaranteeing three disjoint links, a network diameter of less than seven and a maximum node degree of five. It does this by taking a predefined network graph and finding three optimal disjoint paths for every traffic requirement. The resultant sub-graph has minimal total link cost and will define the required capacity for every link. The chromosome is encoded by storing three groups of node numbers that define a path between source and destination. The number of nodes in a group is determined by the network diameter. An entry of 0 in the group indicates that the node number can be disregarded, thus allowing routes of less than the network diameter. No nodes can be repeated so the paths are guaranteed to be disjoint. This encoding enforces the critical diameter and connectivity constraints. Two-point crossover is used with a repair operator that swaps back individual nodes that are duplicated in one chromosome of the offspring. Fitness is evaluated by calculating the cost of carrying the extra capacity along with an offset cost for the first time a link is utilised. The offset cost encourages the GA to choose solutions with fewer links than would otherwise be the case. The costs (C_{max} to C_{min}) are then linearised between 0 and 1 to produce fitness values. This approach can be adapted to different network diameters and connectivity constraints by increasing the number of nodes per route and the number of groups of nodes. The authors claim that for large networks the GA outperforms a method based on Dijkstra's algorithm and compares favourably with the approach taken by Davis et al. (1993). Although the routes are defined using this method, since networks change slowly but traffic changes frequently, this may more fairly be considered network topology design using a routing approach.

Davis et al. (1993) have combined design and routing into a three part chromosome which they label (X, W', W''). The first parameter, X, stores the capacities of each link. W' encodes the order in which traffic demands are considered for routing. W'' encodes the order in which traffic is routed for a network whose links have been reduced in capacity by a factor ($0 \leq s \leq 1$). Crossover is performed by applying uniform crossover to X and by applying uniform order-based crossover to W' and W''. Mutation randomly selects a link to have its capacity changed. A local mutation called 'creep' has a 0.3 probability of modifying a link capacity value, x_i, within the range $x_i \pm 5$. Finally, 'zero-link' sets the value of the capacity of a link in X to zero. Ranking is used to set the fitness value, then a roulette wheel procedure is used for selection with probabilities of 0.4 for crossover, 0.15 for mutation, 0.4 for creep and 0.05 for zero-link

The decoder takes each demand in W' and attempts to route it down the k-shortest paths (typically 10). Routes that have used up their available bandwidth are not included. If there are no paths available the constraint violation parameter C is set to 1,000,000 and the demand is left unrouted. Otherwise as much demand as possible is routed down the available paths. If this still leaves unrouted demand, links on the first path are increased in capacity (repaired) until they can carry all the remaining traffic. An identical process is then followed for W''. Fitness is then evaluated by assessing the link cost of the network (proportional to bandwidth + offset) plus the constraint violation penalty. The repaired chromosome has a 10% chance of replacing the original chromosome in the population; according to the authors this value may be too high. This method proved to be successful when compared with other methods on large problems. Mixed Integer Programming techniques were better for small problems. However, care had to be taken in controlling the number of shortest paths, and the number of links, available to the program.

A very different three segment crossover technique has been proposed by Ko et al. (1997; 1997a). The chromosome is split into topology, routing and capacity assignment. The topology section simply uses a bit per node-pair to indicate if they are connected. The method of obtaining a set of routes for each node-pair is not defined. The routing section holds a list of indices to paths for each source/destination pair. Finally the capacity segment defines the capacity of each link. This is done by having a set of integers determining the number of each type of line for every link. Rules are used to ensure large numbers of low capacity lines are not used in preference to a single large line for a particular link. One-point crossover and random mutation are used on the topology and capacity segments. When assigning the capacity, the link must be able to carry the full flow requirement. Time delays are discouraged by the fitness function. Crossover between the routing segments involves copying the shortest path for every requirement from either parent to the offspring, given the capacities defined in the capacity segment. Mutation randomly selects an alternate path with a probability of 5%. The authors demonstrate the results on a ten node problem that represents one of China's major networks.

The methods considered so far tend to use well understood crossover techniques with an unusual encoding. Cuihong (1997) has taken an entirely different approach to crossover. Each chromosome comprises two sections; the first section defines the capacity (type) of each link, the second has an index for each source/destination pair that points to one of a set of predefined legal routes. The initial population is assigned randomly. Fitness is calculated by subtracting the cost of the proposed network from the maximum feasible cost. The key parameters include time delay assuming an M/M/1 queue, a fixed cost per link and a cost per bit using a link. Each parameter is weighted to give an appropriate cost in the summation. Crossover is done by utilising a novel 'orthogonal crossover' operator (OCX). In OCX a series of random numbers are generated that define where to cut a chromosome into a number of segments (c). Each segment from two parents cut in this manner can be in one of two children. Clearly, the number of ways of arranging two sets of c segments to create two children is 2^c. This problem is well known in manufacturing and other areas where the number of combinations is too large to permit a full search of possibilities. Instead the best result can be found by doing a small sub-set of experiments by carefully choosing the combinations. Designing such experiments has been subject to a lot of mathematical analysis which is based on orthogonal arrays. However the analysis will not be strictly correct if any two parameters are coupled (epistasis). As a result of this problem some orthogonal arrays have been designed to allow certain parameters in the experiment to be coupled and still produce valid results. The author refers readers to an appropriate text (Montgomery, 1991). In practice, orthogonal arrays have proved to be effective even if the experimenter has not ensured the parameters are truly independent. Once the 'experiments' have been done, the best two combinations can be inserted as children into the new population. Mutation randomly chooses a link and calculates the link size by doubling the present size and subtracting a random link size in the legal range. The path is identified by doubling the index number for the existing path and then subtracting a randomly generated index number from the legal range. If the result is beyond the maximum legal value, it is lowered to the legal value; similarly, if it is below zero it is raised to zero. Roulette wheel selection and an elitist strategy select the individual chromosomes.

10.3.6 A Two-dimensional Encoding for Restoration

The most common method for evolving routes is a simple index per traffic requirement to an alternate route, however as mentioned earlier, there are limitations to this approach. In this section, a detailed account of a two-dimensional order-based genetic algorithm. The time required is much greater than that used by the previously mentioned techniques, but the results which were obtained are potentially better due to the larger search space and enhanced use of genetic material. As this algorithm is intended for restoration in a heavily loaded ATM network the authors have assumed that all traffic cannot be re-routed. In network design this is often the case because low priority traffic does not need to be guaranteed in the event of link failure. By using effective bandwidths for reserving bandwidth the time delay issue is largely circumvented. However the number of connections permitted is critical. The following description is largely based on work published at GALESIA'97 (Bentall *et al.*, 1997).

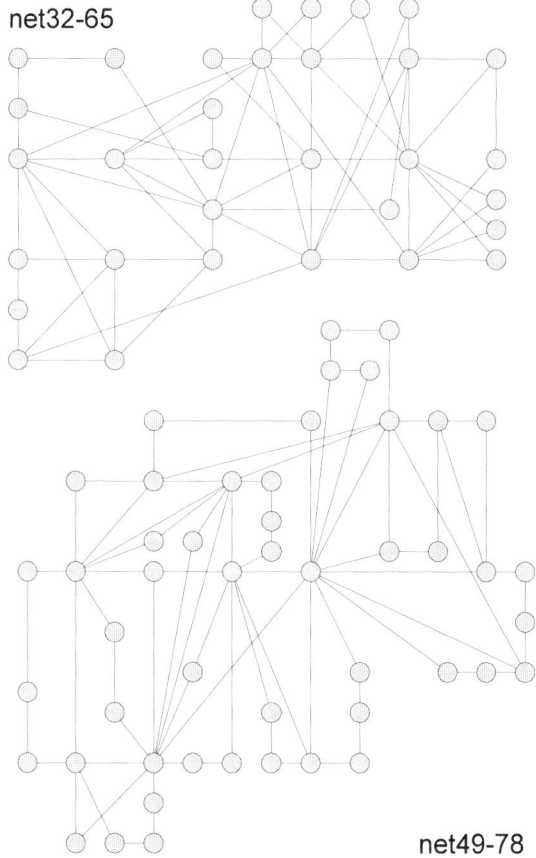

Figure 10.1 Example test networks based on real networks.

Test Problem

For each path over a failed link many alternative paths exist. The number of alternative paths is highly dependent on the maximum number of hops each alternative path is allowed to take. For example, in test network 'net32-65' (see Figure 10.1) and with a maximum hop count of 5 the number of alternative paths for one particular link is, on average, 14.71. However if the maximum hop count is increased to 8, then the average number of alternative paths increases to 733.65. Within the test network, over 2000 virtual paths are randomly generated resulting in between 30 and 250 virtual paths using any particular link. Even for a link with a below average number of paths, for example 80, utilising its capacity, and a maximum hop count of 7 (average of ~280 virtual path alternatives), contains 280^{80} possible solutions.

A Two-dimensional Structure for Path routing

The algorithm is based on an order-based genetic algorithm with each chromosome represented in two dimensions, allowing failed paths and path sets to be evaluated.

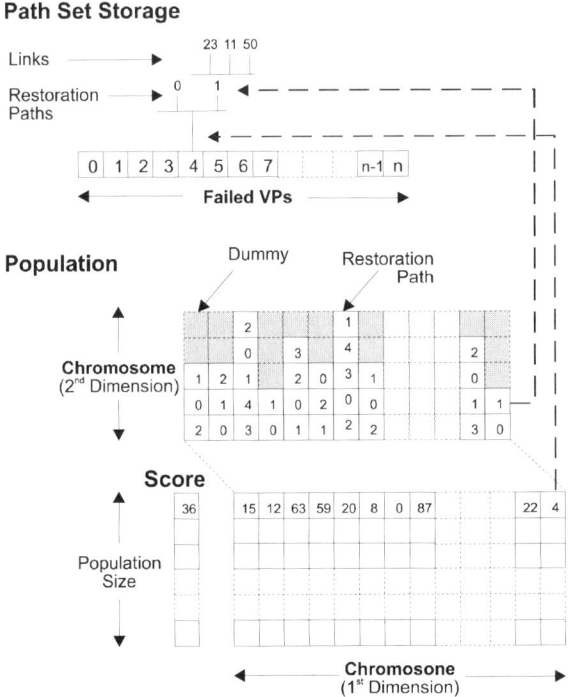

Figure 10.2 GA population structure.

Data Structure

Figure 10.2 shows the two-dimensional structure of the population as implemented in the simulator. The restoration order of the failed paths is represented in the first dimension of the chromosome. For each restoration path there exists a number of alternative paths. Each alternative must be tried until an acceptable path is found. The order in which an alternative path will be selected from the set of alternatives is represented in the second dimension.

Fitness Function

The original purpose of this work was specifically for restoration in heavily loaded networks where some paths would be lost. As it is also connection oriented, effective bandwidths must be assumed for each requirement rather than actual traffic levels. The fitness is based on the success of the chromosome in finding acceptable routes to restore all failed paths. For each chromosome there are two elements representing its fitness. The first element is a direct representation of the number of paths restored. The second element represents the efficiency of the chromosome in keeping spare capacity for future use. The second element can only be fairly compared with other solutions that restore the same number of paths. Consequently, the second element is linearised between 0 and 0.9 for a particular generation and number of restored paths. This figure is then added to the number of paths restored to produce a final fitness value. Figure 10.3 provides an example of a typical score. Note that the second element is only used in the parent selection process and has no meaning outside the current population.

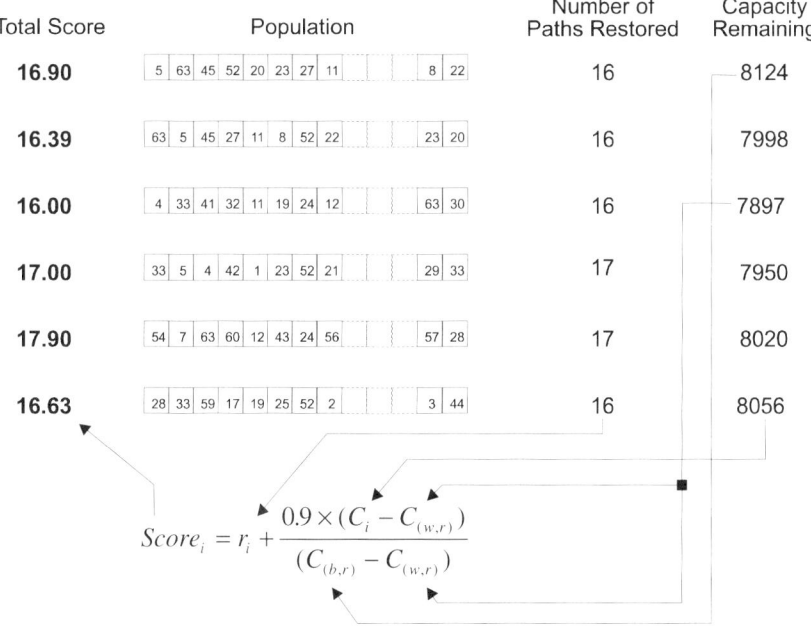

Figure 10.3 Scoring function.

The full fitness function is given in equation 10.1, where i is the chromosome under consideration, r is the number of paths successfully restored, C is the capacity remaining, and $C_{(b,r)}$ and $C_{(w,r)}$ represent the capacity remaining for the best and worst case chromosomes respectively with r failed paths restored within this current population.

$$\text{Fitness}_i = r_i + \frac{0.9 \times (C_i - C_{(w,r)})}{(C_{(b,r)} - C_{(w,r)})} \quad (10.1)$$

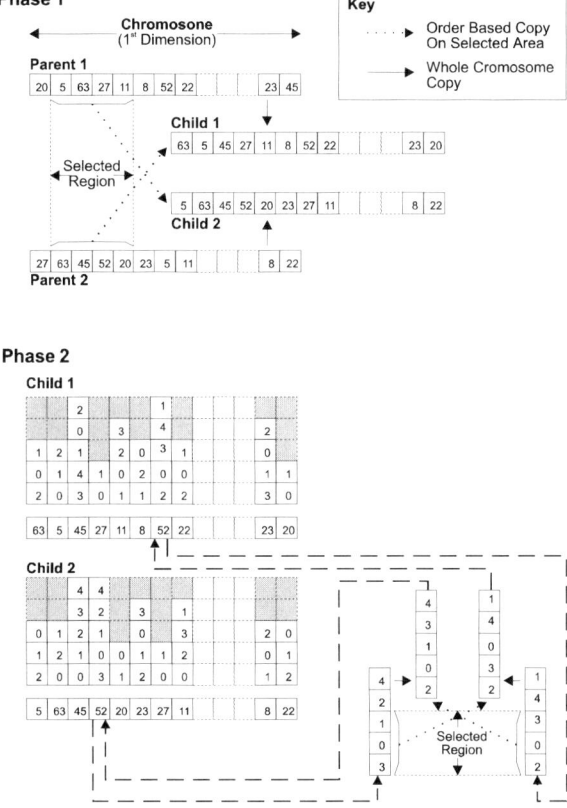

Figure 10.4 Example crossover.

This function enables us to ensure that chromosomes restoring a greater number of paths will always have a greater chance of selection. However, chromosomes that have equal scores are also graded according to the spare capacity remaining within the network after restoration. Chromosomes with the most capacity remaining are fitter, and therefore have a better chance of parent selection.

Evolution Operators

The genetic algorithm is constructed from three evolutionary operations, each fed by a linearised roulette wheel parent selection process. The parent selection process linearises the scores between 0 and 1 before the roulette method is used to select the parent(s). The operators used are order-based mutation, and order-based, 2-dimensional, 2-point crossover.

Mutation

The mutation process selects two genes within the parent and exchanges their order, producing a single mutated child. The mutated genes' counterpart chromosomes are exchanged to ensure that each gene maintains its associated path set.

Crossover
The crossover operator is completed in two phases using a standard two-point order-based crossover on two parents, producing two children. The first phase applies order-based crossover to the first dimension and thus acts on the failed paths' restoration order (see Figure 10.4). When the children of parents 1 and 2 have been evaluated in the first dimension, second dimension crossover is performed (phase 2). Each gene represents a particular path that should be restored. If the gene is selected for crossover then the corresponding gene, same path, in the other child is used. This ensures that a path set (gene) only evolves with other instances of the same path set within different chromosomes. The probability values used in this implementation are 2% mutation and 70% crossover, with a 50% chance for each gene performing crossover in the second dimension. The remaining 18% constitute reproduction.

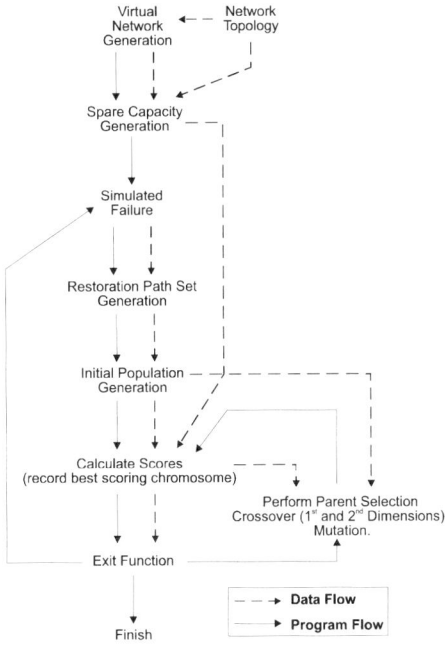

Figure 10.5 Functional and data flow diagram of network generation and GA operators within the simulator.

The simulator is also used to evaluate the genetic algorithm with respect to other techniques. To calculate restoration benchmarks for comparison with restoration algorithms, the topology of the test networks and the assignment of capacity must be decided. The networks selected for testing combine a selection of pre-published, real and evaluation networks. For commercial reasons, the real networks include in this text have slight alterations. The networks' names represent the number of nodes and the number of links, for example 'net49-78' has 49 nodes and 78 links. Two of the test networks are shown in Figure 10.1. Random virtual networks of a controlled fitness are created within these networks during each simulation group.

Once the topology has been established, the quantity of spare capacity must be decided for each link. This is critically different to all other simulations to date. For network implementation the simplest spare capacity allocation is formed from assigning a minimum load to each link. Therefore, the authors have set a flat rate of 90% load to all links. Using this assignment technique produces a network containing links with differing capacities, therefore allowing the algorithms to be tested under a variety of different conditions in one network.

The simulator has several control parameters for the virtual network structure algorithm. The virtual network parameters include Virtual Path (VP) length control, VP capacity control, Number of VPs and Network Load. Figure 10.5 provides a functional and data-flow picture of the operation of the simulator. Further details of the simulator can be found in Bentall *et al.* (1997).

An example of the results can be seen in Figure 10.6, showing the genetic algorithm results compared with random search for a single-link failure scenario. The two-dimensional encoding of this genetic algorithm performs well in terms of the routes restored but is computationally expensive. At present, the GA is too slow to be used as a real-time restoration algorithm because of the time it takes to perform the calculations. At present, the technique is thus best used in establishing a benchmark for other algorithms.

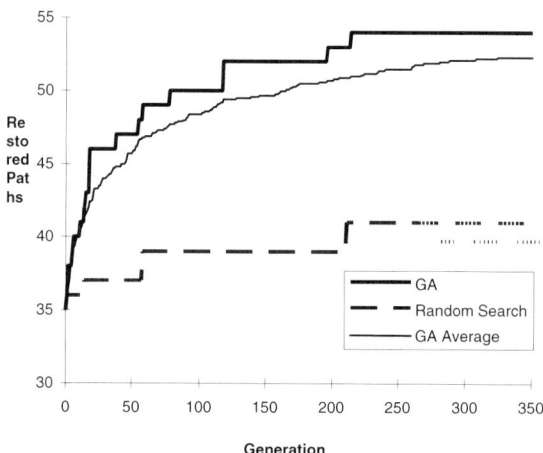

Figure 10.6 GA vs. random search for a single link failure scenario.

10.4 Future for Evolutionary Algorithms in Restoration/Routing

A wide variety of approaches have been tried over the last decade. Most of them have shown great promise but are limited by the time it takes to process the information.

However, given the rapid improvements in processing power and the accelerated performance possible with parallel genetic algorithms, this may not be such a large problem in the future. Work on parallel genetic algorithms is continuing and hardware implementations are being studied. Both ordinary and permutation-based crossover techniques have been designed for hardware implementation (Turton and Arslan, 1995; 1995a). The hardware systems envisaged at present have the potential to work over six orders of magnitude faster than a conventional general purpose computer, and when mass-produced would be a tenth of the cost. There are some who keep predicting that soon there will be more bandwidth available than can conveniently be used and therefore routing may not be a problem. However, all the signs at present indicate that greater capacity generates more traffic. In addition, users like to get the greatest possible use from their leased lines. New challenges will appear as mobile communication systems become more prevalent. Routing a call when the destination may suddenly change to a different part of the network, with different and variable error rates is going to cause far more fluctuations in traffic and link failures. Even traditional networks are growing so large that congestion and flow control are proving to be a problem as Olsen's paper indicates. There is no doubt that evolutionary algorithms hybridised with other techniques and supported by advances in electronics will continue to generate both interesting problems and solutions.

11

GA-Based Verification of Network Protocols Performance

M. Baldi, F. Corno, M. Rebaudengo, Matteo Sonza Reorda, and Giovanni Squillero

11.1 Introduction

Computer networks are gaining increasing importance as they penetrate business and everyday life. Technological evolution results in increased computational power and transmission capacity. These phenomena open the way to the development and exploitation of new applications (e.g. video conferencing) which require demanding services from the network. As a result, computer networks are in continuous evolution and the protocols regulating their operation must be assessed for their suitability to new technologies and ability to support the new applications. This assessment is particularly important since new technologies and applications often differ from the ones the protocol has been designed for.

The effectiveness of a network solution depends on both the specifications of the protocols and their software and hardware implementations. Given leading hardware and software technologies, the protocol specification thus plays a crucial role in the overall network performance. Some aspects are of utmost importance: the *correctness* of the protocol, i.e. the warranty of showing the intended behavior in any specific situation; its *performance*, i.e. the utilization of the available bandwidth it is able to achieve over the physical medium; and its *robustness*, i.e. the property of being able to work correctly under abnormal conditions.

The analysis of these different aspects of communication protocols is usually done according to the following techniques:

Formal modeling: to verify some properties of the protocol. This requires us to model the protocol behavior according to some formal description language, such as LOTOS, and to run some formal verification tool. Description languages are nowadays powerful enough to capture most of the protocol semantics; however, verification tools are often quite limited in the size and complexity of the descriptions they can deal with. Some important aspects of a network protocol that are difficult to verify are those related to different time-out counters, message queues, and transmission strategies based on message contents.

Simulation: to check the protocol in its entire complex behavior and explore the dependencies among the various hardware and software components of a computer network. A typical simulator allows the actions taken by a protocol to be monitored while operating over a computer network with a specified topology and traffic load. The simulator models a number of sources that can be programmed to generate traffic according to statistical models defined after real sources. One or more sources behave according to the protocol under study; the others are considered *background traffic* generators. Poisson statistical processes (Hui, 1990) are often adopted as a model of background traffic.

We feel that neither approach is sufficient to give enough information about the protocol performance. While the formal approach delivers a mathematically proven answer, it is forced to work on an over-simplified view of the protocol. Traditional simulation techniques, on the other hand, rely on a clever choice of the traffic pattern to yield useful results, and worst-case analysis is often performed by hand-designing the traffic.

We propose the adoption of a mixed simulation-based technique that rely on a Genetic Algorithm (GA) (Holland, 1975) to explore the solution space and look for an inconsistency in the verification goal. As a case study problem, we chose the performance verification of the Transmission Control Protocol (TCP) (Postel, 1981), a widely used computer network protocol. The TCP protocol was chosen because its characteristics and behavior are already well known to the computer network research community, and our goal is to better focus on the verification methodology and exploitation of the GA, and not on the protocol itself.

In this work, we show how to integrate a GA with a network simulator to drive the generation of a critical background traffic. The GA aims at generating the worst-case traffic for the protocol under analysis, given some constraints on the traffic bandwidth. Errors and weaknesses in the network protocol can therefore be discovered via simulation of this worst-case traffic. We expect the GA to drive the network to conditions where the protocol performance (e.g. the effective transfer rate) is low, and we can study the behavior of the protocol under extreme load conditions.

In previous work (Alba and Troya, 1996), GAs were applied to the analysis of communication protocols by checking the correctness of their Finite State Machine (FSM). Communicating protocol entities were expressed as a pair of communicating FSMs, a GA generates communication traces between the two protocol entities, and a FSM simulator executes them. The GA aims at detecting deadlocks and useless states. This approach showed several limitations, mainly due to over-simplifications necessary to express a *real* protocol as a mathematical model such as a finite state machine.

By contrast, our approach, being based on a network simulator, is able to model most of the aspects of the protocol stack and the network on which it is employed: not just the state machine, but also counters, message queues, routing tables, different re-transmission speeds, and so on.

The GA generates the worst operating conditions of the protocol (by defining the load in the network) and finds the traffic configuration that minimizes its performance. During this analysis, the protocol is stressed enough to achieve high degrees of confidence, and informally, we can say that both its functionality and its performance are *verified*. Some preliminary experimental results we gathered show that traffic generated by the GA is able to expose protocol shortcomings that were impossible to reveal with formal techniques, and that were not found by means of statistical simulation.

11.2 The Approach

In this work we choose to examine the TCP protocol, whose characteristics and behavior are already well known to the computer network research community, in order to better focus on the verification methodology and exploitation of the GA, and not on the protocol itself. Hence, we do not expect to discover any new information about TCP behavior and performance, but to find in an automated way well known information that, with traditional approaches, requires skills and in-depth knowledge of the protocol.

We aim at studying the TCP protocol in real operating conditions. We set up an IP (Internet Protocol) network (Stevens, 1994) and a number of TCP *probe connections* (i.e. sender-receiver pairs). The network is loaded with a *background traffic* generated by User Datagram Protocol (UDP) (Postel, 1980) sender-receiver pairs. We chose the UDP protocol for the sake of simplicity, but other kinds of background traffic sources can be modeled too.

During the analysis process, the GA provides a *pattern* of the traffic generated by background sources and a network simulation is run on the given topology. During this simulation, relevant data are gathered from probe connections by the simulator program and provided to the GA, which uses them to estimate the 'damage' that the background traffic made. Such information is then used to drive the generation of traffic patterns to be used in subsequent steps of the algorithm.

As for statistical methods, our analysis requires a large number of simulation runs to obtain a significant result, therefore the choice of the simulation program is critical in order to keep the computational time small. In our work, simulations were run using a publicly available simulator called *insane* (Internet Simulated ATM Networking Environment – Mah (1996). *Insane* adopts an event-driven mechanism for the simulation, thus being reasonably fast. Moreover, it is able to deal with an arbitrary network topology described as a TCL (Welch, 1995) script and many characteristics of network nodes and links (such as processing speed, buffering capacity, bandwidth and latency) can be controlled.

Figure 11.1 shows the architecture of the proposed verification environment, called Nepal (Network Protocol Analysis Algorithm), which integrates the GA and the simulator. The approach is quite general and can be applied to different protocols using different simulators with limited effort.

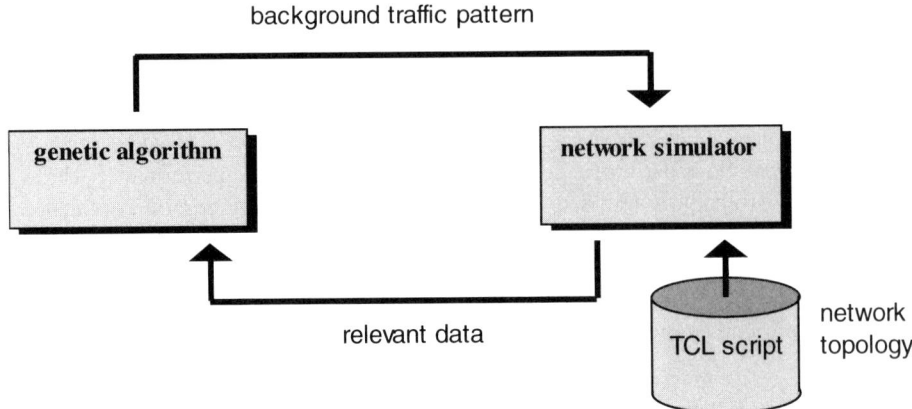

Figure 11.1 Architecture of the Nepal system.

11.3 Genetic Algorithms

The field of Genetic Algorithms (GAs) has been growing since the early 1970s, but only recently were GAs applied to real-world applications which demonstrate their commercial effectiveness.

Holland (1975) invented GAs to mimic some of the processes observed in natural evolution: *evolution, natural selection, reproduction*. GAs have been investigated as a possible solution for many search and optimization problems in the way nature does, i.e., through *evolution*. Evolution operates on *chromosomes*, i.e. organic elements for encoding the structure of live beings. Processes of natural selection cause the chromosomes that encode successful individuals to reproduce more frequently than those less fit. In the process of biological reproduction, chromosomes coming from parents are combined to generate new individuals through *crossover* and *mutation* mechanisms.

In the GA field, biological concepts are mapped as follows. Each solution (*individual*) is represented as a string (*chromosome*) of elements (*genes*); each individual is assigned a *fitness value*, based on the result of an *evaluation function*. The evaluation function measures the chromosome's performance on the problem to be solved. The way of *encoding* solutions to the problem on chromosomes and the definition of an evaluation function are the mechanisms that link a GA to the problem to be solved. As the technique for encoding solutions may vary from problem to problem, a certain amount of art is involved in selecting a good encoding technique.

A set of individuals constitutes a *population* that evolves from one generation to the next through the creation of new individuals and the deletion of some old ones. The process starts with an initial population created in some way, e.g. randomly. A GA is basically an algorithm which manipulates strings of digits: like nature, GAs solve the problem of finding good chromosomes by blindly manipulating the material in the chromosomes themselves.

11.3.1 The Genetic Algorithm in Nepal

The pseudocode of the GA adopted within Nepal is shown in Figure 11.2. When the GA starts, N_p individuals are randomly generated. In our approach, an individual represents a background traffic pattern. The background traffic corresponding to the initial population is generated according to a Poisson process whose inter-arrival time between packets is exponentially distributed.

At each generation the GA creates N_p new individuals by applying a crossover operator to two parents. Parents are randomly chosen with a probability which linearly increases from the worst individual (smallest fitness) to the best one (highest fitness). A mutation operator is applied over each new individual with a probability p_m. At the end of each generation, the N_p individuals with higher fitness are selected for survival (*elitism*), and the worst N_o ones are deleted from the population.

The fitness function measures the probe connections' throughput, i.e., the performance of the probe TCP connections perceived by end-users during the simulation experiment. All bytes successfully received at the TCP level, but not delivered to end-users, such as duplicated packets received due to the retransmission mechanism of the protocol, are therefore not considered.

```
𝒫₀ = random_population(Np);
compute_fitness(𝒫₀);
i=0; /* current generation number */
while (i < max_generation)
{
        𝒜 = 𝒫ᵢ;
for j=0 to Nₒ
        {       /* new element generation */
                s' = select_an_individual();
                s" = select_an_individual();
                sⱼ = cross_over_operator(s', s");
if(rand()≤pₘ)
                        sⱼ = mutation_operator(sⱼ);
𝒜 = 𝒜 ∪ sⱼ;
        }
compute_fitness(𝒜);
i++;
        𝒫ᵢ = {the Np best individuals ∈ 𝒜 };
}
return( best_individual(𝒫ᵢ) )
```

Figure 11.2 Pseudocode of the genetic algorithm in Nepal.

The fitness function should increase with the increasing goodness of a solution, and a solution in Nepal is good when the background traffic pattern is critical; therefore, the fitness function we defined is *inversely proportional* to the total number of bytes perceived by the end-users. Hence, fitness is $1/B_{TCP}$ where B_{TCP} is the total number of bytes delivered to the TCP users on all probe connection.

This simple measure already delivers satisfactory results on simple network topologies. For different problems and different goals, additional parameters can be easily included into the fitness function.

11.3.2 Encoding

Each individual encodes the description of the traffic generated by all the background connections for the whole duration of the simulation. A connection is specified by a UDP source and destination pair, while the route followed by packets depends upon the network topology and on the routing mechanism. We assume that all the packets sent on the network are of the same size, hence individuals do not need to encode this information for each packet.

Individuals are encoded as strings of N genes. Each gene represents a single background packet and N is constant during the whole experiment. Genes are composed of two parts: **TAG,** that indicates which background connection the packet will be sent on; **DELAY,** that represents how long the given source will wait before sending a new packet after sending the current one.

A common model for background traffic commonly adopted in statistical network analysis is a Poisson process with negative exponentially distributed arrival times (Hui, 1990), and we want delays in randomly generated individual to be exponentially distributed. For this reason the **DELAY** field represent an index in an array of values exponentially distributed from 0 to K seconds with an average value of $0.4*K$. Where K is called the *Poisson distribution's parameter*. Please note that we make no assumptions on the number of genes with a given **TAG**; however, N is usually large enough to make the initial random background traffic almost equally distributed between sources.

The crossover operators must have the ability to combine *good* property from different individuals and in Nepal each individual embeds two different characteristics: the background traffic as a function of time (*time domain*) and the background traffic as a function of the background connection (*space domain*). It is important to maintain the conceptual separation of these two characteristic, because, for instance, an individual could represent *good* background traffic during a given period of a simulation experiment, or it can represent a *good* traffic for a given background connection, or both.

It is possible that the genes with a given **TAG** are not enough to generate traffic for the whole experiment, in this case the connection remains idle in the final part of the experiment. On the other hand, it is also possible that the simulation ends before all the packets are sent, and final genes are ignored.

Figure 11.3 shows, on the left-hand side, a possible individual and, on the right-hand side, the corresponding background traffic generated during the simulation experiment. The table mapping the **DELAY** field to exponential delays is reported in the middle of the figure.

In the individual shown in Figure 11.3, packet number 9 is not sent because the simulation is over before the delay associated with gene number 6 expires. On the other hand, one gene with the third tag is missing, since the delay associated with the last packet on that connection (packet number 7) expires before simulation's end.

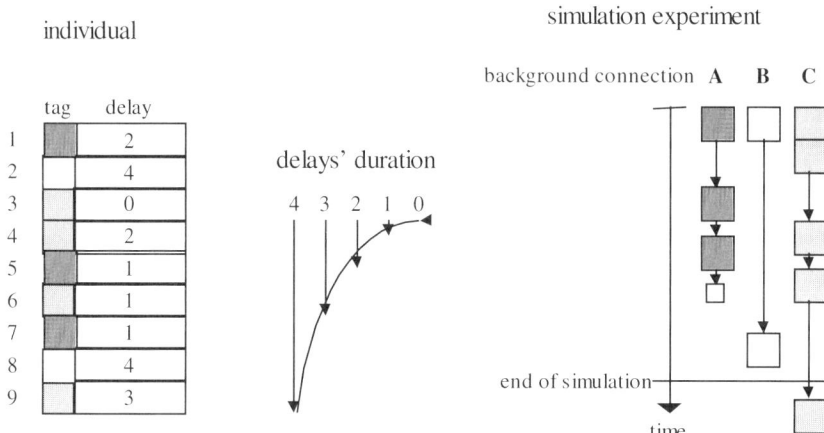

Figure 11.3 Individual encoding.

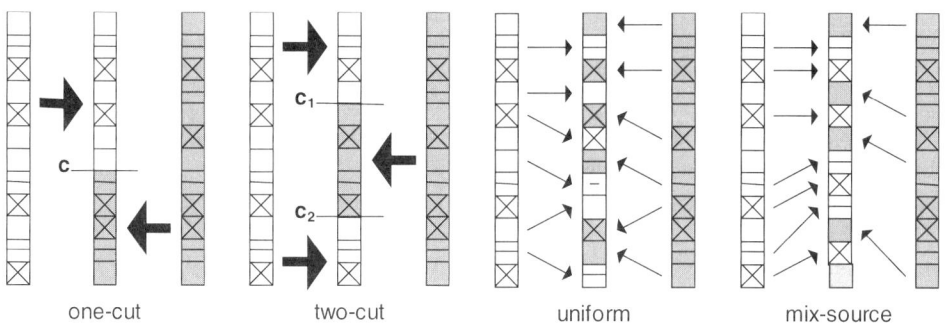

Figure 11.4 Crossover operators

11.3.3 Crossover Operators

The GA in Nepal implements four types of crossover: the widely adopted *one-cut*, *two-cuts* and *uniform*, plus a peculiar one called *mix-source* (Figure 11.4):

One-cut crossover: let one cut point c be a randomly selected point between 1 and N. Nepal generates a new individual copying the first c genes from the first parent and the subsequent $N - c$ ones from the second parent.

Two-cuts crossover: let two cut points c_1 and c_2 be randomly selected between 1 and N with $c_1 < c_2$. Nepal generates a new individual copying the first c_1 and the last $N - c_2$ genes from the first parent. The middle $c_2 - c_1$ genes are copied from the second parent.

These two crossover operators generate a background traffic that is equal to the traffic generated by one parent for a given time, and then suddenly changes and follows the traffic generated by the other parent (strictly speaking, it is possible that the cut points do not correspond exactly to the same time instants in the two parents, however, N is large enough for this to be neglected).

Uniform crossover: Nepal generates a new individual taking each gene from the first or the second parent with the same probability.

Mix-source crossover: each TAG (i.e. each background source) is randomly associated with one of the two parents. Nepal generates a new individual copying each genes with a specific TAG from the associated parent. The new individual can be longer or shorter than N: in the former case the exceeding genes are dropped; in the latter one the individual is filled with random genes.

The *mix-source* crossover aims at putting together good background sources from different individuals.

We can say that the *mix-source* crossover acts in the space (or source) domain while the *one-cut* and *two-cut* ones act in the time domain. The *uniform* crossover is hybrid and acts in both domains.

11.3.4 Mutations

Nepal implements three different mutation operators: *speed-up*, *slow-down* and *random*. Mutation applies on $N_m < N$ contiguous genes, where N_m is a constant. The first gene is randomly selected between 1 and N and the gene list is considered circular.

Speed-up mutation: decreases the DELAY field by one, without modifying the TAG field. DELAY values are considered circular.

Slow-down mutation: increases the DELAY field by one, without modifying the TAG field. DELAY values are considered circular.

Random mutation: substitutes each gene with a new random one.

After each mutation, genes in the individual need to be sorted to fulfill the 'sorted in time' property.

11.3.5 Univocal Representation

A general guideline for GA states that different chromosomes should represent different solutions. Diversely, a GA would waste some of the time in the generation and evaluation identical solutions, and even worse, the population could prematurely converge to a state where chromosomes are different but solutions are equal, and where genetic operators have no effect.

To ensure that a given background traffic is determined by an unique chromosome, genes are sorted to guarantee that the packet associated to gene $i+1$ is never sent before the packet associated with gene i. We say that genes in the individual are 'sorted in time'. Otherwise, it would be possible to exchange adjacent genes with different TAGs without modifying the represented background traffic (e.g. packet number 2 and packet number 3 in Figure 11.3).

Moreover, some crossover operators aim at combining good characteristic in the *time domain*. If an individual does not fulfil the 'sorted in time' property, it would be hard for such operators to preserve property related to such domain, and they would act almost in the same way as the common *uniform* crossover. Our experiments show that the performance of the GA would be significantly reduced.

Chromosomes need to be 'sorted in time' after crossover and mutation operators.

11.4   Experimental Results

We developed a prototypical implementation of the algorithm using the *Perl 5* (Wall et al, 1996) language. The implementation of the GA is less than 500 lines long and, with a limited effort, can be ported to different simulators and different platforms.

The source of the simulator consists of 38,370 lines of C++ code. We needed to modify *insane* to handle an externally generated background traffic; modifications of the original code are limited to about 100 lines. Such modifications allow the traffic source to be completely defined by the user.

11.4.1   Network Topology

Figure 11.5 shows the topology of the IP network exploited in the experiments. Three TCP connections span from the transmitters TXi to the receivers RXi through three IP routers. Each TCP connection performs long file transfers generating a series of 1024 Kbyte messages at a maximum mean rate of 1.33 Mb/s. Thus, the three TCP sources can generate an overall load of 4 Mb/s, if this is allowed by the TCP flow control mechanism. Acknowledgments from each transmitter are carried by messages flowing in the reverse direction. These TCP connections represent the *probe connections* of the experiment.

Two sources (BSi) generate *background* UDP traffic directed to their respective destinations (BDi) over the same links traversed by the TCP connections. The timing of background packets is controlled by the GA, as described earlier. The background traffic is generated only in the forward direction of the TCP connections and thus only data messages can be delayed and discarded due to network overload, not acknowledgments.

In this topology, the link between the right-most two routers represents a bottleneck for the network since it is traversed by all the traffic generated by both the background sources and the TCP connections.

Each link shown in Figure 11.5 has a capacity of 10 Mb/s in each direction and introduces a fixed 10 μs delay. For example, such links can be considered as dedicated 10 Mb/s Ethernet trunks, the 10 μs latency accounting for the propagation delay on the wires and the switching delay introduced by an Ethernet switch.

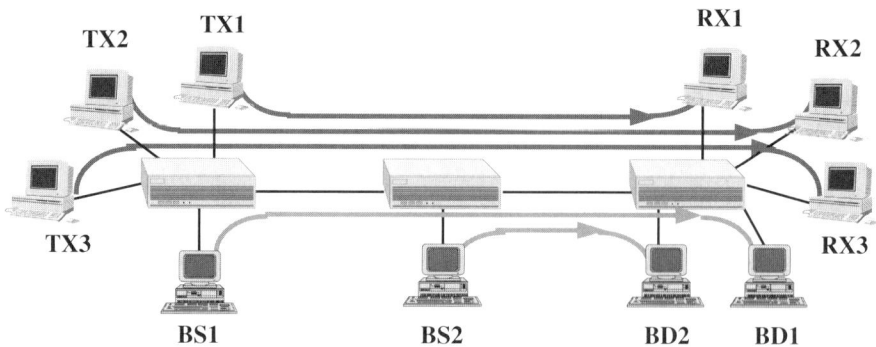

Figure 11.5 Topology of the network.

Routers introduce a fixed 0.1 ms delay component which accounts for the processing required on each packet and adds to the queuing delay. The size of output queues modeled by *insane* is given in terms of number of packets (we used 64 packet queues in our experiments). The buffering capacity of real routers is given in terms of byte volume, i.e. the number of packets that can be stored in a buffer depends upon their size. We choose to deal with fixed size packets to work around this limitation of the simulator program.

11.4.2 Parameter Values

All the parameters used by the GA are summarized in Table 11.1, and the value used in our experiments is reported. The parameters K, N and T are tuned in order to model a 6 Mb/s background traffic level, because, in this situation, the probe connections' traffic (4 Mb/s) plus the background traffic is nearly equal to the links' bandwidth (10 Mb/s). With a smaller load, the background traffic would be unable to interfere significantly with probe connections; while with a larger one, the whole network would be congested and the TCP throughput would be very small, even with a completely random traffic.

Table 11.1 Genetic algorithm parameter values.

PARAMETER	Name	Value
simulation duration [s]	T	3
max number of generations	G	500
population size	N_p	50
offspring size	N_o	20
number of genes in each individual	N	4,500
Poisson distribution's parameter [ms]	K	6.67
mutation's probability	p_m	50%
number of genes affected by a mutation	N_m	100

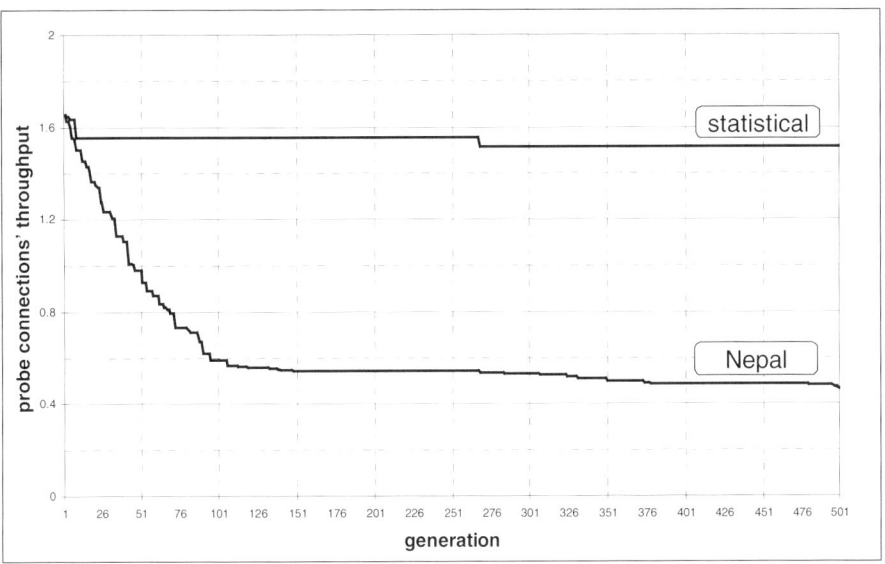

Figure 11.6 Throughput of probe connections.

11.4.3 Results

We let the GA run for G generations on a Sun SPARC Station 5 with 64 Mb of memory. The GA required about 17,000 seconds of CPU time (mainly spent for sorting in time genes of newly generated individuals), and *insane* employed about 53,000 seconds of CPU time for running all the required simulations.

Figure 11.6 shows the throughput of the probe connections, i.e., the connection bandwidth delivered to users. The X axis reports the generation number of the GA. These values are compared to the ones of a standard statistical approach, where the background traffic is randomly generated according an equivalent Poisson process. For the statistical approach, we report the lowest throughput obtained after simulating a number of random patterns equal to the number of simulations required by Nepal until the given generation. This value does not change significantly as new traffic patterns are randomly generated.

During the experiment, Nepal managed to dramatically degrade the probe connections' throughput from 1.66 Mb/s to 0.48 Mb/s with a small increment of the background traffic bandwidth. Thus, the genetic algorithm leads the background traffic sources to generating a traffic pattern that the TCP protocol cannot easily deal with.

Due to the fact that some genes may result unuseful at the end of the simulation run, the bandwidth of a critical background traffic pattern generated by Nepal is slightly larger than the starting one. Thus, in order to eliminate this bias from the results, we defined a *disturbance efficacy* parameter at generation i as $DE_i = (T^* - T_i)/B_i$, where T^* is the throughput of the TCP probe connections without the background noise traffic, T_i is the lowest throughput of the TCP probe connections archived until generation i and B_i is the corresponding background traffic bandwidth. In DE_i, the effects of the traffic are normalized with respect to the varying background traffic bandwidth B_i. We experimentally examined the DE of the statistical approach with different background traffic loads, and we found that the GA reaches the higher DE even when tuning the statistical traffic to provide the same load as the GA.

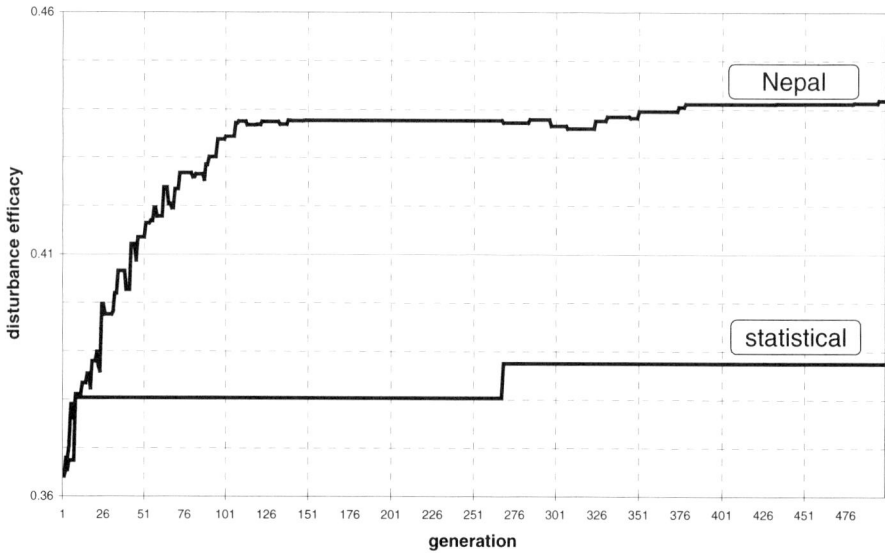

Figure 11.7 Disturbance efficacy.

Figure 11.7 plots DE_i as a function of the GA generation i and clearly shows that a critical traffic pattern is identified by the genetic evolution, and cannot be easily found with standard statistical methods. Performance evaluated with statistical methods can be much larger than the one experienced on the field in equivalent operating conditions. Thus, the GA allows the performance of the protocol to be assessed without making unrealistic assumptions about the probabilistic distribution of the traffic generated by background sources.

Moreover, the GA proves itself able to find a critical pattern examining the whole network almost as a black box, with little information of what is happening inside, and only a few *control knobs* for influencing the system. This is an important point, because Nepal does not rely on the knowledge of a specific problem, but on the strength of a potentially generic GA.

11.4.4   Analysis

In our case study, the genetic algorithm exposes the weaknesses of the TCP congestion reaction mechanism. TCP transmitters infer network congestion from missing acknowledgments and react to congestion by shrinking the transmission control window to

1 TCP message. This forces the transmitter to slow down the transmission rate, thus contributing to eliminate network congestion. The transmission control window is doubled each time an acknowledgment is received and the transmission rate is gradually increased accordingly.

The time required to reach the transmission speed that was devised before shrinking the window depends on the round trip delay of a packet on the connection. As a consequence, the longer the connection, the smaller the average throughput the connection achieves during congested periods.

The background traffic generated by the GA causes routers to discard packets in a way that makes TCP transmitters shrink the window before it can grow large enough to actually exploit all the available bandwidth. As a consequence, TCP connections achieve low throughput even though the network is just slightly overloaded. When the traffic is generated according to a statistical model, the performance limitations due to the shrinking of the control transmission window are less evident. This is shown by Figure 11.8, which plots the dimension of the control window of a TCP transmitter over time when the background traffic is created by the GA and when an equal amount of noise is generated statistically. The graph shows that, in the former case, the maximum dimension reached by the window when the network is overloaded is smaller. Thus, the throughput of the TCP connections with GA controlled background traffic is smaller than with statistical background traffic.

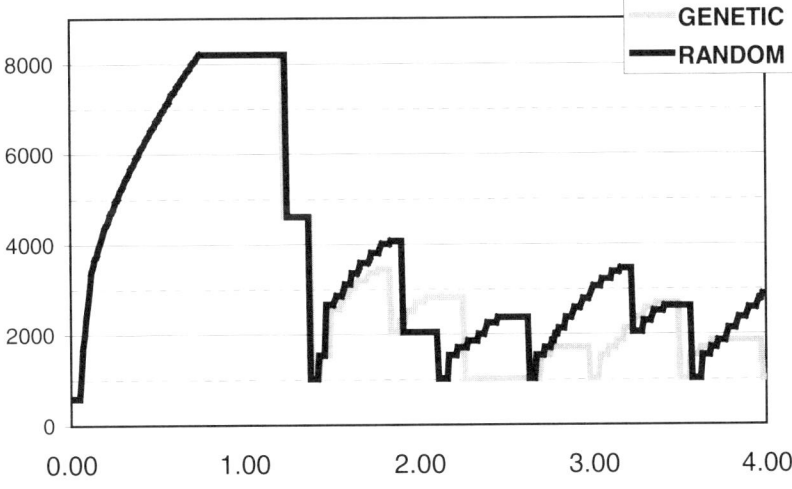

Figure 11.8 Dimension of the control window of a TCP source.

As a side result, our case study shows that in a congested network TCP connections are easily shut up by other traffic not reacting to congestion, like UDP traffic.

11.5 Conclusions

We have presented a new heuristic methodology based on a genetic algorithm, for evaluating the performance of a communication protocol in a realistic environment and possibly finding its weaknesses. We developed a prototypical implementation of the tool called Nepal, and ran some experiments, examining TCP connections in a simple IP network topology.

The results proved that, when the background traffic generation is driven by the genetic algorithm, the TCP performance is much lower than when traffic is generated by statistical methods: Nepal is able to identify traffic patterns particularly critical to the protocol that were not found by means of traditional techniques, such as statistical simulation.

Moreover, Nepal is able to deal with real network topologies and with the real implementation of protocols, instead of a mathematical approximated model. Thus, it can discover problems related to all the features of protocols, including those features that were very hard or impossible to model with formal techniques.

We feel that our approximate methodology could be effectively used in conjunction with traditional methods of verification, giving useful information about the behavior of a transmission protocol when used with a given network topology.

12

Neural Networks for the Optimization of Runtime Adaptable Communication Protocols

Robert S. Fish and Roger J. Loader

12.1 Introduction

The explosive growth of distributed computing has been fuelled by many factors. Applications such as video conferencing, teleoperation and most notably the World Wide Web are placing ever more demanding requirements on their underlying communication systems. Having never been designed to support such diverse communication patterns these systems are failing to provide appropriate services to individual applications.

Artificial neural networks have been used in areas of communication systems including signal processing and call management (Kartalopolus, 1994). This chapter suggests a further use of neural networks for the maintenance of application tailored communication systems. In this context, a neural network minimises the difference between an applications required **Q**uality **o**f **S**ervice (QoS) and that provided by the end-to-end connection.

12.1.1 Problem area

Communication systems based on the ISO Open System Interconnection (OSI) model historically suffered inefficiencies such as function duplication and excessive data copying.

However, a combination of modern protocol implementation techniques and an increase in the power and resources of modern computers has largely eliminated these overheads. Zitterbart (1993) defines the characteristics of various distributed applications and identifes four possible classes based on their communication requirements. Table 12.1 illustrates these classifications and highlights the broad range of transport services required by modern distributed applications. In the face of such diversity, the challenge of optimizing performance shifts from the efficiency of individual mechanisms to the provision of a service that best satisfies the broad range of application requirements. Providing such a service is further complicated by external factors such as end-to-end connection characteristics, host heterogeneity and fluctuations in network utilization. Traditional protocols, such as TCP/IP do not contain the broad functionality necessary to satisfy all application requirements in every operating environment. In addition, the QoS required by an application may change over the lifetime of a connection. If a protocol provides a greater QoS than is required then processor time and network bandwidth may be wasted. For these reasons, applications that use existing protocols do not necessarily receive the communication services they require.

Table 12.1 Diversity of application transport requirements.

Transport service class	Example applications	Average throughput	Burst Factor	Delay sens.	Jitter sens.	Order sens.	Loss Tol.	Priority Delivery
Interactive Time Critical	Voice	Low	Low	High	High	Low	High	No
	Tele conf	Mod	Mod	High	High	Low	Mod	Yes
	Motion video							
Distributed Time Critical	Compressed Motion video	High	High	High	Mod	Low	Mod	Yes
	raw	Very high	Low	High	High	Low	Mod	Yes
Real Time Time Critical	Manufacture Control	Mod	Mod	High	Var	High	Low	Yes
Non Real Time	File transfer	Mod	Low	Low	N/D	High	None	No
	TELNET	Very low	High	High	Low	High	None	Yes
Non Time Critical	Trans. process	Low	High	High	Low	Var	None	No
	File service	Low	High	High	Low	Var	None	No

Configurable protocols offer customised communication services that are tailored to a particular set of application requirements and end-to-end connection characteristics. They may be generated manually, through formal languages or graphical tools, or automatically with code scanning parsers that determine application communication patterns.

12.1.2 Adaptable Communication Systems

Whilst configurable communication systems provide a customized service, they are unable to adapt should the parameters on which they were based change. *Adaptable protocols* support continuously varying application requirements by actively selecting internal protocol processing mechanisms at runtime. There are several advantages in this:

1. **Application QoS:** it is not uncommon for an application to transmit data with variable QoS requirements. For example, a video conferencing application may require different levels of service depending upon the content of the session. Consider a video sequence that consists of a highly dynamic set of action scenes followed by a relatively static close-up sequence. The first part, due to rapid camera movement, is reasonably tolerant of data loss and corruption, but intolerant of high jitter. In contrast, the static close-up scenes are tolerant to jitter but require minimal data loss and corruption.
2. **Connection QoS:** adaptable protocols are able to maintain a defined QoS over varying network conditions. Whilst certain architectures offer guaranteed or statistical services the heterogeneuos mix of interconnection devices that form the modern internet does little to cater for end-to-end QoS. The adverse effects of variables such as throughput, delay and jitter can be minimised by using appropriate protocol mechanisms.
3. **Lightweight:** certain environments are able to support service guarantees such as defined latency and transfer rates. Once these are ascertained an adaptable protocol may remove unnecessary functions to achieve higher transfer rates.

The Dynamic Reconfigurable Protocol Stack (DRoPS) (Fish *et al.*, 1998) defines an architecture supporting the implementation and operation of multiple runtime adaptable communication protocols. Fundamental protocol processing mechanisms, termed *microprotocols* are used to compose fully operational communication systems. Each microprotocol implements an arbitrary protocol processing operation. The complexity of a given operation may range from a simple function, such as a checksum, to a complex layer of a protocol stack, such as TCP. The runtime framework is embedded within an operating system and investigates the benefits that runtime adaptable protocols offer in this environment. Mechanisms are provided to initialize a protocol, configure an instantiation for every connection, manipulate the configuration during communication and maintain consistent configurations at all end points. Support is also provided for runtime adaptation agents that automatically reconfigure a protocol on behalf of an application. These agents execute control mechanisms that optimize the configuration of the associated protocol. The remainder of this chapter will address the optimization of protocol configuration. Other aspects of the DRoPS project are outside the scope of this chapter, but may be found in Fish *et al.* (1998; 1999) and Megson *et al.* (1998).

12.2 Optimising protocol configuration

The selection of an optimal protocol configuration for a specific, but potentially variable, set of application requirements is a complex task. The evaluation of an appropriate configuration should at least consider the processing overheads of all available microprotocols and their combined effect on protocol performance. Additional consideration should be paid to the characteristics of the end-to-end connection. This is due to the diversity of modern LANs and WANs that are largely unable to provide guaranteed services on an end-to-end basis. An application using an adaptable protocol may manually modify its connections to achieve an appropriate service (work on ReSource reserVation Protocols (RSVP) addresses this issue).

Whilst providing complete control over the functionality of a communication system, the additional mechanisms and extra knowledge required for manual control may deter developers from using an adaptable system. History has repeatedly shown that the simplest solution is often favoured over the more complex, technically superior, one. For example, the success of BSD Sockets may be attributed to its simple interface and abstraction of protocol complexities. Manual adaptation relies on the application being aware of protocol specific functionality, the API calls to manipulate that functionality and the implications of reconfiguration. The semantics of individual microprotocols are likely to be meaningless to the average application developer. This is especially true in the case of highly granular protocols such as advocated by the DRoPS framework. As previously stated, protocol configuration is dependent as much on end-to-end connection characteristics as application requirements. Manual adaptation therefore requires network performance to be monitored by the application, or extracted from the protocol through additional protocol specific interfaces. Both approaches increase the complexity of an application and reduce its efficiency. Finally, it is unlikely that the implications of adaptation are fully understood by anyone but the protocol developer themselves. These factors place additional burdens on a developer who may subsequently decide that an adaptable protocol is just not worth the effort. If it is considered that the 'application knows best' then manual control is perhaps more appropriate. However, it is more likely to be a deterrent in the more general case.

It would be more convenient for an application to specify its requirements in more abstract QoS terms (such as tolerated levels of delay, jitter, throughput, loss and error rate) and allow some automated process to optimize the protocol configuration on its behalf.

A process wishing to automate protocol optimization must evaluate the most appropriate protocol configuration with respect to the current application *requirements* as well as end-to-end connection *conditions*. These parameters refer to network characteristics (such as error rates), host resources (such as memory and CPU time) and scheduling constraints for real-time requirements. The complexity of evaluating an appropriate protocol configuration is determined by the number of *conditions* and *requirements*, the number of *states* that each may assume, and the total number of *unique protocol configurations*.

Within DRoPS, a protocol graph defines default protocol structure, basic function dependencies and alternative microprotocol implementations. In practice, a protocol developer will specify this in a custom Adaptable Protocol Specification Language (APSL). Defining such a graph reduces the number of possible protocol configurations to a function of the number of objects in the protocol graph and the number of alternative mechanisms provided by each. This may be expressed as:

$$\prod_{k=1}^{K} F_k \quad (12.1)$$

where, F_k is the number of states of configuration k and K the total number of functions in the protocol graph. The automated process must therefore consider N combinations of requirements, conditions and configurations, which is defined as:

$$N = \prod_{i=1}^{I} C_i \cdot \prod_{j=1}^{J} R_j \cdot \prod_{k=1}^{K} F_k \quad (12.2)$$

where C_i is the number of states of condition i and R_j the number of states of requirement j, and where I and J are the total number of conditions and requirements. This represents the total number of evaluations necessary to determine the most appropriate configuration for each combination of requirements and conditions. The complexity of this task increases relentlessly with small increases in the values of I, J and K; as illustrated in Figure 12.1. Part (a) shows the effect of adding extra protocol layers and functions, and part (b) the effect of increasing the condition and requirement granularity.

12.2.1 Protocol Control Model

The runtime framework supports mechanisms for the execution of protocol specific *adaptation policies*. These lie at the heart of a modular control system that automatically optimises the configuration of a protocol. The methods used to implement these policies are arbitrary and of little concern to the architecture itself. However, the integration of DRoPS within an operating system places several restrictions on the characteristics of these policies. The adaptation policy must posses a broad enough knowledge to provide a good solution for all possible inputs. However in the execution of this task it must not degrade performance by squandering system level resources. Therefore, any implementation must be small to prevent excessive kernel code size and lightweight so as not to degrade system performance.

Adaptation policies are embedded within a control system, as depicted in Figure 12.2. Inputs consist of QoS requirements from the user and performance characteristics from the functions of the communication system. Before being passed to the adaptation policy, both sets of inputs are shaped. This ensures that values passed to the policy are within known bounds and are appropriately scaled to the expectations of the policy.

User requirements are passed to the control system through DRoPS in an arbitrary range of 0 to 10. A value of 0 represents a 'don't care' state, 1 a low priority and 10 a high priority. These values may not map 1:1 to the policy, i.e. the policy may only expect 0 to 3. The shaping function normalizes control system inputs to account for an individual policies interpretation.

End-to-end performance characteristics are collected by individual protocol. Before being used by the policy, the shaping function scales these values according to the capability of the reporting function. For example, an error detected by a weak checksum function should carry proportionally more weight than one detected by a strong function. The shaped requirements and conditions are passed to the adaptation policy for evaluation. Based on the policy heuristic an appropriate protocol configuration is suggested.

The existing and suggested configurations are compared and appropriate adaptation commands issued to convert the former into the latter. Protocol functions, drawn from a library of protocol mechanisms, are added, removed and exchanged, and the updated protocol configuration is used for subsequent communication. The DRoPS runtime framework ensures that changes in protocol configuration are propagated and implemented at all end points of communication. The new configuration should provide a connection with characteristics that match the required performance more closely than the old configuration. Statistics on the new configuration will be compiled over time and if it fails to perform adequately it will be adapted.

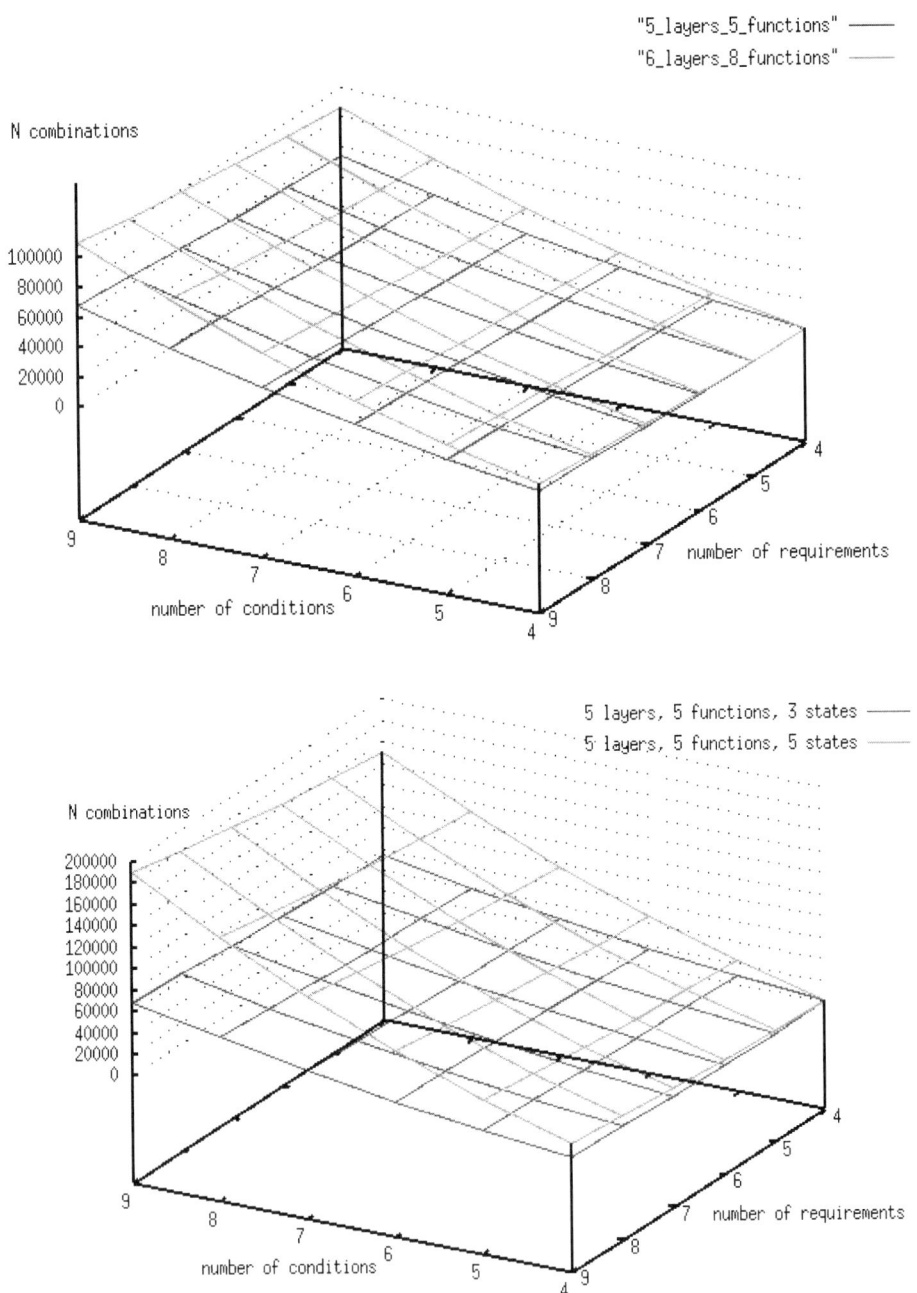

Figure 12.1 Increasing complexity of the configuration task.

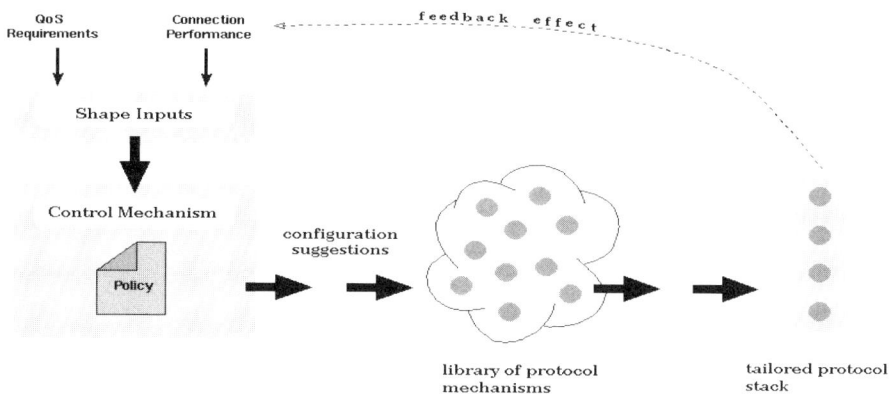

Figure 12.2 Model of automatic adaptation control system.

12.2.2 Neural Networks as Adaptation Policies

Various projects have attempted to simplify the process of reconfiguration by mapping application specified QoS requirements to protocol configurations. Work by Box *et al.* (1992) and Zitterbart (1993) classified applications into service classes according to Table 12.1 and mapped each to a predefined protocol configuration. The DaCaPo project uses a search based heuristic, CoRA (Plagemann *et al.*, 1994), for evaluation and subsequent renegotiation of protocol configuration. The classification of building blocks and measurement of resource usage are combined in a structured search approach enabling CoRA to find suitable configurations. The properties of component functions, described in a proprietry language L, are based on tuples of attribute types such as throughput, delay and loss probability. CoRA configures protocols for new connections at runtime with respect to an applications requirements, the characteristics of the offered transport service and the availability of end system resources. The second approach provides a greater degree of customisation, but the time permitted to locate a new configuration determines the quality of solution found. Beyond these investigations there is little work on heuristics for the runtime optimisation of protocol configuration.

In the search for a more efficient method of performing this mapping, an approach similar to that used in Bhatti and Knight (1998) for processing QoS information about media flows was considered. However, the volume of data required to represent and reason about QoS rendered this solution intractable for fine-grained protocol configuration in an Operating System environment. Although impractical, this served to highlight the highly consistent relationships between conditions, requirements and the actual performance of individual configurations. For example, consider two requirements, bit error tolerance and required throughput, and a protocol with variable error checking schemes. The more comprehensive the error checking, the greater the impact it has on throughput. This is the case for processing overhead (raw CPU usage) and knock-on effects from the detection of errors (packet retransmission). As emphasis is shifted from correctness to throughput, the selection of error function should move from complete to non-existent, depending on the level of error in the end-to-end connection.

12.2.3 Motivation

If requirements and conditions are quantized and represented as a *vector*, the process of mapping to protocol configurations may be reduced to a pattern matching exercise. Initial interest in the use of neural networks was motivated by this fact, as pattern is an application at which neural networks are particularly adept. The case for neural network adaptation policies is strengthened by the following factors:

1. **Problem data:** the problem data is well suited to representation by a neural network. Firstly extrapolation is never performed due to shaping and bounding in the control mechanism. Secondly, following shaping the values presented at the input nodes may not necessarily be discrete. Rather than rounding, as one would in a classic state table, the networks ability to interpolate allows the suggestion of protocol configurations for combinations of characteristics and requirements not explicitly trained.

2. **Distribution of overheads:** the largest overhead in the implementation and operation of a neural network is the training process. For this application the overheads in off line activities, such as the time taken to code a new protocol function or adaptation policy, do not adversely effect the more important runtime performance of the protocol. Thus, the overheads are being moved from performance sensitive online processing to off line activities, where the overheads of generating an adaptation policy are minimal compared to the time required develop and test a new protocol.

3. **Execution predictability:** the execution overheads of a neural network are constant and predictable. The quality of solution found does not depend upon an allotted search time and always results in the best configuration being found (quality of solution is naturally dependent on the training data).

12.2.4 The Neural Network Model

The aim of using a neural network is to capitalise on the factors of knowledge representation and generalisation to produce small, fast, knowledgeable and flexible adaptation heuristics. In its most abstract form, the proposed model employs a neural network to map an input vector, composed of quantized requirements and conditions, to an output vector representing desired protocol functionality.

A simple example is illustrated in Figure 12.3. Nodes in the input layer receive requirements from the application and connection characteristics from the protocol. The values presented to an input node represents the quantized state (for example low, medium or high) of that QoS characteristic. No restrictions are placed on the granularity of these states and as more are introduced the ability of an application to express its requirements increases. Before being passed to the network input node, values are shaped to ensure they stay within a certain range expected by the policy. It should be noted that this process does not round these values to the closest state as would be required in a state table. The networks ability to generalise allows appropriate output to be generated for input values not explicitly trained.

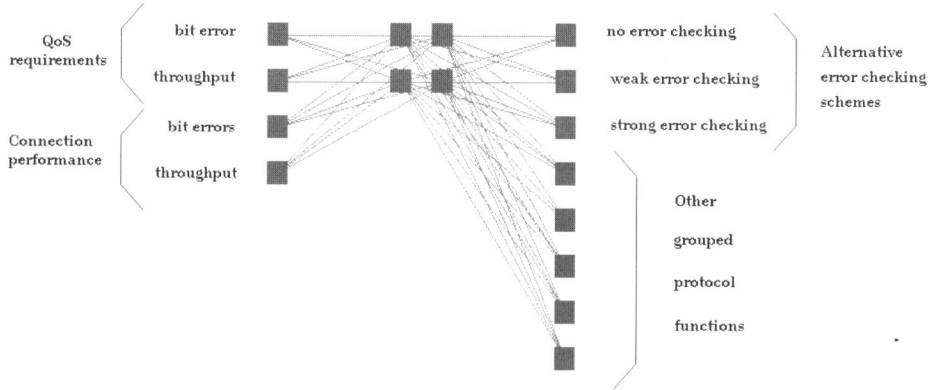

Figure 12.3 Mapping QoS parameters to protocol configuration.

When the network is executed the values written in the nodes of the output layer represent the set of functions that should appear in a new protocol configuration. To achieve this, output nodes are logically grouped according to the class of operation they perform; individual nodes represent a single function within that class. Output nodes also represent non-existent functions, such as that representing no error checking in the example. This forms a simple YES / NO pattern on the output nodes, represented by 1 and 0 respectively. For example, if error checking is not required, the node representing *no error checking* will exhibit a YES whilst the other nodes in this logical class with exhibit NO.

In many cases, the values presented at the output nodes will not be black and white, 1 or 0, due to non-discrete input values and the effect of generalisation. Therefore the value in each node represents a degree of confidence that the function represented should appear in any new configuration. When more than one node in a logical group assumes a non-zero value, the function represented by the highest confidence value is selected. To reduce processing overhead, only protocol functions that have alternative microprotocols are represented in the output layer.

12.3   Example Neural Controller

This section outlines the steps taken to implement a neural network based adaptation policy for the Reading Adaptable Protocol (RAP). RAP is a complete communication system composed of multiple microprotocols; it contains a number of adaptable functions, summarised in Table 12.2, a subset of which are used in the example adaptation policy.

Table 12.2 Adaptable functionality of the Reading Adaptable Protocol.

Protocol mechanism	Alternative implementations
Buffer allocation	preallocated cache, dynamic
Fragmentation and reassembly	stream based, message based
sequence control	none, complete
flow control	none, window based
acknowledgement scheme	IRQ, PM-ARQ
checksums	none, block checking, full CRC

12.3.1 Adaptation Policy Training Data

A neural network gains knowledge through the process of learning. In this application the training data should represent the most appropriate protocol configuration for each combination of application requirements and operating conditions. The development of a neural network adaptation controller is a three stage process:

1. **Evaluate protocol performance:** this process determines the performance of each protocol configuration in each operating environment. Network QoS parameters are varied and the response of individual configurations logged.
2. **Evaluate appropriate configurations:** the result of performance evaluation is used to determine the most appropriate configuration for each set of requirements in each operating environment. This requires development of an appropriate fitness function.
3. **Generate a policy:** having derived an ideal set of protocol configurations a neural network must be trained and embedded within an adaptation policy.

The result of these three stages is an adaptation policy that may be loaded into the DRoPS runtime framework and used to control the configuration of a RAP based system.

12.3.1 Evaluating Protocol Performance

The evaluation of protocol performance is performed by brute force experimentation. During protocol specification, a configuration file is used to identify microprotocol resources and default protocol configurations. Using this file it is possible for the APSL parser to automatically generate client and server applications that evaluate the performance characteristics of all valid protocol configurations.

Evaluating every protocol configuration in a static environment, where connection characteristics remain fixed, does not account for the protocols performance over real world connections in which connection characteristics are potentially variable. To function correctly in such circumstances an adaptation policy requires knowledge of how different configurations perform under different conditions. To simulate precisely defined network characteristics, a traffic shaper is introduced. This intercepts packets traversing a host's network interface, incurs bit and block errors, introduces variations in delay, loses packets and restricts throughput. The shaper is contained within a microprotocol module that may be dynamically loaded into the runtime framework. In concept this module lies on the connection between the client and server evaluation applications. In fact it is located in the runtime framework at the server side node. A basic overhead of 0.0438 microseconds is incurred for each invocation (measured on a Pentium II 300MHz based machine) with additional undefined overheads for the shaping code. The use of a modular structure permits protocol designers to add their own code for the shaping of network characteristics. Each microprotocol posses a control interface allowing instructions to be passed from the runtime framework. This permits an application to simulate numerous network conditions by scaling parameters, such as residual error rate, used by the shaper mechanisms.

Figure 12.4 presents a a pseudo-code algorithm that forms the core functionality of the evaluation application. To evaluate the performance of a configuration, two individual tests are performed; a ping test and a raw throughput test. The former is used to judge latency and jitter whilst the latter determines throughput and error detection capabilities. Rather than burden an application with detecting loss and corruption, the error checking functions themselves are required to post notification of any observed losses or errors to a shaper structure. If the function is not included in the protocol configuration the error will not be detected and will go unreported. In addition, mechanisms within the shaper keep track of the number of losses and errors that are caused. After the ping and throughput tests these statistics are combined to determine the configurations performance within the simulated environment. To obtain accurate statistics, each configuration must be evaluated several hundred times for each condition. Depending upon the actual condition, each evaluation can take several seconds to perform. The minimal number of evaluations that must be performed for each condition are:

$$\prod_{k=1}^{K} F_k \cdot Eval$$

where $Eval$ is the number of evaluations performed for each configuration and the rest of the notation is consistent with equation 12.1. Even in the simple case the evaluation of all configurations can take tens of hours to complete.

The result of evaluation is a performance profile for every protocol configuration in each operating environment. This takes the form of a simple report for each combination of configuration and connection characteristic. Figure 12.5 illustrates such a report for a RAP configuration operating in a connection with a latency of 300 microseconds and a bit error occurring for approximately every 8 megabytes of data. The performance of this configuration occupies the remainder of the report. For the example protocol, 1943 reports were generated. These were generated by considering the protocol functions introduced in Table 12.2 with only four QoS characteristics (each assuming three states).

12.3.3 Evaluating Fitness of Configuration

The second stage attempts to determine the most appropriate configuration for each combination of requirements and conditions. This relies on the report file generated as output from configuration performance evaluation. Unlike the previous phase, where the evaluation applications are automatically generated by the APSL parser, the responsibility for creating the fitness evaluation application falls to the protocol designer. For each entry in the report file, the fitness evaluator has to determine how well every combination of application requirements is served. Figure 12.6 presents a pseudo code example of the core fitness evaluator function for requirements of loss, error, delay and jitter. The fitness evaluation function reads individual records from the report file described in the previous section. For each report, the configurations performance is evaluated for every combination of application requirements. Evaluation is performed by a fitness function that must be defined by the protocol developer. When considering how well a particular configuration satisfies an applications requirements in a particular operating environment each combination is assigned a fitness value. The mechanism used to generate this value depends upon the desired objectives of the adaptation policy. The fitness function used by the example adaptation policy is based on a weighted sum with two objectives:

```
/* initialise protocol configuration */

initialise_environmental_conditions();

do {
   /* initialise environmental conditions */

   initialise_protocol_configuration();

    /* evaluate every protocol configuration in this
     * environment */

   do {
      /* PERFORMANCE TEST THE CURRENT PROTOCOL
       * CONFIGURATION */

      evaluate_configuration();

      /* next possible configuration */

      cycled_through_all = increment_configuration();

   } while( !cycled_through_all );

   /* increment environmental conditions */

   cycled_through_all_conditions = increment_conditions();

} while( !cycled_through_all_conditions );
```

Figure 12.4 Pseudocode for the generation of protocol performance characteristics.

```
264 ------------------------------------------------
   conditions:
   latency 300 jitter 0 bit errors 8000000 loss 0

   configuration:
   buffer 1  frag 1  addr 1  seq ctrl 0  flow ctrl 0  csum 0  ack 3

   results:
   throughput: ave 2.361896 MB/s
   delay      : min 2000.885010, max 2009.724976, ave 2008.575073
   jitter     : ave 4.419983 microseconds
   loss       : caused 0, observed 0
   error      : caused 539, observed 264
```

Figure 12.5 Performance report generated by performance evaluation.

$$W_0 \cdot \left(1 - \frac{\text{Throughput}_{actual}}{\text{Throughput}_{max}}\right) \\ + W_1 \cdot \left(\text{Quality}_{error} + \text{Quality}_{loss} + \text{Quality}_{delay} + \text{Quality}_{jitter}\right) \quad (12.3)$$

where Quality_x is the normalized satisfaction rating of requirement x, and W_0 and W_1 are weights used to trade off between satisfaction of requirements and maximum throughput. The primary objective of this function is to provide a protocol configuration that satisfies the specified QoS requirements and the secondary objective is to maximize throughput. The components of this function determine the runtime objectives of the adaptation policy.

It should be noted that the significance of certain microprotocols can not be determined solely through performance evaluation. These atomic functions are either required or not. Functions such as message and stream based fragmentation, encryption and compression can only be evaluated by the above fitness function in terms of their effect on observable characteristics. A fitness function developer may wish to place additional clauses in the fitness function to favour message-based fragmentation if it is explicitly required by the application. Perhaps these should be placed after the result of the neural controllers suggestion. In this way, suggestions by the controller may be overridden by application specifications for particular functions. Unlike the protocol performance evaluator, there is currently no mechanism for generating the fitness evaluation application. Fitness evaluation results in a data set containing the most appropriate protocol configurations for each requirement in each operating environment. The configurations suggested by this set are determined by the objectives of the fitness function and may be used to train the neural network adaptation policy.

```
for every report in the report file
 while( read_item_from_report_file ) {
        foreach (loss_requirement) {
           foreach (error_requirement) {
              foreach (delay requirement) {
                 foreach (jitter requirement) {

                    /*
                     * how well does current configuration
                     * suit these requirements in current
                     * environment
                     */
                    fitness = evaluate_fitness_of_configuration();

                    /*
                     * update most appropriate configuration
                     */
                    if ( fitness > best_fitness ) {
                        best_fitness = fitness;
                    }
                 }
              }
           }
        }
    }
write_training_set();
```

Figure 12.6 Pseudocode for the evaluation of configuration fitness.

12.3.4 The Neural Network

Training data for a neural network is generated from the fitness evaluation of each protocol configuration. The remainder of this subsection describes the training process using an example policy for the optimization of RAP.

A simple feed forward MultiLayer Perceptron (MLP) was created using the Stuttgart Neural Network Simulator (SNNS). The SNNS is a software simulator for neural networks developed at the Institute for Parallel and Distributed High Performance Systems (IPVR) at the University of Stuttgart. The projects goal is to create an efficient and flexible simulation environment for research on and application of neural networks. The SNNS consists of two main components; a simulator kernel and a graphical user interface. The kernel operates on the internal network data structures and performs all operations of learning and recall. The user interface provides graphical representations of a neural network and controls the kernel during the simulation run. In addition, the user interface has an integrated network editor which can be used to directly create, manipulate and visualise neural nets in various ways. Choosing the optimal number of hidden layer nodes is an *ad hoc* process best determined through experimentation. An excess number of nodes can lead to large runtime execution overheads and too few nodes can lead to to poor classification and generalisation. The example MLP has nine input nodes, six hidden layer nodes (in a 2×3 arrangement) and nine output nodes. Four of the input nodes represent conditions whilst the remaining five represent user requirements. The topology of the example network, including logical partitioning of input and output nodes, is shown in Figure 12.7. A network is trained using the SNNS backpropagation algorithm with the patterns generated by the fitness function. The x-axis denotes the number of epochs, where each epoch represents the presentation of all training patterns to the network. The y-axis represents the sum of the squared differences at each output neuron between actual and required, referred to as the Sum Squared Error (SSE). The eagerness with which the network learns highlights how the mappings generated by the evaluation application are consistent enough to be successfully learnt by a neural network. Figure 12.8 visualizes the error development observed during the training of the example network. Various network topologies were implemented but no benefit was noticed in increasing the number of hidden layer nodes above 6, which is surprisingly small.

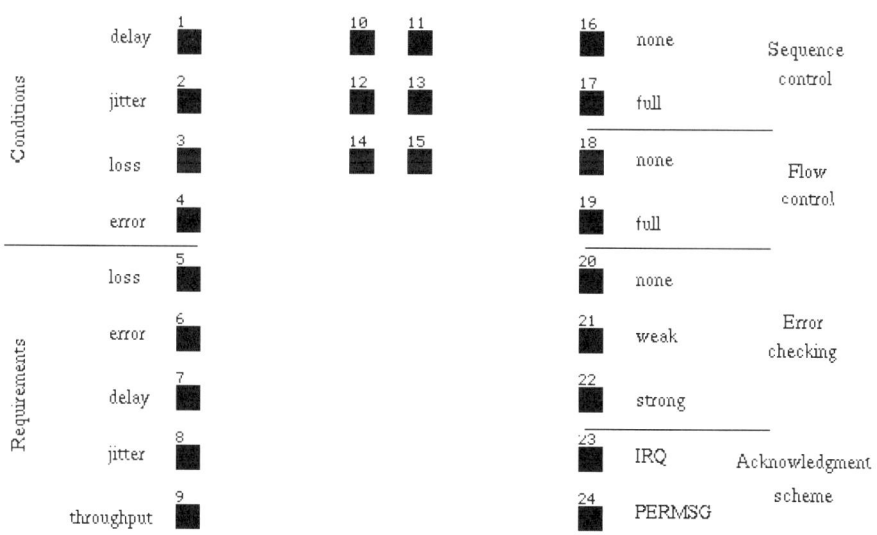

Figure 12.7 Mapping and logical grouping in example neural network.

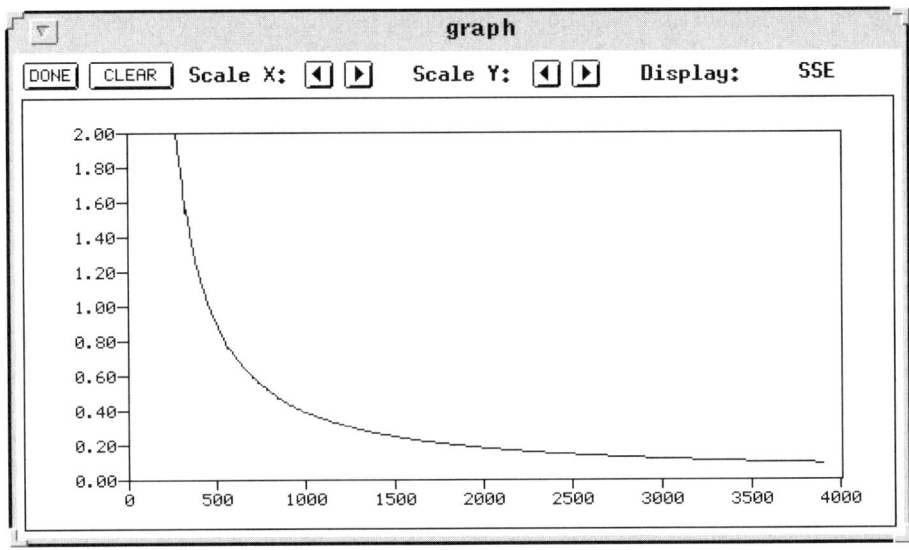

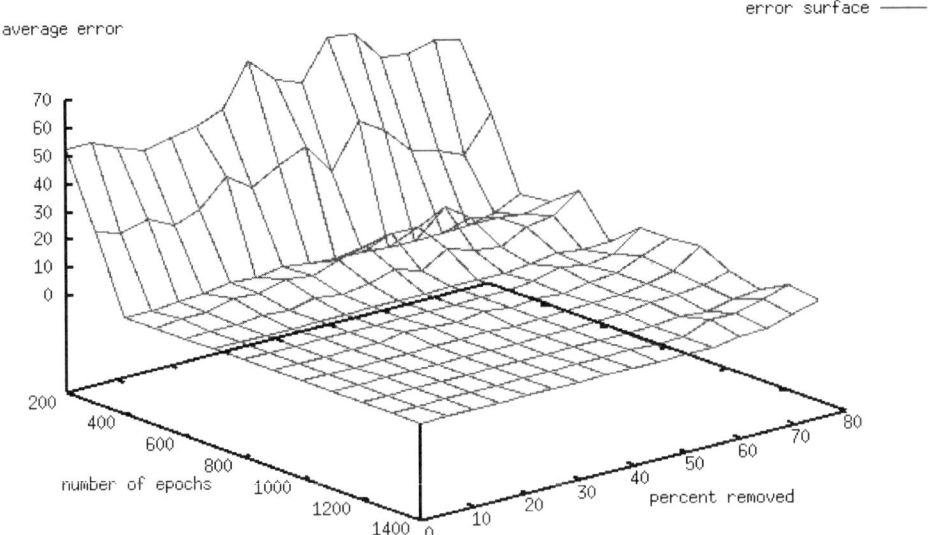

Figure 12.8 Progression of training (upper) and error surface (lower).

The SNNS provides tools to visualise various aspects of an implemented neural network. Figure 12.9 illustrates two such visualisations, presenting the networks' suggested configuration for two extreme scenarios. The nodes in this figure correspond to those in Figure 12.7. The upper example represents a network with all requirements and conditions set to 0, and the lower example demonstrates the other extreme. The former suggests a

Neural Networks for the Optimization of Runtime Adaptable Communication Protocols

protocol configuration with no sequence control, no flow control, no checksum function and the most lightweight acknowledgement scheme. The latter is operating in a noisy, highly lossy environment, and so suggests a more conservative configuration.

The SNNS allows a trained network to be exported as an independent C program. The first noticeable feature of this code is its size in comparison to the raw set of training data. For this problem about 3Mb is required to store the raw performance data as compared to 7Kb for a neural network trained on this data. The runtime performance of the network is minimal due to the small number of nodes required. Averaged over a million iterations the network takes circa 40 microseconds to execute. Not only is this time small, it also remains constant.

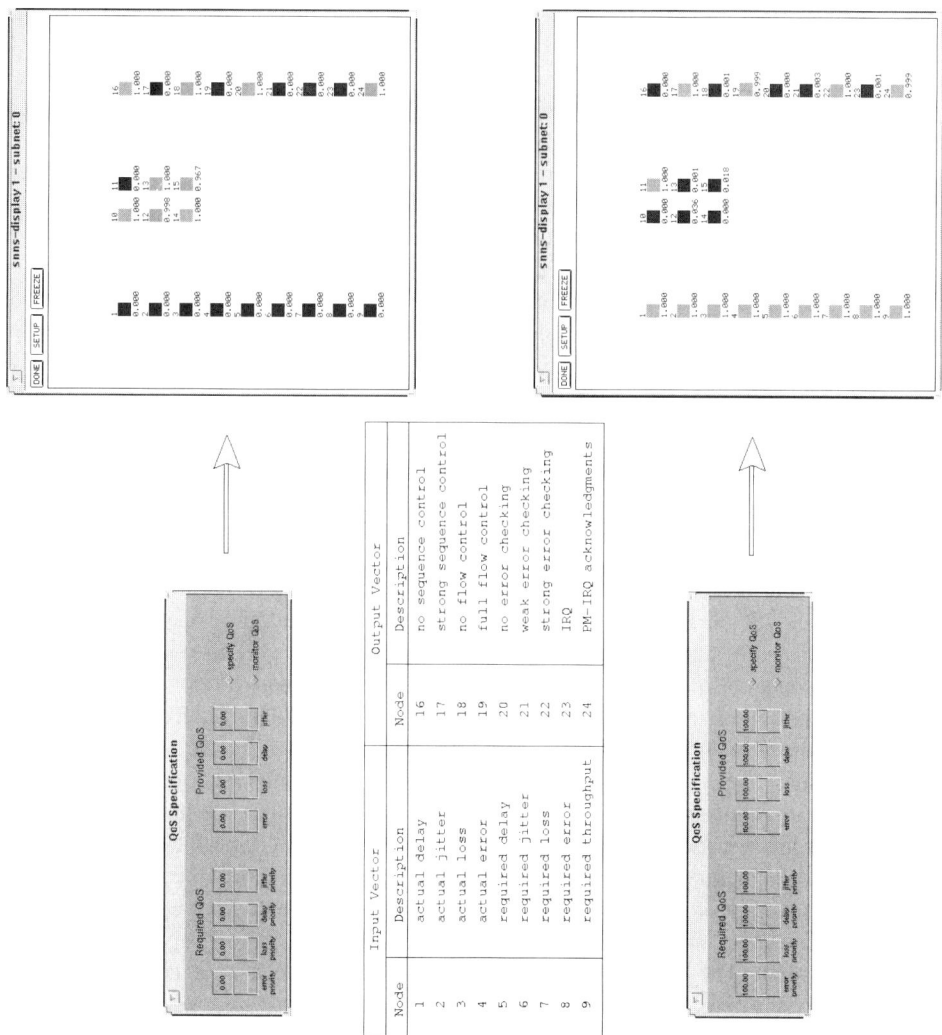

Figure 12.9 Mapping combinations of requirements and end-to-end conditions.

12.4 Implications of Adaptation

The previous sections introduce the notion of neural networks as control heuristics for the optimisation of protocol configuration. Whilst a model is presented, no reference is made to the effect of using these controllers. To evaluate the impact of protocol adaptation on real world applications, a distributed video tool has been implemented. Based on the MPEG TV Player decoding engine the tool is composed of a file server and video player. An MPEG encoded data stream is sent from the server to player using a DRoPS adaptable protocol. Various network conditions are simulated using the same technique as in the generation of the neural network training data. In addition, the tool supports mechanisms to manually define a protocol configuration or specify required QoS levels that are passed to an adaptation policy. The interface to this tool, shown in Figure 12.10, is composed of several windows. The main window contains typical video controls such as Play, Pause and Rewind. In addition, it provides controls to create and break connections with an MPEG server. The QoS specification window allows a user to define their required QoS and either set or monitor the provided QoS. These values are passed to the runtime framework where the former is passed to the adaptation policy and the latter to the traffic shaping function. The Adaptation Control window presents a graphical interface to simplify manual reconfiguration. The current configuration is displayed in the two large canvas objects that occupy the body of the window. The individual objects shown in these canvases represent the component functionality of protocol that is being used to transfer the MPEG stream. This window provides a number of controls to perform the common tasks of adaptation allowing functions to be enabled, disable, added, removed and exchanged. A fourth window supporting the specification of Quality of Perception (QoP) is supported but not shown here. QoP specifies the quality of a media presentation in terms of more abstract notions such as satisfaction and understanding. In co-operation with another project the DRoPS architecture is being used to experiment with the interactions of QoS and QoP. The interested reader may find more information in Fish *et al.* (1999).

As an example of the implications of protocol reconfiguration, a video sequence containing of highly dynamic action and relatively static scenes is considered. Such an example was considered earlier as one of the motivating factors for protocol adaptation. In this example the subject is trying to describe certain features of a set of suspension forks for a mountain bike. The sequence is composed of action scenes, cycling over rough terrain, with large amounts of movement and static where the components of the cycle are studied in detail. In the action scenes, the effect of data corruption is tolerable. Even where whole blocks of quantized data are lost, the visible impact is minimised by the fact that so much else is happening in the video clip. Whilst these anomalies are tolerable, the effects of jitter in the datatream, causing the clip to pause momentarily once in a while, are much more noticeable. In this situation there is a trade-off between errors and the fluidity of the video clip. Rather than maintaining absolute correctness in the data flow, the jitter caused by retransmitting corrupt or missing data should be minimised by simply not retransmitting that data. In the relatively static part of the video sequence, the effects of lost and corrupt data are more apparent. They may also be more problematic as the loss or corruption of this content may result in the loss of critical information. For example, over an extremely lossy link, individual characters of text may become unreadable. When this occurs, the jitter caused by retransmission correcting the corrupt or lost data become more acceptable.

Neural Networks for the Optimization of Runtime Adaptable Communication Protocols 217

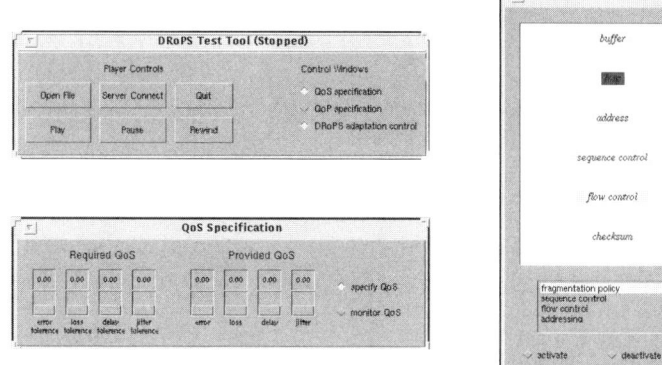

Figure 12.10 Interface of multimedia evaluation tool.

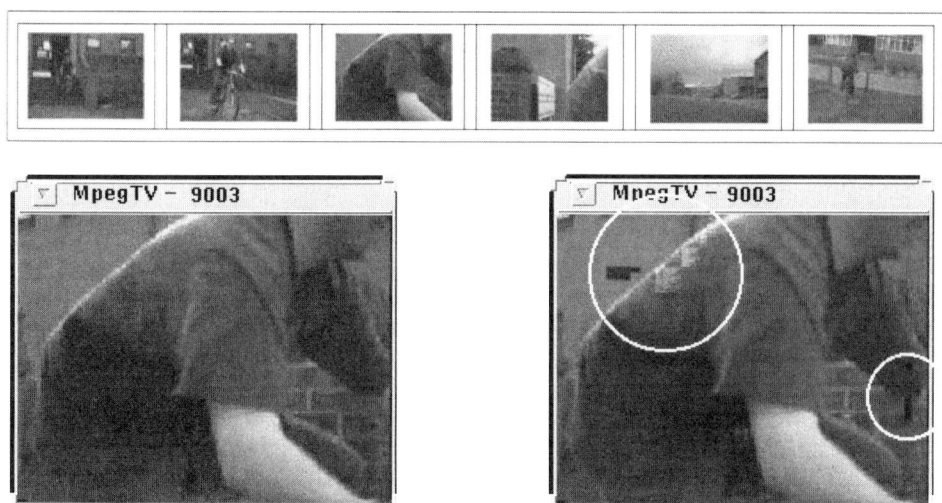

Figure 12.11 Video sequence with large degree of movement.

Figure 12.11 demonstrates the effect of losing data. The strip of images at the top of this Figure represents snapshots at arbitrary points in the video sequence. The duration of the clip is approximately 8 seconds, and the cycle passes the camera going 12 mph at a distance of 3 meters. The larger frames shown below this strip show the same image. The first does not suffer any corruption whilst the second suffers from lost data. Two circular rings highlight the effect of this loss. At this speed and distance from the camera, the corrupted frame is highly active and the effect of the error is likely to escape largely unnoticed.

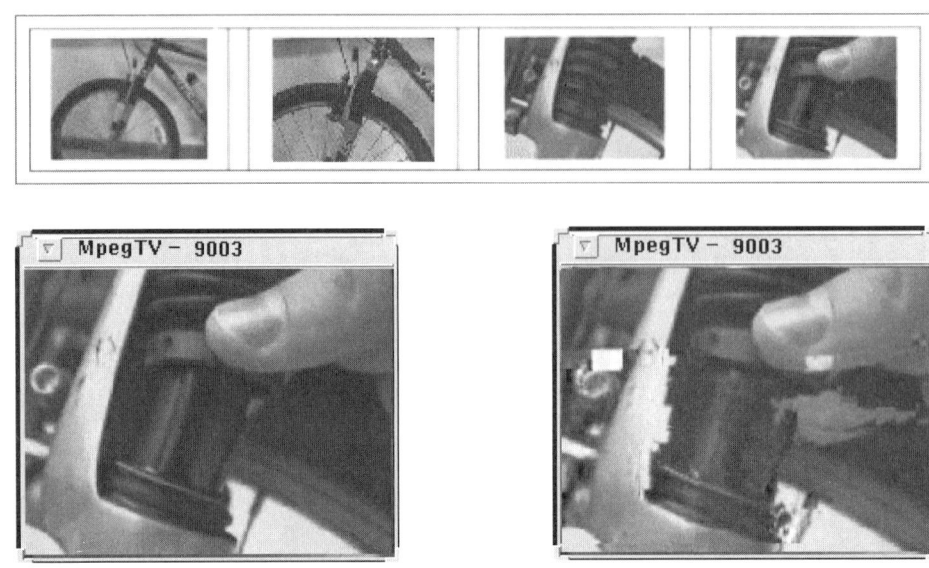

Figure 12.12 Video sequence with small degree of movement.

After demonstrating the operation of the suspension forks the video clip effectively changes its QoS requirements by playing a rather static set of images that show close-up images of the components. Figure 12.12 shows a similar set of images to the previous example and again data is lost. However, in this example the result of loss is more noticeable as the background is itself less changeable. In contrast, the action segment of the video this segment is trying to communicate detailed information about specific components. It therefore cannot tolerate the same degree of loss as its predecessor. In addition, which jitter should still be minimized, it is more important for the contents of this clip to remain 100% correct. In addition to the control model, a selection of graphical examples demonstrate the benefits of using protocol adaptation. These enforce the notion that the required QoS within an application is potentially changeable. They also reinforce the argument that the communication systems serving these applications should be aware of these potentially changing requirements and be able to respond appropriately. In the given example, this is simply a case of the application respecifying its communication requirements. The mapping of these requirements to an appropriate protocol configuration is performed automatically by the neural network.

Using the test tool introduced earlier, the QoS sliders may be set to reflect an applications requirements at each stage in the lifetime of the video. These values are passed to the control system which then evaluates a protocol configuration which best suits the requirements in the given environment. In the case of these clips, the first (action) corresponds to the first configuration of Figure 12.9 whilst the second clip (static) is represented by the second configuration.

12.5 Conclusions

This chapter has presented a model for the use of neural networks in the optimisation of highly granular communication protocols for the DRoPS architecture. The application of a neural network to this problem allows such optimization without the use of computationally expensive search based techniques. In addition, the execution time of the neural network is constant and the quality of solution is not dictated by the amount of time permitted to find a configuration. These benefits are achieved at the expense of a many hours evaluating all available protocol configurations. However, within an operating system architecture, such as DRoPS, the small runtime overhead and minimal resource are worth the offline overheads.

The functionality of RAP is continually expanding. Whilst increasing the ability of the protocol to accommodate a greater set of application requirements, it increases the complexity of the configuration process. The current set of configurations, conditions and requirements form only a small subset of those that would be found in a production system. It is therefore important to emphasize that whilst this model has been seen to work for all our current protocol implementations, issues of how well it scales have yet to be addressed. Future work will continue to study the effect of added functionality and attempt to identify other evolutionary methods for reducing the time taken to generate training data. At present, the generation of a neural network adaptation policy is performed by hand, a long and tedious process. The data and mechanisms exist to automate this process such that a user may define the functionality of an adaptable protocol, have it validated and an adaptation automatically generated.

Part Three

Software, Strategy and Traffic Management

A critical and increasingly problematic bottleneck in the growth of telecommunications firms is the software development process. New services need new, fast, flexible and complex software, but the speed with which it needs to be developed causes considerable extra cost. One of the chapters in this part looks at this problem, focussing on the software testing and validation process, and how this can be effectively automated with evolutionary computation.

Strategy refers to some of the more complex and dynamic decision making processes which occur in network management and service provision. An emerging idea to cope with this, and to make reasonable and effective decisions in highly complex, competitive scenarios, is to use game theory. Two chapters in this part look at game theory and its use, in conjunction with genetic algorithms, to support decisions about service provision and traffic control.

More on traffic control and management is addressed by four other chapters in this part; two concerning static networks and two concerning mobile networks. In static networks, the traffic management problem amounts to a load-balancing problem in which resources, be they servers, links, or switches, need to be effectively used in such a way that perceived quality of service is maximized. The 'static network' chapters concern evolutionary algorithm and heuristic approaches to this issue in the conetext of distributed database service provision. In mobile networks, traffic management amounts to good channel assignment strategies, which negotiate often complex constraints on bandwiths and relative frequencies in order to maximize, for example, the number of calls that can be concurrently handled by a base station. Again, modern optimization techniques, artfully designed, are found to be highly effective at such tasks.

13

Adaptive Demand-based Heuristics for Traffic Reduction in Distributed Information Systems

George Bilchev and Sverrir Olafsson

13.1 Introduction

As Internet connectivity is reaching the global community, information systems are becoming more and more distributed. Inevitably, this overnight exponential growth has also caused traffic overload at various places in the network. Until recently, it was believed that scaling the Internet was simply an issue of adding more resources, i.e. bandwidth and processing power could be brought to where they were needed. The Internet's exponential growth, however, exposed this impression as a myth. Information access has not been and will not be evenly distributed. As it has been observed, user requests create 'hot-spots' of network load, with the same data transmitted over the same network links again and again. These hotspots are not static, but also move around, making it impossible to accurately predict the right network capacity to be installed. All these justify the requirement to develop new infrastructure for data dissemination on an ever-increasing scale, and the design of adaptive heuristics for traffic reduction.

In this chapter, we develop a distributed file system model and use it as an experimental simulation tool to design, implement and test network adaptation algorithms. Section 13.2 describes in detail the distributed file system model and explains the implemented simulation environment. Two adaptation algorithms are developed in section 13.3. One is based on the 'greedy' heuristic principle and the other is a genetic algorithm tailored to handle the constraints of our problem. Experiments are shown in section 13.4, and section 13.5 gives conclusions and discusses possible future research directions.

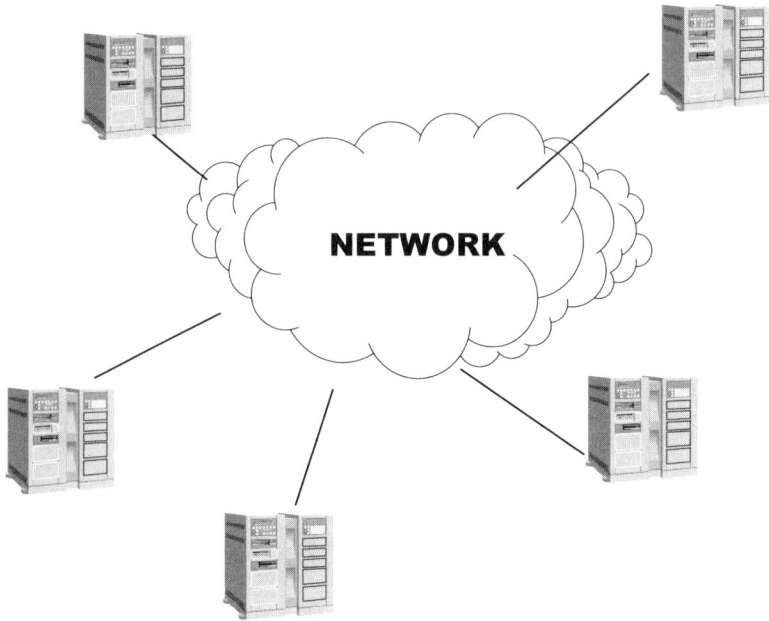

Figure 13.1 A schematic representation of the network and the distributed file system.

13.2 The Adaptation Problem of a Distributed File System

The World Wide Web is rapidly moving us towards a distributed, interconnected information environment, in which an object will be accessed from multiple locations that may be geographically distributed worldwide. For example, a database of customers' information can be accessed from the location where a salesmen is working for the day. In another example, an electronic document may be co-authored and edited by several users.

In such distributed information environments, the replication of objects in the distributed system has crucial implications for system performance. The replication scheme affects the performance of the distributed system, since reading an object locally is faster and less costly than reading it from a remote server. In general, the optimal replication scheme of an object depends on the request pattern, i.e. the number of times users request the data. Presently, the replication scheme of a distributed database is established in a static fashion when the database is designed. The replication scheme remains fixed until the designer manually intervenes to change the number of replicas or their location. If the request pattern is fixed and known *a priori*, then this is a reasonable solution. However, in practice the request patterns are often dynamic and difficult to predict. Therefore, we need an adaptive network that manages to optimize itself as the pattern changes. We proceed with the development of a mathematical model of a distributed information/file system.

A distributed file system consists of interconnected nodes where each node i, $i = 1, N$ has a local disk with capacity d_i to store files – see Figures 13.1 and 13.2. There is a collection of M files each of size s_j, $j = 1, M$. Copies of the files can reside on any one of the disks provided there is enough capacity. The communication cost $c_{i,k}$ between nodes i and k (measured as transferred bytes per simulated second) is also given.

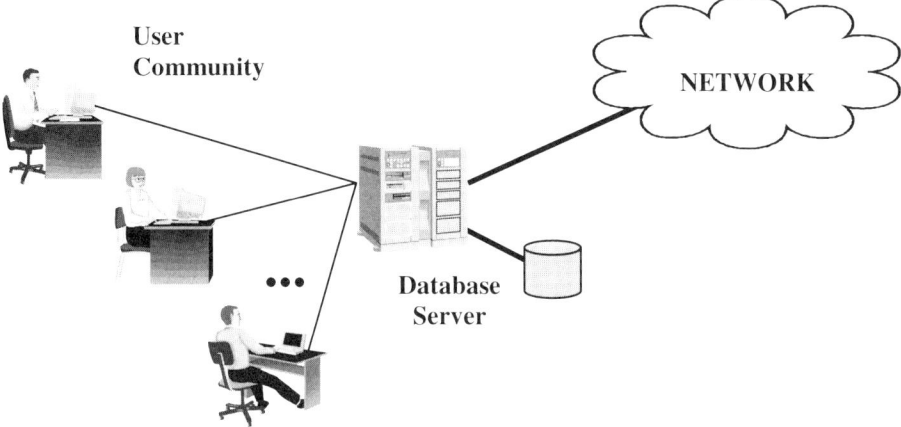

Figure 13.2 Users connect to each node from the distributed files system and generate requests.

In our model each node runs a file manager which is responsible for downloading files from the network (Figure 13.3). To do that, each file manager i maintains an index vector l_{ij} containing the location where each file j is downloaded from. User applications running on the nodes generate file requests the frequency of which can be statistically monitored in a matrix $\{p_{i,j}\}$.

To account for contention and to distribute the file load across the network it has been decided to model how busy the file managers are at each node k as follows:

$$b_k = \frac{\sum_{\substack{i,j \\ l_{i,j}=k}} p_{i,j} \cdot s_j}{\sum_{\substack{n=1,N \\ m=1,M}} p_{i,j} \cdot s_j}$$

Thus, the response time of the file manager at node k can be expressed as waiting time in a buffer (Schwartz, 1987):

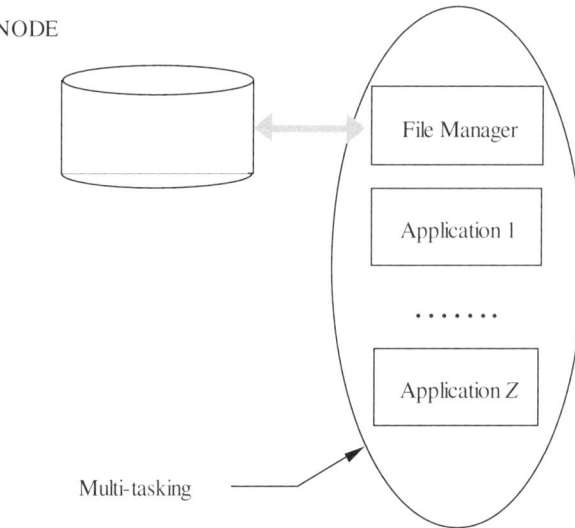

Figure 13.3 Each node runs a file manager (responsible for allocating files on the network) and a number of user applications which generate the requests.

$$r_k = \begin{cases} \dfrac{1}{\tau_k - b_k} & \tau_k > b_k \\ \infty & \text{otherwise} \end{cases}$$

where τ_k reflects the maximum response capacity of the individual servers. The overall performance at node i can be measured as the time during which applications wait for files to download (i.e. response time):

$$O_i = \sum_{j=1}^{M} \left(\frac{s_j}{c_{i,l_{i,j}}} + r_{l_{i,j}} \right) \cdot p_{i,j}$$

The first term in the sum represents the time needed for the actual transfer of the data and the second term reflects the waiting time for that transfer to begin. The goal is to minimize the average network response time:

$$\min_{\{l_{i,j}\}} \frac{1}{N} \sum_{i=1}^{N} \sum_{j=1}^{M} \left(\frac{s_j}{c_{i,l_{i,j}}} + r_{l_{i,j}} \right) \cdot p_{i,j}$$

The minimization is over the index matrix $\{l_{i,j}\}$. There are two constraints: (1) the available disk capacity on each node should not be exceeded; and (2) each file must have at least one copy somewhere on the network.

13.2.1 The Simulation Environment

In our distributed file system model the users generate file requests and the network responds by allocating and downloading the necessary files. The file requests can be statistically monitored and future requests predicted from observed patterns. The simulation environment captures these ideas by modeling the user file requests as random walks:

$$p_{i,j}(t) = p_{i,j}(t-1) + \gamma$$

where γ is drawn from a uniform distribution $U(-r,r)$. The parameter r determines the 'randomness' of the walk. If it is close to zero then $p_{i,j}(t) \approx p_{i,j}(t-1)$.

During the simulated interval $[t, t+1]$ the model has information about the file requests that have occurred in the previous interval $[t-1, t]$. Thus the dynamics of the simulation can be formally defined as:

```
For t=1,2,3,…
    generate new file requests:   P(t) = P(t-1) + {γ_{i,j}}
    simulate network:
```

$$O(t) = \frac{1}{N}\sum_{i=1}^{N} O_i(t) = \frac{1}{N}\sum_{i=1}^{N}\sum_{j=1}^{M}\left(\frac{s_j}{c_{i,l_{i,j}}} + r_{l_{i,j}}\right) \cdot p_{i,j}(t)$$

An adaptive distributed file system would optimize its file distribution according to the user requests. Since the future user requests are not known and can only be predicted the optimization algorithm would have to use an expected value of the requests derived from previous observations:

$$\tilde{P}(t) = \text{Prediction}(P(t-1))$$

Thus an adaptive distributed file system can be simulated as follows:

```
For t=1,2,3,…
    file requests prediction:       P̃(t) = Prediction(P(t-1))
    optimization:                   L(t) = Optimize(P̃(t))
    generate new file requests:     P(t) = P(t-1) + {γ_{i,j}}
    simulate network:
```

$$O(t) = \frac{1}{N}\sum_{i=1}^{N} O_i(t) = \frac{1}{N}\sum_{i=1}^{N}\sum_{j=1}^{M}\left(\frac{s_j}{c_{i,l_{i,j}}} + r_{l_{i,j}}\right) \cdot p_{i,j}(t)$$

The next section describes the developed optimization algorithms in detail.

13.3 Optimization Algorithms

13.3.1 Greedy Algorithm

The 'greedy' principle consists of selfishly allocating resources (provided constraints allow it) without regard to the performance of the other members of the network (Cormen *et al.*, 1990). While greedy algorithms are optimal for certain problems (e.g. the minimal spanning tree problem) in practice they often produce only near optimal solutions. Greedy algorithms, however, are very fast and are usually used as a heuristic method. The greedy approach seems very well suited to our problem since the uncertainties in the file request prediction mean that we never actually optimize the real problem, but our expectation of it.

The implemented greedy algorithm works as follows:

```
For each file j check every node i to see if there is enough
space to accommodate it and if enough space is available
calculate the response time of the network if file j was at
node i:
```

$$\sum_{k=1}^{N}\left(\frac{s_j}{c_{k,i}} + r_i\right) \cdot p_{k,j}$$

```
After all nodes are checked copy the file to the best found
node.
```

The above described algorithm loads only one copy of each file into the distributed file system. If multiple copies are allowed, then add copies of the files in the following way:

```
For each node i get the most heavily used file
(i.e., max(p_{i,j} · s_j)) which is not already present.
       j

Check if there is enough space to accommodate it.
If yes, copy it. Continue until all files are checked.
```

13.3.2 Genetic Algorithm

Genetic Algorithms (GAs) are very popular due to their simple idea and wide applicability (Holland, 1975; Goldberg, 1989). The simple GA is a *population-based* search in which the *individuals* (each representing a point from the search space) exchange information (i.e. *reproduce*) to move through the search space. The exchange of information is done through *operators* (such as mutation and crossover) and is based on the 'survival of the fittest' principle, i.e. better individuals have greater chance to reproduce.

It is well established that in order to produce good results the basic GA must be tailored to the problem at hand by designing problem specific representation and operators. The overall flow of control in our implemented GA for the distributed files system model is similar to the steady state genetic algorithm described in Chapter 1.

In order to describe the further implementation details of our GA we need to answer the following questions: How are the individuals represented? How is the population initialized? How is the selection process implemented? What operators are used?

Individuals representation: each individual from the population represents a distribution state for the file system captured by the matrix $\{l_{i,j}\}$.

Initialization: it is important to create a random population of feasible individuals. The implemented initialization process randomly generates a node index for each object and tries to accommodate it on that node. In case of a failure the process is repeated for the same object.

Selection process: the individuals are first linearly ranked according to their fitness and then are selected by a roulette-wheel process using their rank value.

Operators: the main problem is the design of operators which preserve feasibility of the solutions (Bilchev and Parmee, 1996). This is important for our problem since it intrinsically has two constraints: (i) the disk capacity of the nodes must not be exceeded; and (ii) each file must have at least one copy somewhere on the network. (If feasibility is not preserved by the operators, then the fitness function would require to be modified by an appropriate penalty function in order to drive the population into the feasible region.) We have developed two main operators both preserving feasibility:

The new operators developed in this work are called Safe-add and Safe-delete. Safe-add works as follows:

```
For each node randomly select and copy a file which is not
already locally present and whose size is smaller than the
available disk space.

Check to see if any of the nodes would respond faster by
downloading files from the new locations and if yes, update
the matrix {l_{i,j}}.
```

Safe-delete is as follows:

```
For each node randomly select and delete a file provided it
is not the last copy.

Update the matrix {l_{i,j}} to reflect on the above changes.
```

In our experiments we have used a population size of 70 individuals for 30 generations. During each generation 50 safe-add and three safe-delete operators are applied. During the selection process the best individual has 5% more chances of being selected as compared to the second best, and so on.

13.4 Simulation Results

In our experiments we start with an *offline* simulation during which the optimization algorithms are run when the file system is not used (i.e. overnight, for example). In this scenario, we assume that both algorithms have enough time to finish their optimization before the file system is used again.

A typical simulation is shown in Figure 13.4. Tests are done using seven nodes and 100 files. All simulation graphs start from the same initial state of the distributed file system. Then the two optimization algorithms are compared against a non-adaptive (static) file system (i.e. when no optimization is used). The experiments undoubtedly reveal that the adaptive distributed file system produces better results as compared to a static file system. The graphs also clearly indicate the excellent performance of the GA optimizer, which consistently outperforms the greedy algorithm.

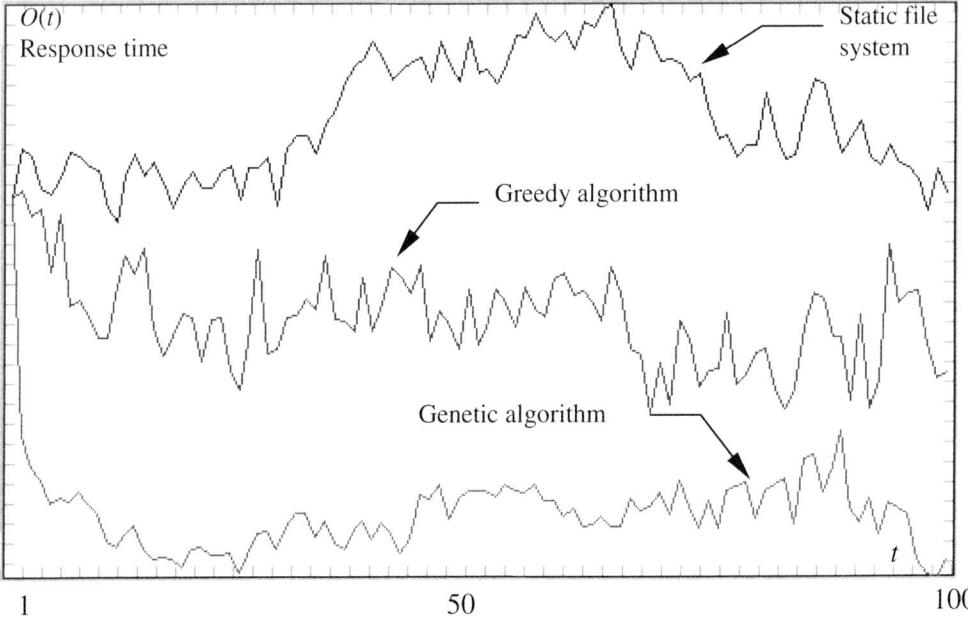

Figure 13.4 An offline simulation. Adaptive distributed file systems utilizing a genetic algorithm and a greedy algorithm respectively are compared against a static distributed file system. The experiments use seven nodes and 31 files.

To show the effect of delayed information, we run the greedy algorithm once using the usage pattern collected from the previous simulation step $P(t-1)$ (which are available in practice) and once using the actual $P(t)$ (which is not known in practice). The difference in performances reveals how much better we can do if perfect information were available (Figure 13.5).

Adaptive Demand-based Heuristics for Traffic Reduction in Distributed Information Systems

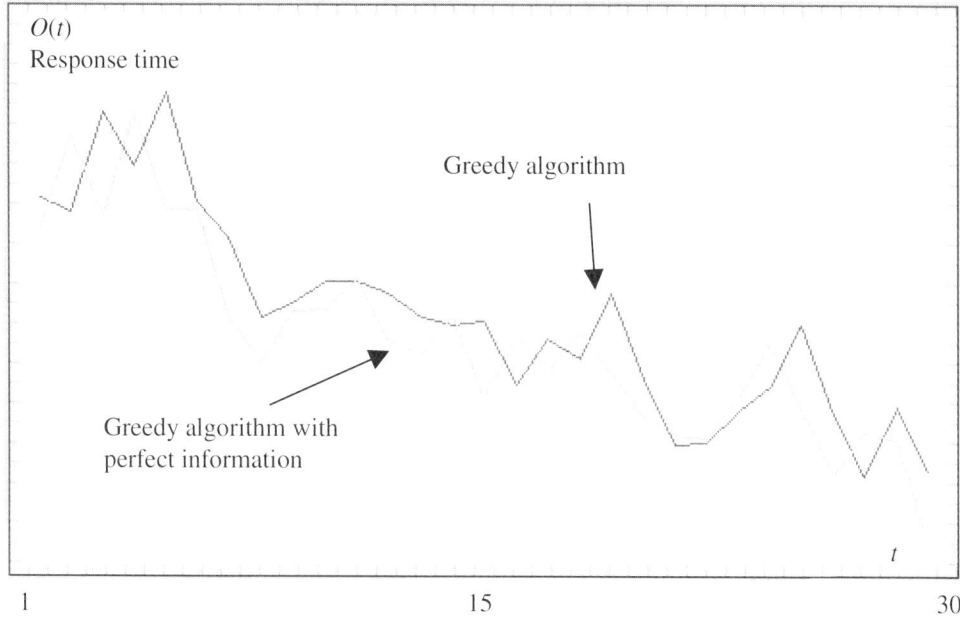

Figure 13.5 A greedy algorithm with perfect information is compared to a greedy algorithm with delayed information. In practice, we only have delayed information.

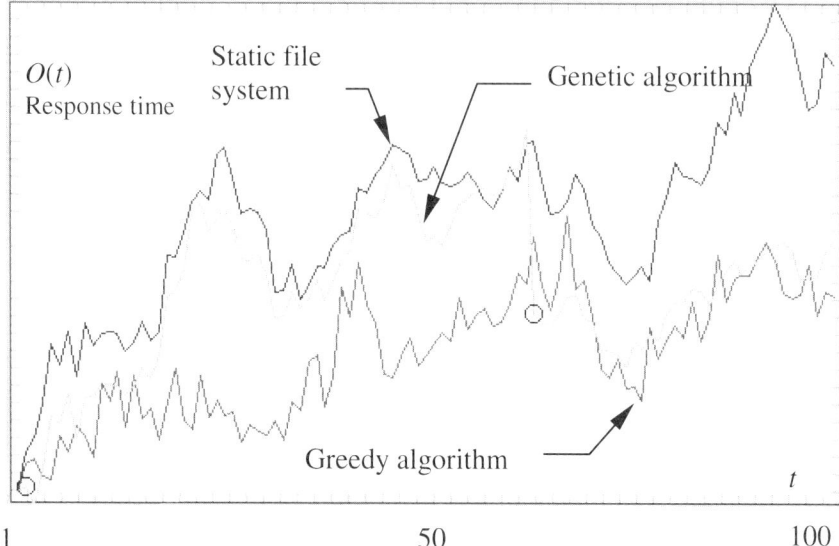

Figure 13.6 Online simulation. The circles indicate when the GA optimization takes place. The GA/greedy algorithm ratio is 60 (i.e. the GA is run once for every 60 runs of the greedy algorithm).

232 Telecommunications Optimization: Heuristic and Adaptive Techniques

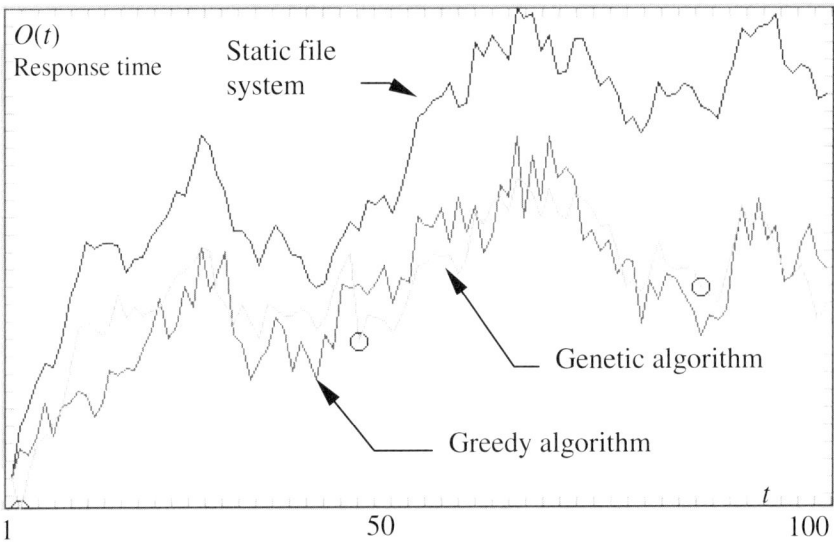

Figure 13.7 Online simulation. The GA/greedy algorithm ratio is 40. This is the critical ratio where the average performance of both algorithm is comparable.

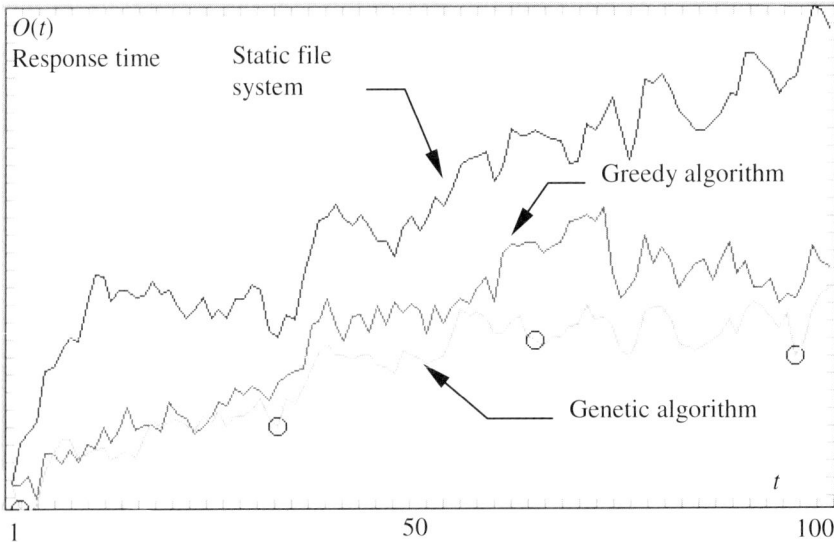

Figure 13.8 Online simulation. The GA/greedy algorithm ratio is 30. The GA manages to maintain its performance advantage.

Next we consider online simulations. In an *online* simulation a faster optimization algorithm would be executed at a greater rate than a slower algorithm. To account for the fact that the GA is much slower than the greedy algorithm, we introduce a new simulation variable, namely the GA/greedy algorithm ratio. It shows how many greedy optimizations can be achieved during the time it takes for one GA optimization to finish.

Figure 13.6 shows a typical online simulation. Here the GA is run every 60 steps while the greedy algorithm is run every step. The GA optimizes the file system at the beginning of the simulation. Then for the next 60 steps the file system is static. This allows the greedy algorithm to take over and outperform the GA.

It is interesting to note the critical GA/greedy algorithm ratio after which the greedy algorithm outperforms the GA. For our simulation model, this happens to be around 40. Figure 13.7 shows such a simulation. When the ratio drops even further then the GA is capable of maintaining its performance advantage (Figure 13.8).

The simulation results suggest that the GA is more suitable for situations where the file request pattern does not change very quickly, while the greedy algorithm is more appropriate for rapidly changing request patterns.

13.5 Conclusions and Future Work

Our investigations showed that adaptation is crucial for network performance optimization. Its importance would be even more evident in a real network where there are limited resources and slow communication links.

The simulation results confirmed our initial hypothesis that a tailored genetic algorithm would outperform 'classical' heuristics such as the greedy algorithm. However, while being able to find better solutions, the GA is considerably slower and doesn't scale as well as the greedy algorithm. Saying that, it is important to mention that both algorithms are centralized. They run on a single node and use global information. Centralized algorithms in general suffer from scalability problems. Operations such as collecting global information can be prohibitive for large networks. This evidence suggests a direction for future studies, namely distributed adaptation algorithms where only local information is used and optimization decisions are executed locally on each node. While it is difficult to predict how global network behavior emerges out of the many local interactions, when designed such an algorithm would be intrinsically scaleable.

Another important aspect requiring future studies is network resilience. User requests (file requests in our model) should not fail if a node is temporary disconnected. While designing resilient networks is a huge problem in itself, it is even more important to design resilient networks which are optimal and adaptable.

Acknowledgements

We would like to thank Brian Turton (University of Wales), Dave Corne (Reading University), Marc Wennink (BT) and Alan Pengelly (BT) for useful comments.

14

Exploring Evolutionary Approaches to Distributed Database Management

Martin J. Oates and David Corne

14.1 Introduction

Many of today's data intensive applications have the common need to access exceedingly large databases in a shared fashion, simultaneously with many other copies of themselves or similar applications. Often these multiple instantiations of the client application are geographically distributed, and therefore access the database over wide area networks. As the size of these 'industrial strength' databases continue to rise, particularly in the arena of Internet, Intranet and Multimedia servers, performance problems due to poor scalabilty are commonplace. Further, there are availability and resilience risks associated with storing all data in a single physical 'data warehouse', and many systems have emerged to help improve this by distributing the data over a number of dispersed servers whilst still presenting the appearance of a single logical database

The Internet is a large scale distributed file system, where vast amounts of highly interconnected data are distributed across many number of geographically dispersed nodes. It is interesting to note that even individual nodes are increasingly being implemented as a cluster or 'farm' of servers. These 'dispersed' systems are a distinct improvement over monolithic databases, but usually still rely on the notion of fixed master/slave relationships (mirrors) between copies of the data, at fixed locations with static access configurations. For 'fixed' systems, initial file distribution design can still be complex and indeed evolutionary algorithms have been suggested in the past for static file distribution by March and Rho (1994, 1995) and Cedano and Vemuri (1997), and for Video-on Demand like services by Tanaka and Berlage (1996). However as usage patterns change, the efficiency of the original distribution can rapidly deteriorate and the administration of such systems, being mainly manual at present, can become labour intensive as an alternative solution, Bichev and Olafsson (1998) have suggested and explored a variety of automated evolutionary caching techniques. However, unless such a dispersed database can dynamically adjust which copy of a piece of data is the 'master' copy, or indeed does away with the notion of a 'master copy', then it is questionable whether it can truly be called a 'distributed' database.

The general objective is to manage varying loads across a distributed database so as to reliably and consistently provide near optimal performance as perceived by client applications. Such a management system must ultimately be capable of operating over a range of time varying usage profiles and fault scenarios, incorporate considerations for multiple updates and maintenance operations, and be capable of being scaled in a practical fashion to ever larger sized networks and databases. To be of general use, the system must take into consideration the performance of both the back-end database servers, and the communications networks, which allow access to the servers from the client applications.

Where a globally accessible service is provided by means of a number of distributed and replicated servers, accessed over a communications network, the particular allocation of specific groups of users to these 'back-end' servers can greatly affect the user perceived performance of the service. Particularly in a global context, where user load varies significantly over a 24 hour period, peak demand tends to 'follow the sun' from Europe through the Americas and on to the Asia Pacific region. Periodic re-allocation of groups of users to different servers can help to balance load on both servers and communications links to maintain an optimal user-perceived Quality of Service. Such re-configuration/re-allocation can also be usefully applied under server node or communications link failure conditions, or during scheduled maintenance.

The management of this dynamic access configuration/load balancing in near real time can rapidly become an exceedingly complex task, dependent on the number of nodes, level of fragmentation of the database, topography of the network and time specific load characteristics. Before investigation of this problem space can be contemplated, it is essential to develop a suitable model of the distributed database and network, and a method of evaluating the performance of any particular access and data distribution given a particular loading profile is required. This model and evaluation method can then be used for fitness function calculations within an evolutionary algorithm or other optimisation technique, for investigating the feasibility and effectiveness of different access configurations based on sampled usage and other data. Armed with such a 'performance predicting' model, an automated load balancing system can be devised which uses an optimiser to determine ideal access configurations based on current conditions, which can then be used to apply periodic database self-adaption in near real time.

14.2   An Overview of the Model

Figure 14.1 shows a block diagram of such an automated, self adapting, load balancing, distributed database. The system employs a performance predicting model of the servers

Exploring Evolutionary Approaches to Distributed Database Management

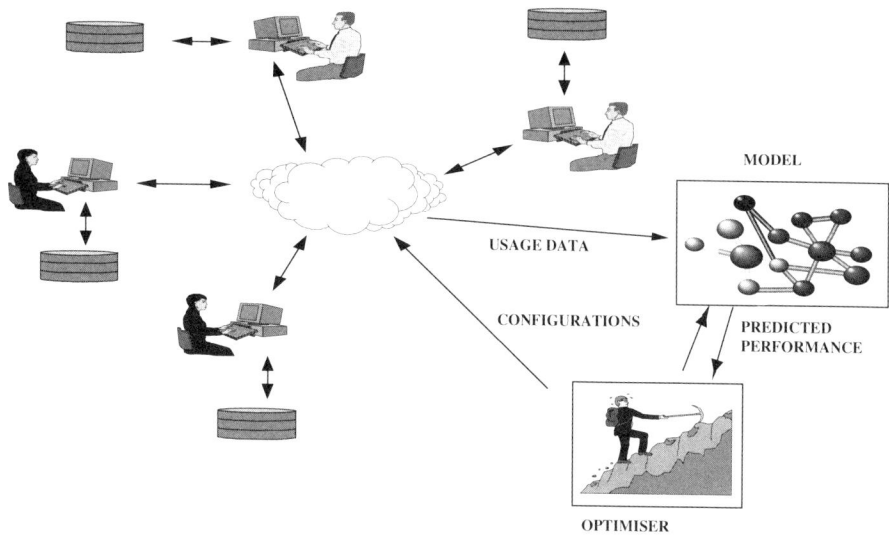

Figure 14.1 Schematic of an automated, self-adapting, load-balancing distributed database.

and communication links, and an optimiser which produces possible allocations of groups of users to 'back-end' servers. These 'allocations' (solution vectors) are evaluated by the model, which uses them to determine how to combine respective workloads onto selected servers and predicts the degraded performance of each server and communication link using two key formulae based on the principles of Little's Law and MM1 queuing. These are:

$$\text{Degraded Response Time} = \frac{1}{((1/\text{BTT}) - \text{TAR})} \qquad (14.1)$$

where BTT is the server Base Transaction Time and TAR is Transaction Arrival Rate, and:

$$\text{CTR} = \text{CV} \cdot \left(\left(\sum_{i \in S} \text{TR}_i \right) - \max_{i \in S} \text{TR}_i \right) + \max_{i \in S} \text{TR}_i \qquad (14.2)$$

where CTR stands for Combined Transaction Rate, taking into account individual transaction rates TR from a range of sources S, and where CV is a Contention Value representing a measure of the typical degree of collision between transactions.

Each node can be considered to be both a client (a source of workload) and a potential server. As a client, the node can be thought of as a 'Gateway' or 'Portal' aggregating user load for a particular geographical sub-region or interest group. This is referred to as the 'Client node' loading and is characterised for each node by a Retrieval rate and Update rate together with a transaction overlap factor. As a server, each node's ability to store data and/or perform transactions is characterised by its Base Transaction Time (the latency experienced by a solitary transaction on the server – this then degrades as work load increases) and a resource contention factor. Workload retrievals from a particular node are performed on the server, specified in a solution vector supplied by the optimiser, with updates applied to all active servers. Each nodal point-to-point communications link is also characterised by a Base Communications Time which deteriorates with increased load. Specified as a matrix, this allows crude modelling of a variety of different interconnection topologies.

The optimiser runs for a fixed number of evaluations in an attempt to find a configuration giving the least worst user transaction latency, moderated by a measure of overall system performance (variants of this will be described in due course). As the system is balancing worst server performance, communications link performance and overall system performance, this effectively becomes a multi-objective minimisation problem which can be likened to a rather complex bin-packing problem. Experiments described here utilise 10 node 'scenarios' for the problem space which are described later.

A typical solution vector dictates for each client node load, which server node to use for retrieval access as shown below :

Client	1	2	3	4	5	6	7	8	9	10
Server to use	1	3	3	4	1	3	3	4	1	3

This solution vector is generated by the optimiser using a chromosome of length 10 and an allelic range of the integers 1 through 10 – and is manipulated as a direct 10-ary representation rather than in a binary representation more typical of a cannonical genetic algorithm (see Bäck, 1996; Goldberg, 1989; Holland, 1975). Previous publications by the author and others have demonstrated differential algorithmic performance between HillClimbers, Simulated Annealers and differing forms of GA on this problem set (see Oates *et al.* 1998; 1998a; 1998b), under different tuning values of population size and mutation rates (see Oates *et al.*, 1998c), on different scenarios (Oates *et al.*, 1998b) and using different operators (Oates *et al.* 1999). Some of these results are reviewed over the next few pages.

The scenarios investigated typically vary the relative performance of each node within the system and the topography of the communications network. Two such scenarios were explored in (Oates *et al.*, 1998b) where the first, Scenario A, consists of all servers being of similar relative performance (all Base Transaction Times being within a factor of 2 of each other) and similar inter-node communication link latency (again all within a factor of 2). The communications link latency for a node communicating with itself is obviously set significantly lower than the latency to any other node. This scenario is shown schematically in Figure 14.2 and, with the basic 'least worst performing server' evaluation function, is found to have many different solutions with the same globally optimum fitness value.

Scenario B considers the case where the 10 nodes are split into two regions, all nodes in each region being connected by a high speed LAN and the two LANs being interconnected by a WAN, the WAN being 10 times slower than the LANs. This is represented by high communication latencies for clients accessing servers outside their region, medium latencies for access within their region, and the lowest latencies for access to themselves. One node in each region is considered a Supernode, with one tenth the Base Transaction Time of the other nodes in its region. This scenario, shown in Figure 14.3, has only one optimal solution under most load conditions, where all nodes in a region access their own region's supernode.

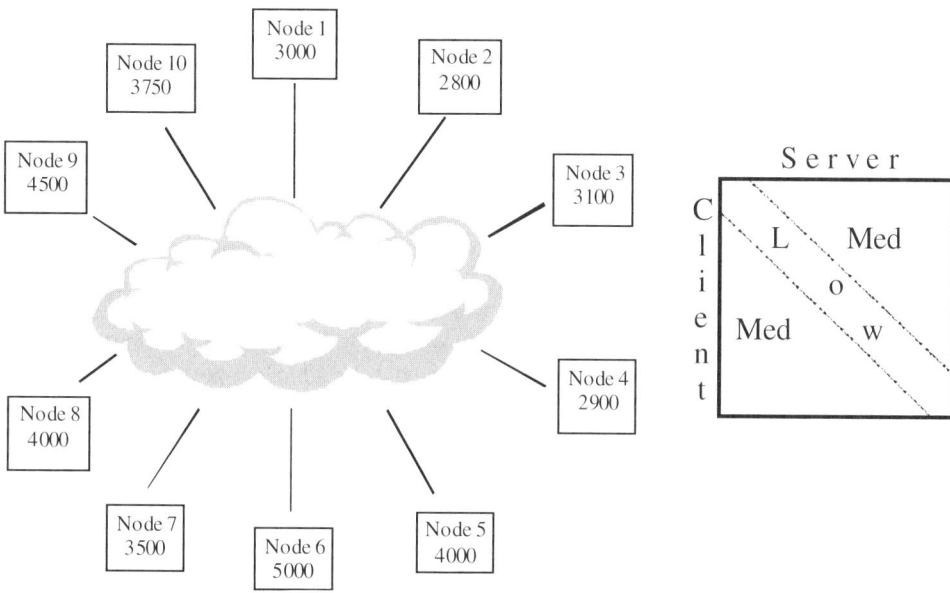

Figure 14.2 Logical topology of Scenario A.

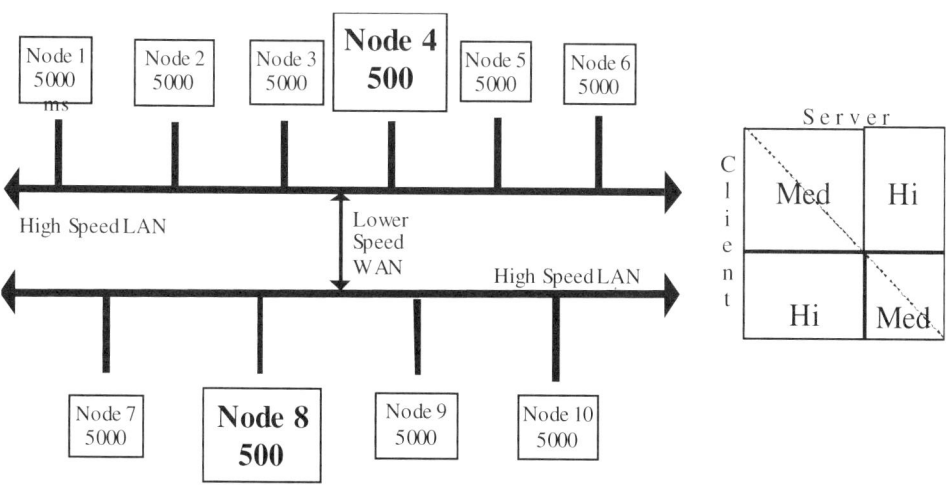

Figure 14.3 Logical topology of Scenario B.

Several different optimisation algorithms have been explored, and selected results from these experiments are presented and compared below. As a baseline, a simple random mutation 'Hill Climber' was used, where the neighbourhood operator changed a single random gene (Client) to a new random allele value (representing the server choice for that client). If superior, the mutant would then become the current solution, otherwise it would be rejected. This optimisation method is later referred to as HC. A Simulated Annealer (SA) was also tried, using the same neighbourhood operator, with a geometric cooling schedule and start and end temperatures determined after preliminary tuning with respect to the allowed number of iterations.

Three types of genetic algorithm were also tried, each of these maintaining a population of potential solution vectors, intermixing sub-parts of these solutions in the search for ever better ones. Firstly a 'Breeder' style GA (see Mühlenbein and Schlierkamp-Voosen, 1994) was used employing 50% elitism, random selection, uniform crossover and uniformly distributed allele replacement mutation. Here, each member of the population is evaluated and ranked according to performance. The worst performing half are then deleted, to be replaced by 'children' generated from randomly selected pairs of parent solutions from the surviving top half of the population. These are created, for each client position, by choosing the nominated server from either of the two parent solutions at random. This process is known as Uniform Crossover (see Syswerda, 1989). These 'children' are then all evaluated and the entire population is re-ranked and the procedure repeated. The population size remains constant from one generation to the next. This is later referred to as 'BDR'.

The results from a simple 'Tournament' GA (Bäck, 1994) were also compared, using three way single tournament selection, where 3 members of the population were chosen at random, ranked, and the best and second best used to create a 'child' which automatically replaces the third member chosen in the tournament. This GA also used uniform crossover and uniformly distributed allele replacement mutation and is later referred to as 'TNT'.

Finally, another 'Tournament' style GA was also used, this time using a specialised variant of two point crossover. With this method the child starts off as an exact copy of the second parent but then a random start position in the first parent is chosen, together with a random length (with wrap-around) of genes, and these are overlaid into the child starting at yet another randomly chosen position. This is then followed by uniformly distributed allele replacement mutation. This gives a 'skewing' effect as demonstrated below and is later referred to as 'SKT'.

```
Gene Position :     1  2  3  4  5  6  7  8  9  10

First Parent :     | A | B | C | D | E | F | G | H | I | J |

Second Parent :    | a | b | C | d | e | f | g | h | i | j |
```

Random start position in second parent   : 8
Random length chosen from second parent   : 5
Random start position in child   : 4

```
Resulting Child    | A | B | C | h | i | j | a | b | I | J |
```

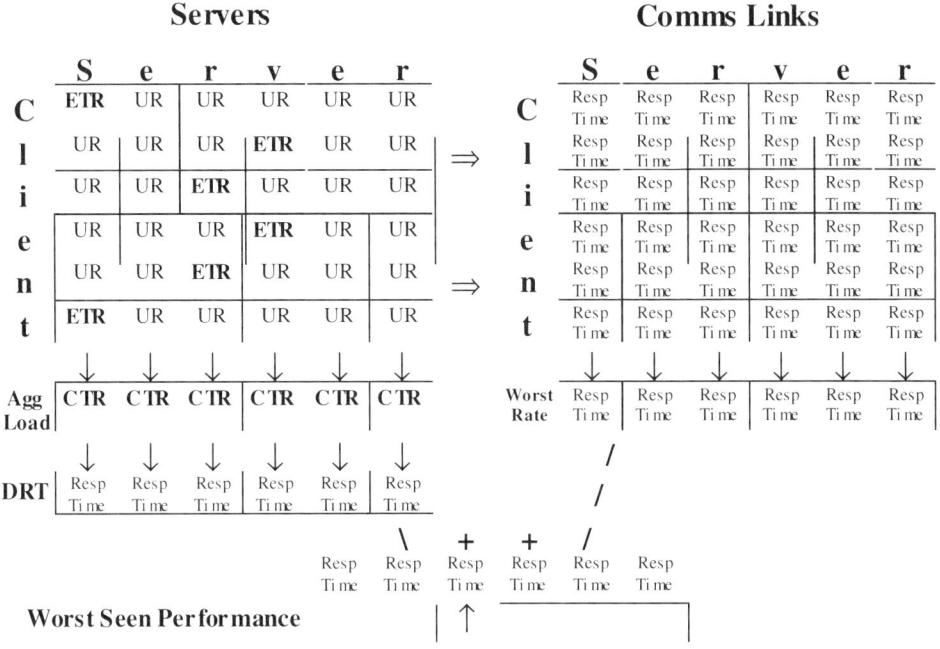

Figure 14.4 The 'Basic' model evaluation function.

14.3 The Model

The basic model was devised by Derek Edwards as part of the Advanced Systems and Networks Project at British Telecommunications Research Labs, and is demonstrated in Figure 14.4. It assumes that all nodes can act as both clients and servers. For each client node, its Effective Transaction Rate (ETR = combined Retrieval and Update rates) is calculated using equation 14.2, and this is entered into the left hand table of Figure 14.4 under the server entry denoted for this client by the solution vector. The update rate from this client is entered into all other server positions in that row. This is then repeated for each client. In the example shown (with only 6 nodes) the solution vector would have been 1, 4, 3, 4, 3, 1. Reading down the columns of the left hand table and using equation 14.2 with the appropriate server resource contention value, the Combined Transaction Rate (or aggregate load) is then calculated for each server. Using equation 14.1 for each server, this is then converted into a Degraded Response Time (DRT) using the server's specified BTT.

Using equation 14.1 the degraded response time for each point-to-point link is now calculated and entered into the right hand table using the appropriate base communications time and the traffic rate specified in the corresponding entry in the left hand table.

The highest entry in each communications table column is now recorded, denoting the slowest response time to that server seen by any client. Each of these communications times is then added to the corresponding server's DRT to produce the worst overall response time as seen by any client to each server. The highest value in this row now represents the worst overall response time seen by any client to any server and it is this value that is returned by the evaluation function. It is the optimisers job to minimise this, leading to the concept of 'least worst' performance. Checks are made throughout to ensure that any infinite or negative response time is substituted by a suitably large number.

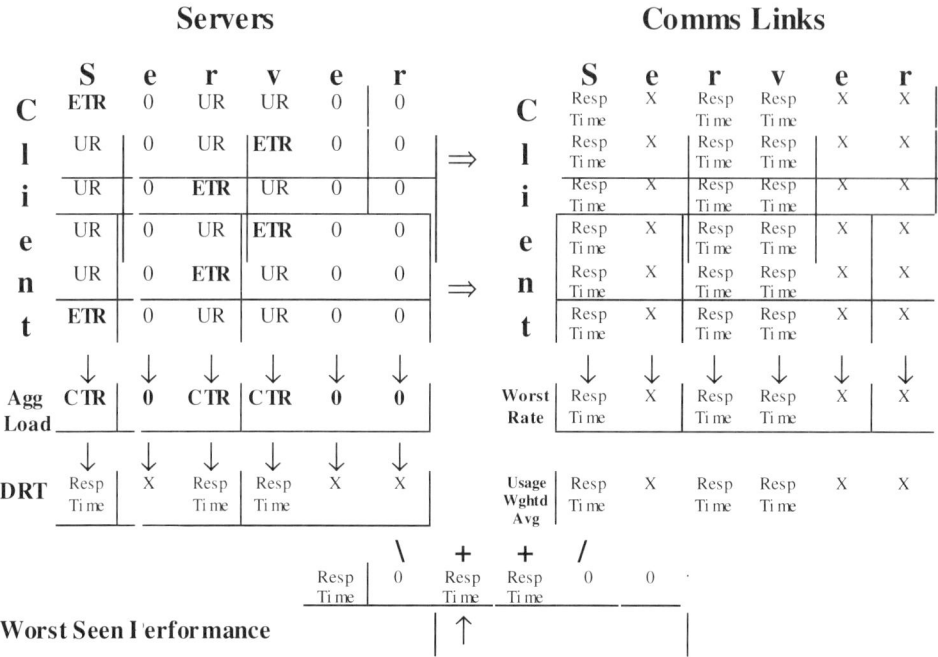

Figure 14.5 The 'Plus Used' model evaluation function.

Several variants of this 'basic' evaluation function have been explored. The first of these (plus avg) again assumes that all nodes are potential servers. It therefore applies updates to all nodes, however this time 10% of the average performance of all nodes is added to the performance of the worst transaction latency seen by any user.

Another variant restricts updates only to those servers considered to be 'active', i.e. appear in the solution vector and are therefore 'in use'. This variant is termed 'just used' and has been investigated but is not reported on here. Yet another variant starts from the 'just used' position but this time adds a usage weighted average to the worst communications time as shown in Figure 14.5. This the 'plus used' variant and is seen as a good overall reflection of user perceived quality of service. It is the basis of many results presented here. Previous publications have shown how different combinations of these scenarios and evaluation functions produce radically different fitness landscapes which vary dramatically in the difficulty they present to Genetic Search (see Oates *et al.*, 1998b; 1999).

14.4 Initial Comparative Results

For each optimiser and each scenario, 1000 trials were conducted, each starting with different, randomly generated initial populations. For each trial, the optimisers were first allowed 1000 and then 5000 iterations (evaluations) before reporting the best solution they had found. For the SA, cooling schedules were adjusted to maintain comparable start and end temperatures between the 1000 iteration and 5000 iteration runs. For the BDR GA, the number of 'generations' used was adjusted with respect to population size.

Of the 1000 trials it is noted how many trials found solutions with the known globally optimal fitness value. These are referred to as being 'on target'. It was also noted how many times the best solution found was within 5% of the known globally optimal fitness value, as this was deemed acceptable performance in a real-time industrial context. Finally it was noted how many times out of the 1000 trials, the best solution found was more than 30% worse than the known globally optimal fitness value – this was deemed totally unacceptable performance. The results of these trials for Scenario A with the 'plus average' fitness model are shown in Figure 14.6.

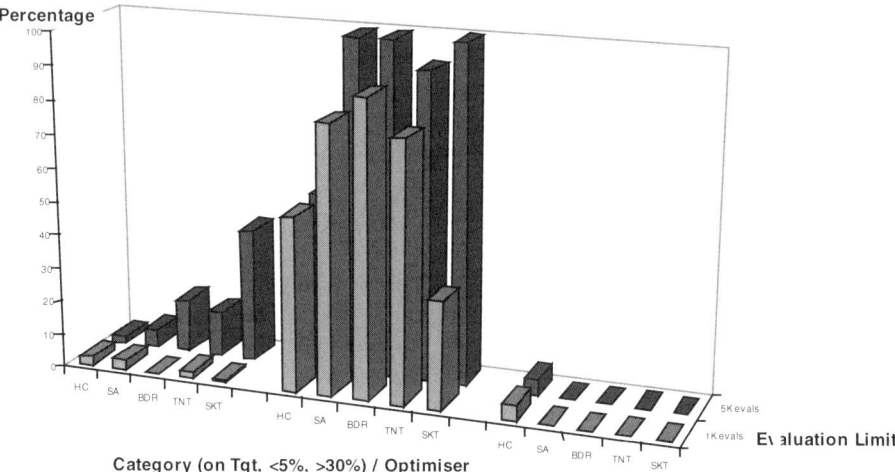

Figure 14.6 Scenario A with the 'plus average' fitness model.

Here it can be seen in the left-hand set of columns that at only 1000 evaluations (the foreground row), very few trials actually found the global optimum solution. The Breeder (BDR) and Skewed Tournament (SKT) genetic algorithms actually perform worst however neither Hillclimber (HC) nor Simulated Annealing (SA) nor Tournament Genetic Algorithm (TNT) deliver better than a 3% success rate. Still at only 1000 evaluations, Hillclimber can be seen to totally fail (right hand set of columns) around 5% of the time with all other techniques never falling into this category. At 5000 evaluations (the background row), the performance of the genetic algorithms improves significantly with Skewed Tournament delivering around 30% 'on target' hits. For best results falling within 5% of the global optimum fitness value (the middle set of columns), there is little to choose between Simulated Annealing, Breeder or Skewed Tournament GA, all delivering success rates above 99%. The third set of columns at 5000 evaluations shows the failure rate where best found solutions were more than 30% adrift of the global optimum fitness value. Only Hillclimber has any significant entry here. Interestingly it is only Hillclimber that fails to show any significant improvement in its performance when given five times the number of evaluations. This implies the fitness landscape must have some degree of multi-modality (or 'hanging valleys') which Hillclimber quickly ascends but becomes trapped at.

Figure 14.7 shows similar performance charts for the five optimisers on Scenario B with the 'plus used' evaluation function. Here it is clear that only the Skewed Tournament Genetic Algorithm gives any degree of acceptable performance, and even this requires 5000 evaluations. In terms of best solutions found being worse than 30% more than the global optimum, even at 5000 evaluations all techniques, with the exception of Skewed Tournament, are deemed to fail over 75% of the time. Skewed Tournament gives on target hits 99.7% of the time with no complete failures.

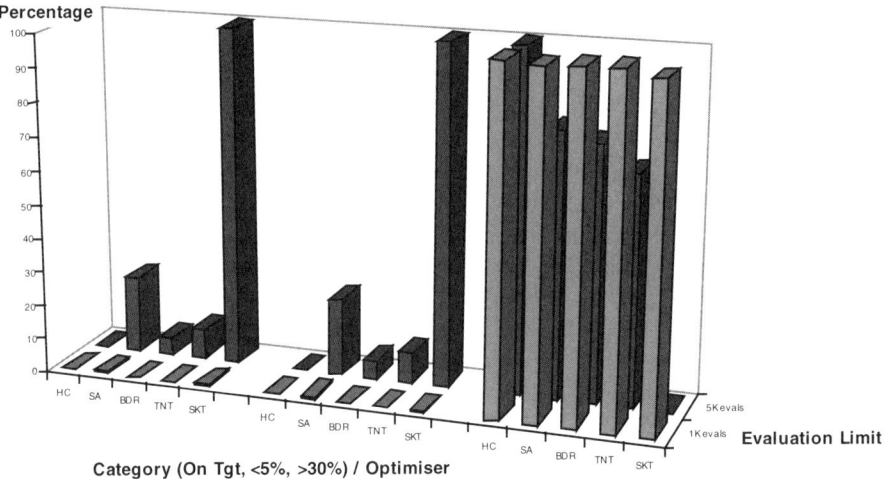

Figure 14. 7 Scenario B with the 'plus used' fitness model.

These results and others are summarised in Table 14.1 with respect to the performance of simulated annealing. In this table, the difficulty with which simulated annealing was able to find the best result on various scenario/evaluation function pairings is classified roughly as either 'Very Easy', 'Easy', 'Moderate', 'Fairly Hard' or 'Very Hard'. One clear trend is that the imposition of the 'plus used' evaluation function on Scenario B produces a landscape that makes optimal solutions particularly difficult to find. However it is intriguing that the 'plus average' model yields an easier problem in the Scenario B case than with Scenario A.

Table 14.1 Summary of search space difficulty.

Model	Scenario A	Scenario B
Basic	Very Easy	Moderate
Just used	Very Easy	Fairly Hard
Plus avg.	Easy	Very Easy
Plus used	Very Easy	Very Hard

14.5 Fitness Landscape Exploration

Wright (1932) introduced the concept of a 'fitness landscape' as a visual metaphor to describe relative fitness of neighbouring points in a search space. To try to discover more about those features of our ADDMP search space landscapes that cause difficulties to evolutionary search, a number of investigations were carried out exploring the characteristics of the landscape around the 'global optimum' solution to Scenario B using the 'plus used' model. This 'neighbourhood analysis' focused on the average evaluation values of 100 'n-distance nearest neighbours' to try and determine whether a 'cusp' like feature existed immediately surrounding the 'globally optimal solution'. Such a feature in the 10 dimensional landscape, would make it difficult to 'home in' on the globally optimal solution, as the nearer the solution got in terms of Hamming distance, the worse the returned evaluation value would be, and this would generate negative selection pressure within the GAs. This technique is similar to that of Fitness Distance Correlation which is described and demonstrated in detail by Jones and Forrest (1995).

The left-hand edge of Figure 14.8 shows the average evaluation value of 100 randomly chosen, single mutation neighbours of the 'globally optimum solution' to both the 'plus average' and 'plus used' models both against Scenario B (the global optimum evaluation value being less than 9000 in both cases). The plot continues from left to right, next introducing the average evaluation value of 100 randomly chosen, dual mutation neighbours. This continues up to the final two points showing the average evaluation value of 100 points in the search space, each different from the globally optimal solution in eight gene positions. It was hoped to see a significant difference between the two plots, but this is clearly not the case. Indeed, in the case of the 'plus used' plot, it was hoped to see a peak value at 1 mutation, dropping off as Hamming distance increased. This would have supported a hypothesis of a 'cusp' in the 10 dimensional search space which would have provided a degree of negative selection pressure around the global optimum solution, hence making it 'hard' to find.

An examination of the distribution of evaluation values of the 100 points at each Hamming distance however, on close examination, does provide some supporting evidence. Figure 14.9 shows the distribution of these points for 'plus avg' on Scenario B. Clearly as Hamming distance increases, evaluation values in excess of 100,000 become more frequent (however it must be borne in mind that each point shown on the plot can represent 1 or many instances out of the 100 samples, all with the same evaluation value. Figure 14.10 gives the same plot for 'plus used'.

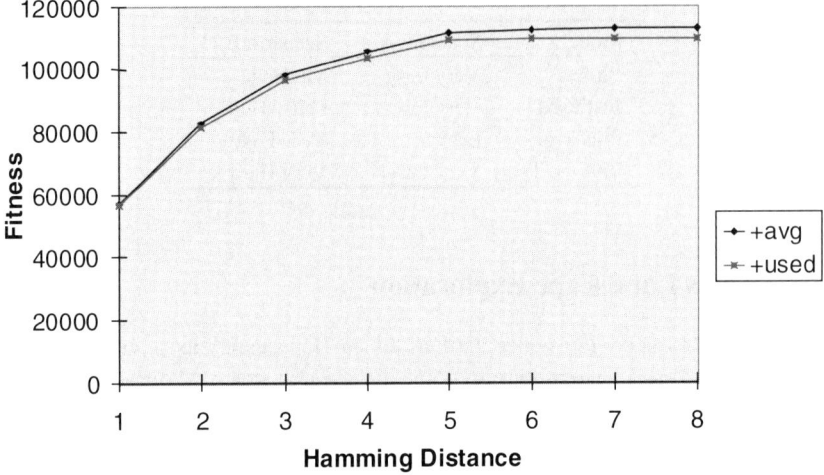

Figure 14.8 Neighbourhood analysis.

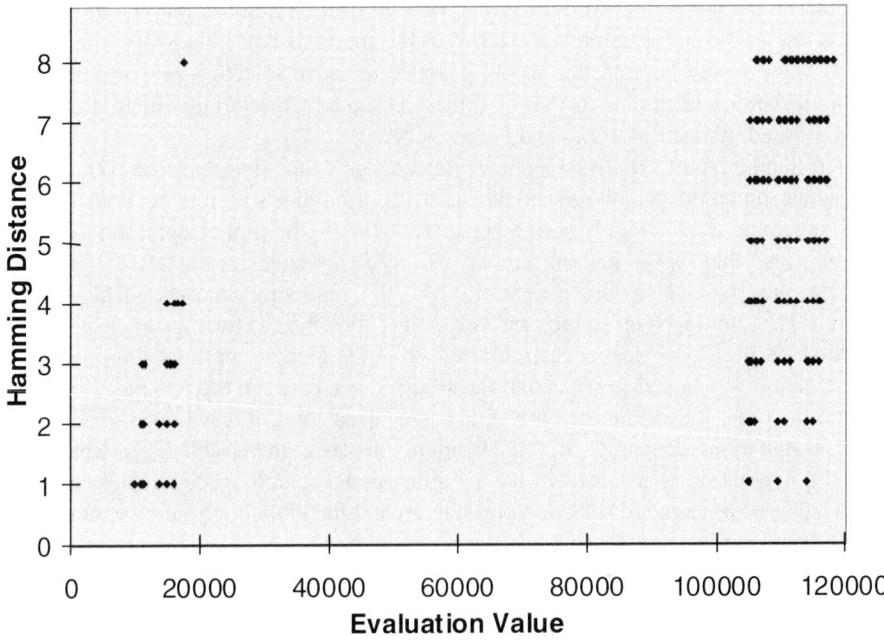

Figure 14.9 Distribution for 'plus avg'.

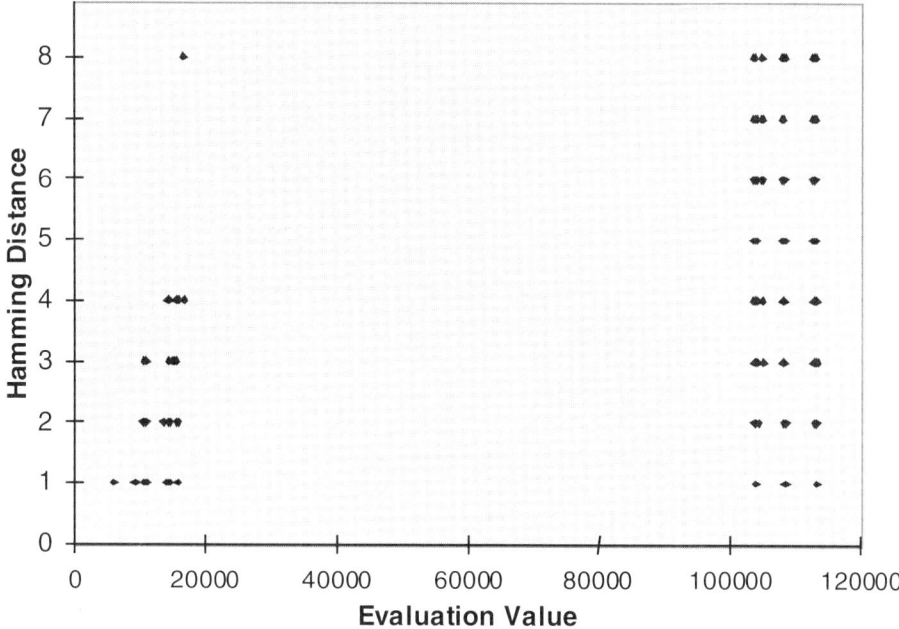

Figure 14.10 Distribution for 'plus used'.

Taking a closer look at the high evaluation value groupings in these figures shows that for 'plus avg' (in Figure 14.11), high evaluation value points decrease in evaluation value as Hamming distance decreases. However, for 'plus used' (in Figure 14.12), there is a repeated trend implying an increase in evaluation value as Hamming distance decreases. Bearing in mind this is a minimisation problem, this feature would act as a deterrent to 'homing in' on the global optimum, providing negative selection pressure the closer the search came to the edge of the 'cusp'. Although this requires many assumptions on the nature of association between points on the plot, it is nonetheless an interesting result which requires further investigation.

The possibility of the 'cusp' is also explainable by examining the evaluation function itself. Considering a single deviation from the global optimum solution for Scenario B using 'plus used' could simply incur a greater communications overhead to access an existing used server (if the deviation simply causes a client to access the wrong region's active server). Alternatively, the deviation could introduce a 'new used server'. This would add to the list of 'used servers' and would mean the application of a single 'retrieval rate' and a combined 'update rate' to an inappropriate node. This will almost certainly give a new 'worst server' result, significantly worse than the global optimum. A second deviation could add another 'new used server' to the list which, whilst probably no worse than the preceding effect, increases the number of 'used servers', and hence reduces the bias, as the evaluation function divides this by the number of used servers which has now increased further. This would cause the first deviation to produce a radically worst first nearest neighbour, but with the effect reducing with increased Hamming distance, and would produce exactly the negative selection pressure postulated. The fact that more 'used servers' are performing worse is irrelevant to this model as it considers the average of client access to the worst server, not the average of the 'used servers'.

By contrast, increasing deviations from the global optimum with 'plus avg' on Scenario B, whilst still likely to introduce 'new used servers', will see an increasingly worsening effect as the 'average server performance' is influenced by an increasing number of poorly performing servers.

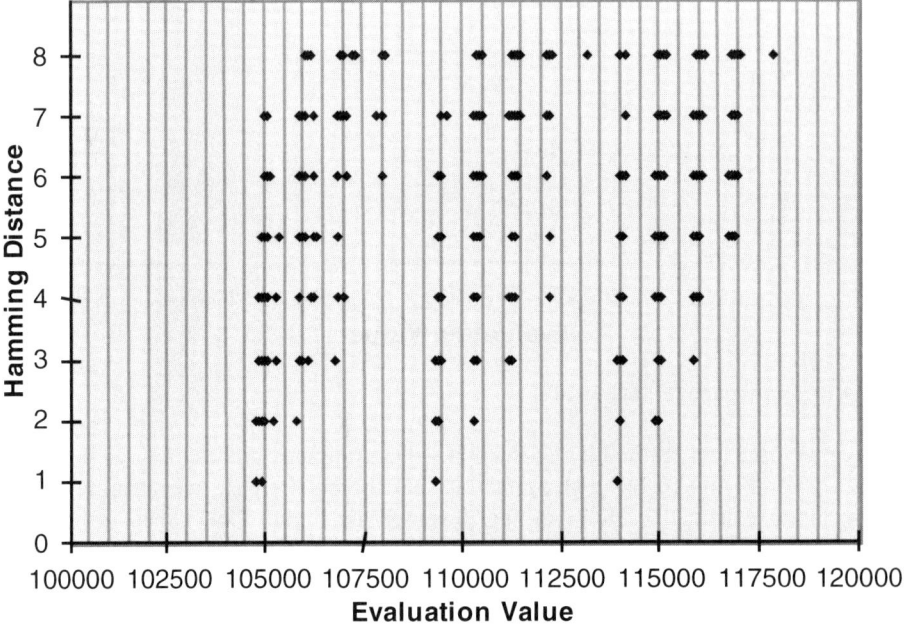

Figure 14.11 'Many evaluation' fitness distribution for 'plus avg'.

The 'cusp' hypothesis is not as directly applicable in the case of Scenario A. In Scenario A, not only are there several solutions attainable which share the 'best known fitness value', but these solutions usually contain a wide genotypic diversity. That is, there are multiple optimal solutions which are quite distinct in terms of the 'servers' used by clients. Deviations from these solutions will have a far less marked effect than in a case when the best known solution is a unique vector, or perhaps a small set containing very little diversity. However, such potential shallow multimodality will produce a degree of ruggedness which, as already demonstrated by Figure 14.6, is seen to be sufficient to prevent a basic hillclimbing algorithm from finding the global optimum.

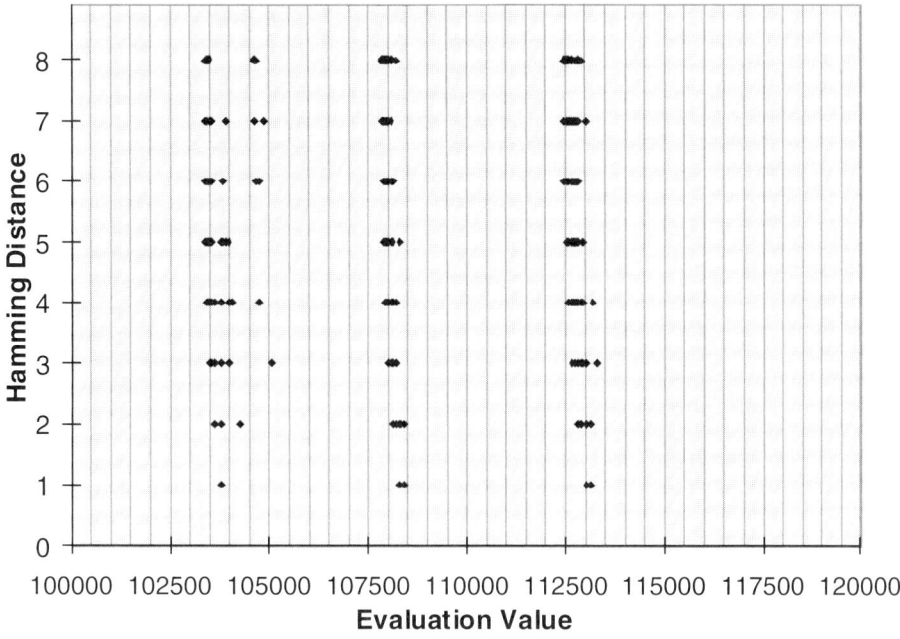

Figure 14.12 'Many evaluation' fitness distribution for 'plus used'.

14.6 Landscape Visualisation

In an effort to visualise the fitness landscape around the global optimum, a sampling technique is required, however there are significant issues to be overcome with such a technique. Firstly, the 'base' axes from which the landscape is to be projected must be chosen. Simply picking two gene positions at random to act as the X and Y base plane is unlikely to be effective. Statistical sampling techniques could be used to try to select 'dominant' gene positions, but there is little guarantee that this would cover any interesting landscape feature. Secondly, even if two gene positions could be determined, the order in which the alleles were plotted would have a significant bearing on the 3D landscape visualised. With our examples from the Adaptive Dynamic Database Management Problem (ADDMP), allele values range as integers from 1 to 10 but with no ordinal significance, i.e. '1' is as different from '2' as it is from say '7'. It is effectively a symbolic representation. As such, a feature in the landscape which for some reason exploited a common characteristic from the allele values '2', '5', '7' and '9' would appear as a rugged zig-zag in a visualisation which plotted allele values in ascending numerical order. In this case, plotting the fitness of solutions with the odd valued alleles followed by the even valued ones might expose more of a 'clustered' feature. Clearly, it would not be practical to explore all possible permutations in both dimensions.

Further, it can be argued that simply plotting fitness values over a range of allele values for two specified genes is not representative of the landscape as 'seen' by the processes of mutation and crossover. At low levels of mutation, it is likely that the GA would approach the global optimum via a series of single allele mutations, approaching with ever decreasing Hamming distance from an '*n*'th nearest neighbour through an '*n*–1'th nearest neighbour to a 1st nearest neighbour before finding the global optimum. This of course assumes at least one path of positive selection pressure exists amidst what could be a multi-modal or deceptive landscape. Crossover complicates this by allowing the GA to make multiple changes to a chromosome in each evolutionary step, therefore potentially jumping from '*x*'th nearest neighbour to '*y*'th nearest neighbour in a single step, where '*x*' and '*y*' may be significantly different values. Nonetheless, the beneficial effects of crossover are often most dominant in the early stages of evolutionary search, when much initial diversity exists in the gene pool. By the time the GA is 'homing in' on the global optimum towards the end of the search, it is likely that significantly less diversity is available to the crossover operator (unless the problem has a high degree of multi-modality, and even in this case, towards the end of the search, hybrids are unlikely to have improved fitness over their parents). Thus in the latter stages of the search, mutation for fine tuning (local search) is more likely to be the dominant operator.

Based on this assumption, examination of solutions neighbouring the global optimum, differing by only one gene/allele combination, can be argued to give an indication of the ruggedness that the GA sees in the final stages of its search. To this end, a technique was suggested in Oates *et al.* (2000), by which a series of increasingly Hamming distant neighbours from the global optimum are sampled and their fitnesses plotted in a concentric fashion around the fitness of the global optimum solution.

In our ADDMP examples we have 10 gene positions each with 10 possible allele values. Hence we have 90 potential first nearest neighbours to the global optimum (10 genes by 9 other allele values). Given that this quantity is quite low, this set can be exhaustively evaluated. If the range were higher, some form of sampling would be necessary. From these 90 variants, four are then chosen. Firstly, the 90 variants are ranked according to fitness and the mutation that created the first nearest neighbour with the lowest fitness is noted (for a minimisation problem such as the ADDMP this would represent the 'best' first nearest neighbour). Then the mutation that created the highest fitness of the 90 first nearest neighbours is noted and labelled the 'worst' first nearest neighbour. The mutation which created the median of the ranked list of first nearest neighbours (here the 45th member) is chosen as the third variant, and the fourth variant is that mutation variant whose fitness is closest to the mean fitness of the 90 sampled points. This last selection is adjusted if necessary to ensure that it is not the same member of the 90 first nearest neighbours as any of the three previously selected members.

We have now chosen four different first nearest neighbours to the global optimum, noting for each both the gene position that was changed and the new allele value substituted. The fitness of these variants is then plotted on a grid as shown in Figure 14.13 using two defined axes of variance from the centrally placed global optimum, named the 'Worst-Best' and the 'Median-Mean' axes respectively. The corners of this centrally placed 3 by 3 grid are then populated by the fitnesses of the 2nd-nearest neighbour hybrid solutions generated by applying the mutations from both of their immediately adjacent first nearest neighbours.

Exploring Evolutionary Approaches to Distributed Database Management

		Hybrid 5th Nearest N	Mean 4th Nearest N	Hybrid 5th Nearest N		
...	Hybrid 5th Nearest N	Hybrid 4th Nearest N	Mean 3rd Nearest N	Hybrid 4th Nearest N	Hybrid 5th Nearest N	...
Hybrid 5th Nearest N	Hybrid 4th Nearest N	Hybrid 3rd Nearest N	Mean 2nd Nearest N	Hybrid 3rd Nearest N	Hybrid 4th Nearest N	Hybrid 5th Nearest N
Hybrid 4th Nearest N	Hybrid 3rd Nearest N	Hybrid 2nd Nearest N	Mean 1st Nearest N	Hybrid 2nd Nearest N	Hybrid 3rd Nearest N	Hybrid 4th Nearest N
Worst 3rd Nearest N	Worst 2nd Nearest N	Worst 1st Nearest N	**Global Optimum**	Best 1st Nearest N	Best 2nd Nearest N	Best 3rd Nearest N
Hybrid 4th Nearest N	Hybrid 3rd Nearest N	Hybrid 2nd Nearest N	Median 1st Nearest N	Hybrid 2nd Nearest N	Hybrid 3rd Nearest N	Hybrid 4th Nearest N
Hybrid 5th Nearest N	Hybrid 4th Nearest N	Hybrid 3rd Nearest N	Median 2nd Nearest N	Hybrid 3rd Nearest N	Hybrid 4th Nearest N	Hybrid 5th Nearest N
...	Hybrid 5th Nearest N	Hybrid 4th Nearest N	Median 3rd Nearest N	Hybrid 4th Nearest N	Hybrid 5th Nearest N	...
		Hybrid 5th Nearest N	Median 4th Nearest N	Hybrid 5th Nearest N		

Figure 14.13 Base axes neighbour selection.

Starting from the first nearest neighbour termed 'best', 81 new variants are generated and evaluated. These variants are allowed to adjust the nine other gene positions available to them (one already having been varied to generate this 'best' first nearest neighbour) to each of nine different allele values. The 'best' of these 81 variants, each of Hamming distance 2 from the global optimum, is then chosen as the 'best' second nearest neighbour, and its mutation specification (gene position and new allele value) is noted. A similar process is repeated starting from the 'worst' first nearest neighbour, generating 81 variants and selecting the worst. Likewise in each direction of the 'Mean' and 'Median' axis. Third and fourth nearest neighbour hybrids can now be generated in the diagonal areas using the combined mutation specifications of the values selected along the axes. This process is then repeated moving out from the global optimum, generating four sets of 72 third nearest neighbours, four sets of 63 fourth nearest neighbours, etc. The fitnesses of these points are then plotted to produce a surface similar to that shown in Figure 14.14, which gives some indication as to the ruggedness or otherwise of the fitness landscape around the global optimum. (Note, however, that the floor of Figure 14.14 is false – see later.)

While this technique is by no means guaranteed to select all critical neighbours of the global optimum, many constraints being arbitrary, we argue that it offers a 'window' onto the ruggedness of a multi-dimensional combinatorial landscape. A further criticism could be raised in that the technique effectively looks 'outwards' from the global optimum rather than moving 'inwards' from a range of sampled points towards the global optimum, as indeed the GA does during its search. Indeed a possible alternative might be to start with four randomly selected 'n' Hamming distant neighbours at the extremes of the axes and progress inwards towards the global optimum. This idea is ongoing work.

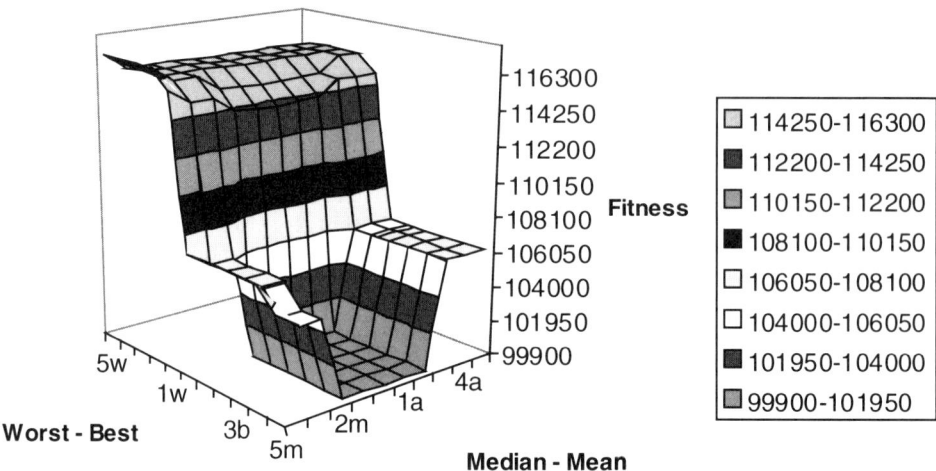

Figure 14.14 3D fitness landscape projection for Scenario B with the 'plus avg' model.

Another potential criticism of the above techniques is the central dependence on the global optimum. For many problems this will not be known, however the technique could be trialled either from an arbitrary point or from known local optima, either of which should still give an indication of the ruggedness of the landscape.

Returning to our two ADDMP examples, Figure 14.14 shows the fitness landscape plot generated by the described technique around the known global optimum for our Scenario B example using the 'plus avg' model. The flat floor depicted at a fitness of 99900 is false, being a limitation of the drawing package. Fitnesses in this region continue downwards (as this is a minimisation problem) to just below 9000 with positive selection pressure focussing on a 'well' with the centrally placed global optimum at its base. What is of most importance in this figure is the positive selection pressure gradients in almost all areas of the surface shown. Consider a ball dropped anywhere on the surface. With the exception of the foremost corner, gravity would force the ball towards the bottom of our 'well' created by our globally optimum solution. Even in the foremost corner, where a sudden drop in fitness occurs as we increase Hamming distance from a solution containing the third best nearest neighbour mutation to the fourth best, selection pressure is still positive as we reduce Hamming distance in the Mean-Median axis.

By contrast, Figure 14.15 has several regions of distinct negative selection pressure. There is a peak in the landscape immediately adjacent to the global optimum in the 'Worst' axis, which immediately provides a deterrent to selections towards the global optimum. More importantly a 'ridge' feature exists along the 'best' axis at solutions containing the third 'average' or 'mean' nearest neighbour mutation (this is the white area along the right hand side of the diagram). This feature alone could provide a significant region of negative selection pressure in the landscape which could divert simple evolutionary search strategies from finding the global optimum if they tried to approach the global optimum from more distant neighbours containing this mutation variant.

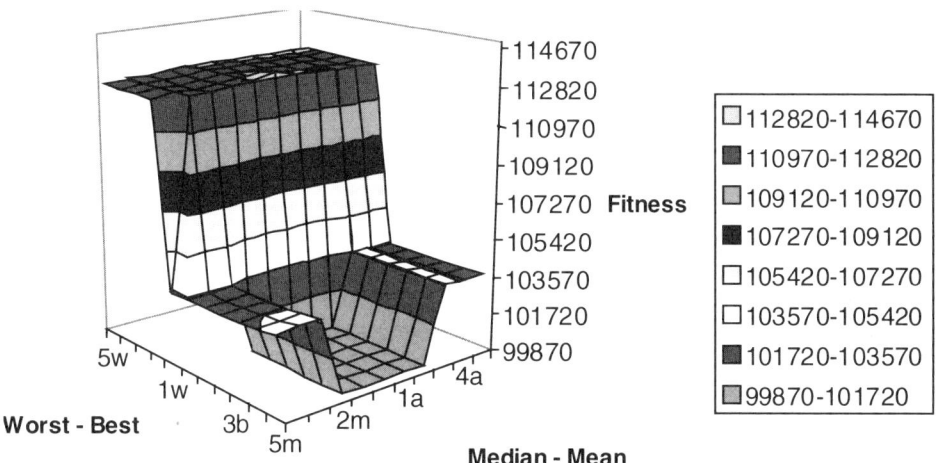

Figure 14.15 3D fitness landscape projection for Scenario B with the 'plus used' model.

Whilst this landscape visualisation technique has been demonstrated here using a non-binary allelic representation, it is easily adapted to the binary case. If this ADDMP were represented via a binary equivalent encoding, it would be likely to require some 33 to 40 bits. This would allow some 32 or 39 first nearest neighbours, etc.

14.7 GA Tuning and Performance on the Easy Problem

For the GA to be an industrially acceptable optimiser in this problem domain, it must be shown to not only be robust to scenario conditions, but also to be near optimally tuned in terms of reliable performance in the minimum number of evaluations with the highest degree of accuracy in finding global optima. Many studies (including Bäck, 1993; Deb and Agrawal, 1998; Goldberg, Deb and Clark, 1992; Muhlenbein, 1992; van Nimwegen and Crutchfield, 1998; 1998a) have shown that population size and mutation rate can greatly affect GA performance, and so it is necessary to explore the effect these parameters have.

Unless stated otherwise, further results presented here are based on a Generational Breeder GA (see Muhlenbein *et al.*, 1994) utilising 50% elitism and random elite parental selection for population regeneration. Uniform Crossover (Syswerda, 1989) is used at 100% probability, followed by uniformly distributed allele replacement mutation at a given rate per gene. A wide range of population sizes and mutation rates have been explored with each result shown being the average of 50 runs, each run starting with a different randomly generated initial population.

The performance of the GA was measured over a range of population sizes (from 10 to 500 members in steps of 10) and over a range of mutation rates starting at 0%, 1E-05% then doubling per point to approximately 83% chance of mutation per gene (this later extreme degenerating the GA virtually into random search). The GA was allowed 5000 evaluations (the number of generations being adjusted with respect to population size) reporting the fitness of the best solution it could find, and the evaluation number at which this was first found. Figures 14.16 and 14.17 show the results of this exercise, each point representing an average of 50 independent runs. Figure 14.16 plots the number of evaluations taken to first find the best solution that the GA could find in 5000 evaluations averaged over the 50 runs. In many cases, this may represent premature convergence on a poor quality solution and so the average difference from the fitness of the known global optimum solution is plotted in Figure 14.17 (note the population axis is deliberately reversed in this figure to aid visibility).

The bi-modal profile seen in Figure 14.16 was first reported in (Oates *et al.*, 1999a) together with the linear feature seen along the left-hand side of the figure, which shows a direct relationship between population size and the number of evaluations exploited. In Oates *et al.*, (1999a) we postulated that this figure represented a characteristic performance profile for genetic search, backed up by, amongst other results, similar evidence from performance evaluations against the simple Unitation problem (where fitness = number of '1's in a binary string) and standard de Jong's test functions (see Goldberg, 1989) with a canonical, binary representation GA.

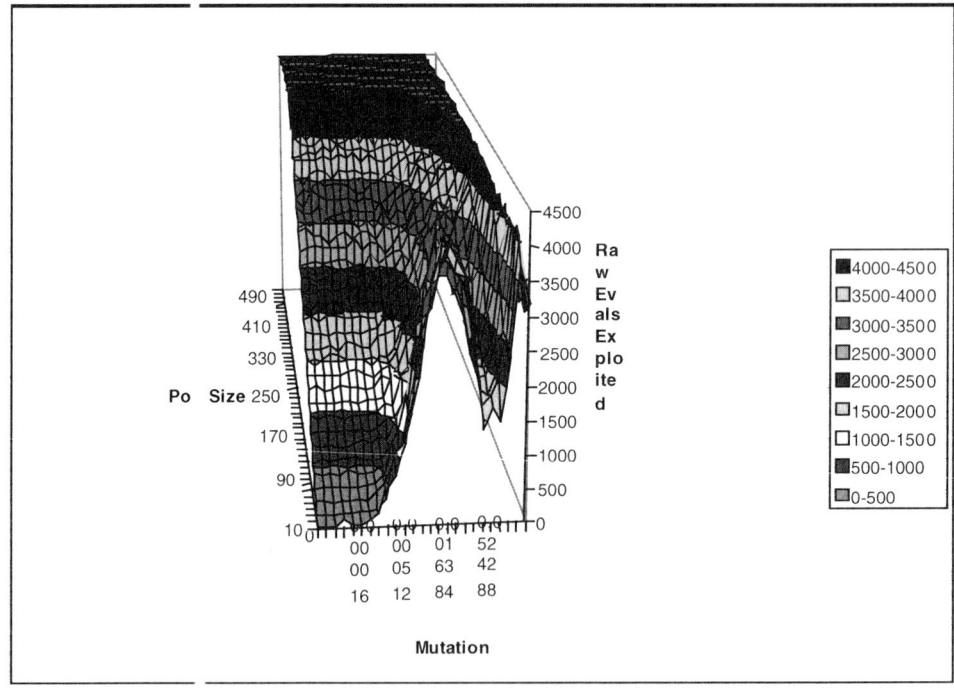

Figure 14.16 Raw evaluations exploited by the breeder GA at 5000 evaluations limit.

It was suggested that this profile was a feature of the 5000 evaluation limit and so subsequent trials were conducted at 1000 and 20000 evaluations, where a similar profile was observed. It was then shown, in Oates *et al.* (1999a), that the position of the first peak and trough varied slightly with evaluation limit – the higher the limit, the lower the mutation level at which the feature was observed. A GA operating at mutation rates and population sizes in this first trough (mutation rates from 2% to 10% in this case) can claim to be optimally tuned, with the GA consistently delivering optimal solutions in a minimal number of evaluations.

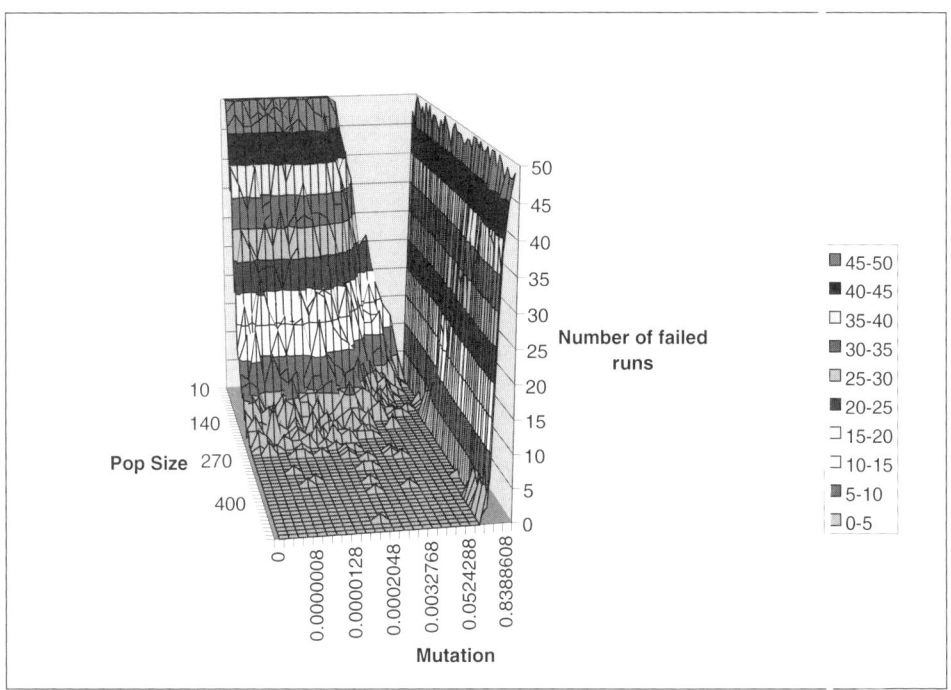

Figure 14.17 Failure rate for the Breeder GA at 5000 evaluations limit.

Whilst Figure 14.16 shows the performance of the GA in terms of speed (i.e. the number of evaluations needed to first find an optimal solution), Figure 14.17 shows the number of times out of the 50 runs that this optimum solution was more than 30% adrift of the known global optimum fitness for this problem. Performance outside this margin was deemed totally unacceptable in an industrial context. The population size axis has been deliberately reversed to make the results more visible.

As can clearly be seen, at very low population sizes there is only a small window of optimal performance with mutation rates ranging from 5% to 20%. However, as population size increases, this range rapidly expands until, with a sufficiently large population (> 200), no mutation at all is required to find consistently excellent performance. Conversely, at the extreme right hand side of Figure 14.17, as population size increases, the GA's ability to exploit very high levels of mutation deteriorates. At low population sizes (up to around 200), the GA can still utilise 40% mutation (although this must be almost random search), whilst as population size exceeds 400, this starts to fall off to 10% mutation. With the Breeder GA being generational and 50% elitist, the amount of evolutionary progress that can be made by recombination at these population sizes, with only 5000 evaluations, must be severely limited (of the order of only 20 generations). This helps to explain the GA's increasing intolerance of high levels of mutation.

The fact that good solutions can be consistently found with moderate population sizes and no mutation emphasises the point that the fitness landscape of this scenario be categorised as 'Easy' to search. However, at low population sizes it is important to note that the range of optimal mutation rates (of the order of 2% to 10%) in terms of consistently finding good solutions, coincide with those of Figure 14.16. That is, where the GA is able to consistently find optimal solutions in a minimum number of evaluations. Here the GA can be said to be optimally tuned.

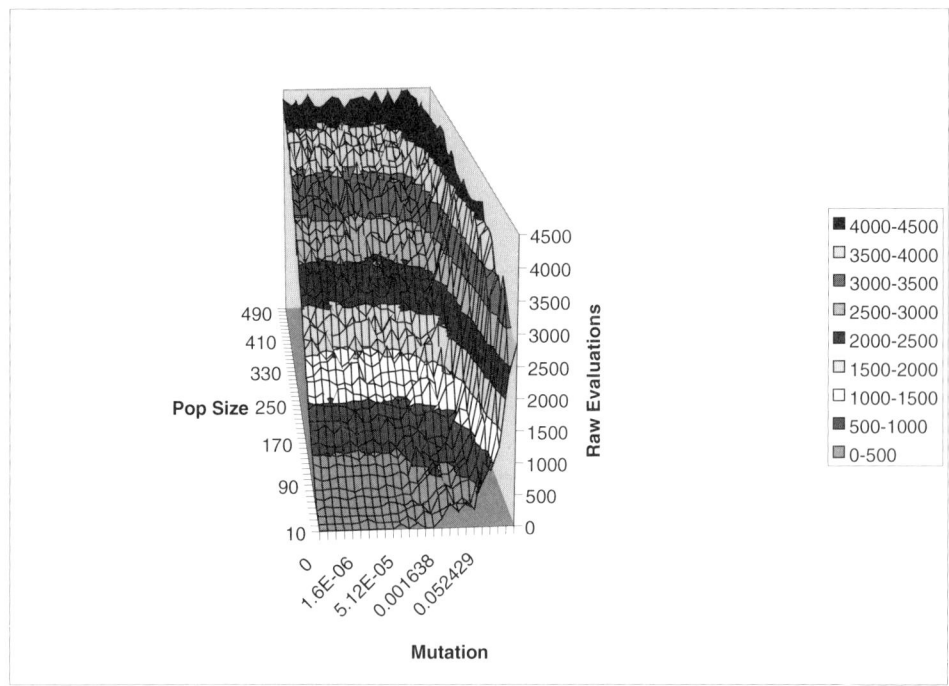

Figure 14.18 Raw Evaluations Exploited by Breeder GA on Scenario B with 'plus used' model at 5000 evaluations.

14.8 GA Tuning and Performance on the Hard Problem

However, this scenario and evaluation function (Scenario A, plus avg. model) has already been summarised as being relatively 'easy' to search (see Table 14.1 and Oates *et al.*

1998b) and so a similar set of runs was conducted against the more difficult problem of Scenario B with the 'plus used' evaluation function. This was first reported on in Oates *et al.* (1999b). Figure 14.18 shows the average raw evaluations exploited for this case. Whilst the linear feature relating number of evaluations exploited to population size is again clearly apparent, the peak and trough features are considerable attenuated. The smaller ridge at about 0.3% mutation is only apparent at low population sizes before being swamped by the general rising feature in the background. Figure 14.19 shows a more detailed expansion of the area of low population size, ranging from two members to 100 members in steps of 2. The peak and trough are more clearly evident here at 0.3% and 1.3% mutation respectively, with the height of both the ridge and the bottom of the trough seen to rise with increased population size.

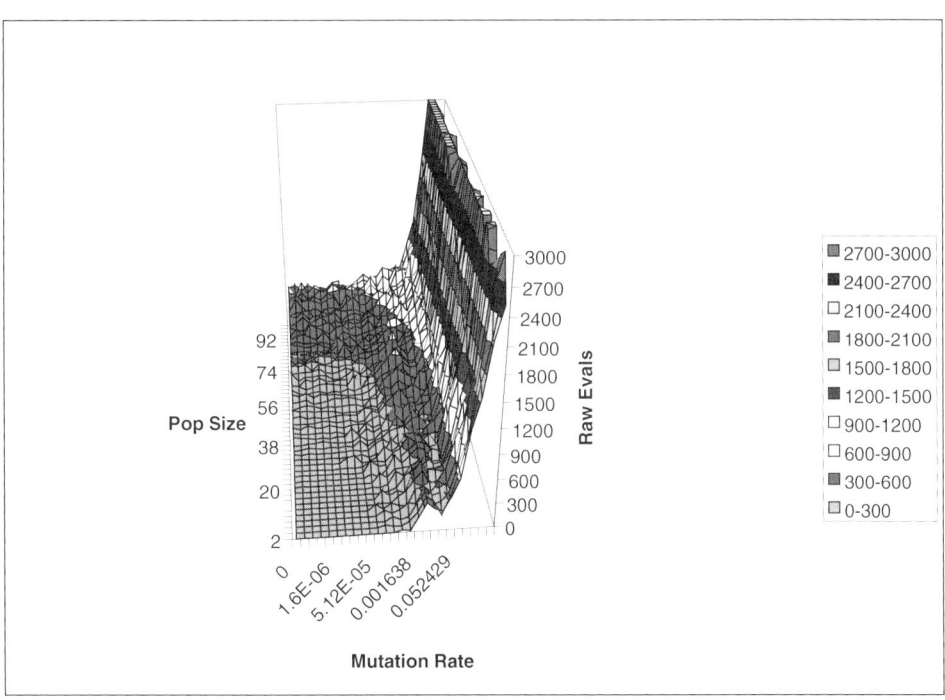

Figure 14.19 Evaluations exploited by breeder GA on hard problem with lower population sizes.

The failure rate for the Breeder GA on the hard problem shown in Figure 14.20 is in stark contrast to Figure 14.17, albeit with a lower range of population. Here it can be seen that for the vast majority of the tuning surface, the GA consistently fails to find good solutions. There is a small 'trough of opportunity' for mutation rates between 10% and 20%, but at best even this still represents a 70% failure rate. The lowest point occurs near population size 40 with 68% of runs unable to find solutions within 5% of the known best.

Figure 14.21 shows the results over the same population range where the GA is allowed 20,000 evaluations. Both the peak and the trough are clearly present but again suffer from considerable attenuation, the peak occurring this time at around 0.08% mutation whilst the trough occurs at around 0.6%. These results (from Oates *et al.*, 1999b) mirror those seen in Oates *et al.* 1999a where the mutation rates for the peaks and troughs were seen to reduce with increased evaluation limit and this is summarised in Table 14.2.

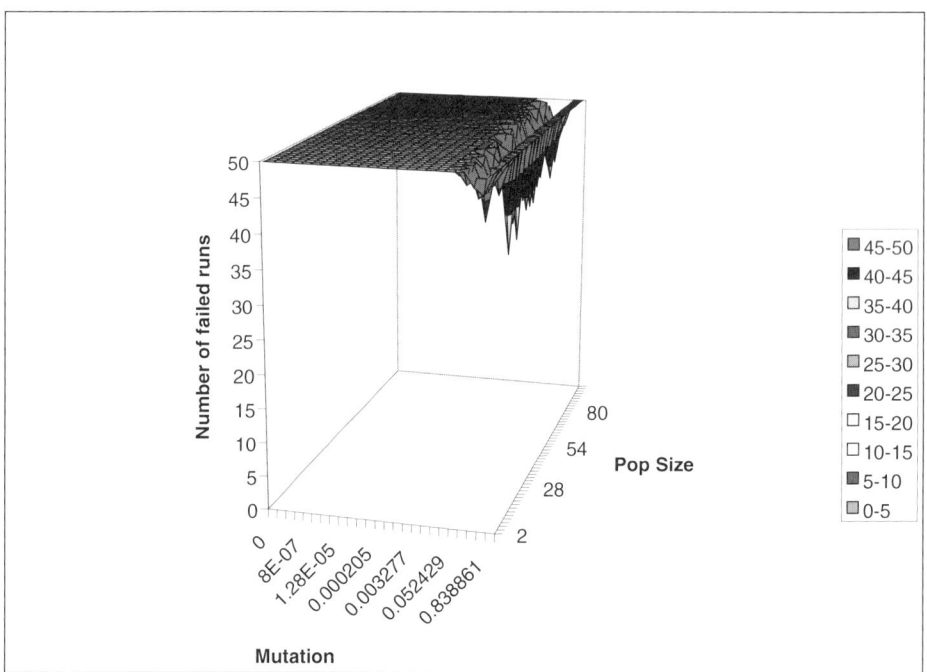

Figure 14.20 Failure rate for breeder GA on hard problem at 5000 evaluations limit.

Table 14.2 Effects of problem complexity and evaluation limit on feature position.

	Easy Problem		Hard Problem	
Eval. limit	Peak	Trough	Peak	Trough
1,000	1.3%	10%	n/a	n/a
5,000	0.3%	5%	0.3%	1.3%
20,000	0.08%	2%	0.08%	0.6%

These results suggest that the mutation rate at which the peaks occur may be independent of problem complexity, but depend upon evaluation limit, while the trough position (the ideally tuned position) varies both with complexity and evaluation limit but by a lesser degree. Where the problem is too complex, or there are too few evaluations allowed, the trough is effectively subsumed into the peak, either cancelling each other out or being attenuated out of detectabilty as shown in Figure 14.22, where the Breeder GA is allowed only 1000 evaluations. Here there is no significant evidence for the initial peak and trough.

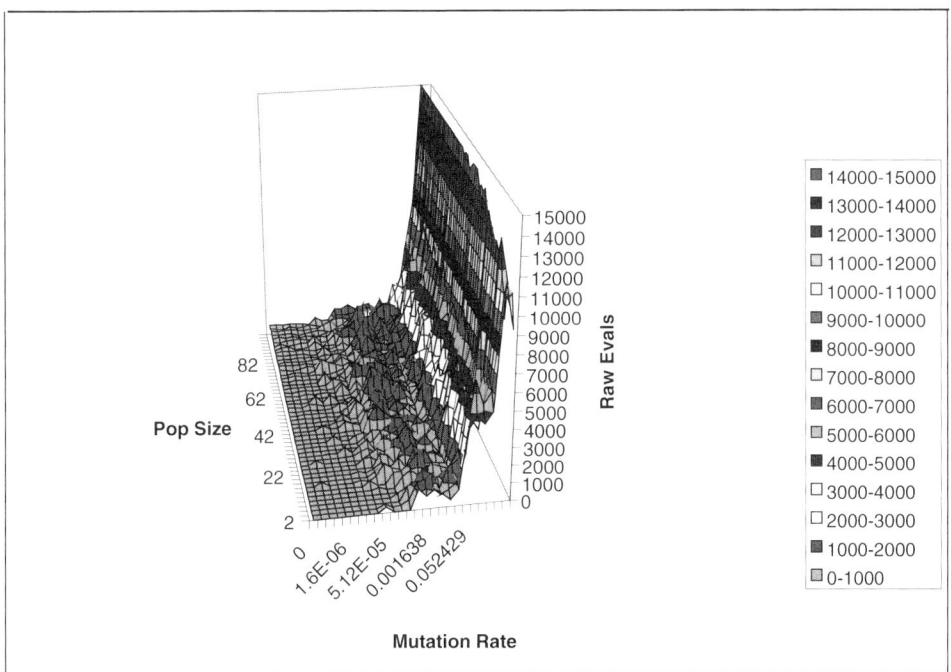

Figure 14.21 Breeder GA on scenario B with 'plus used' model at 20,000 evaluations.

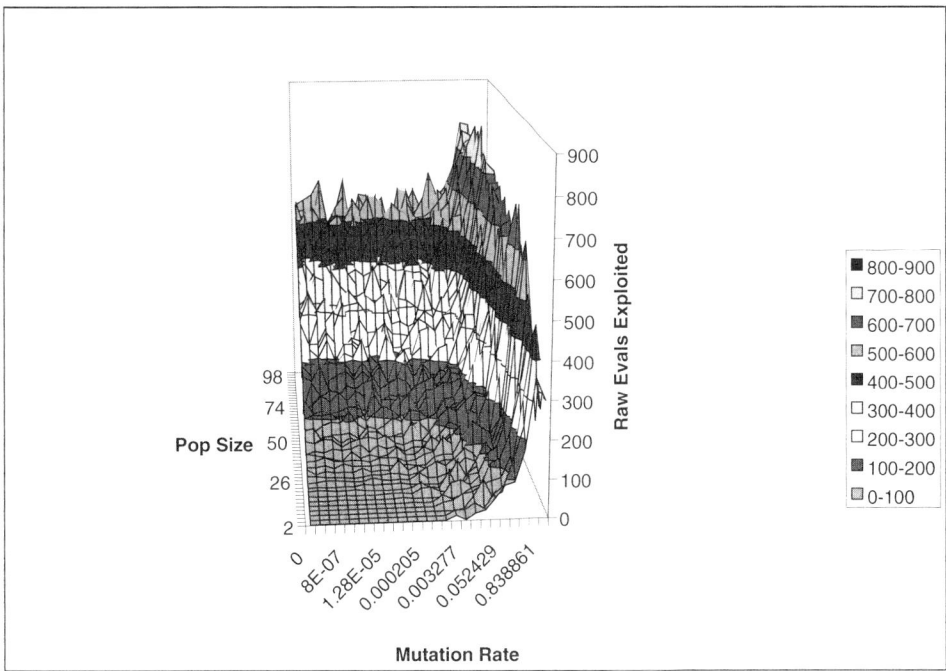

Figure 14.22 Breeder GA on scenario B with 'plus used' model at 1000 evaluations.

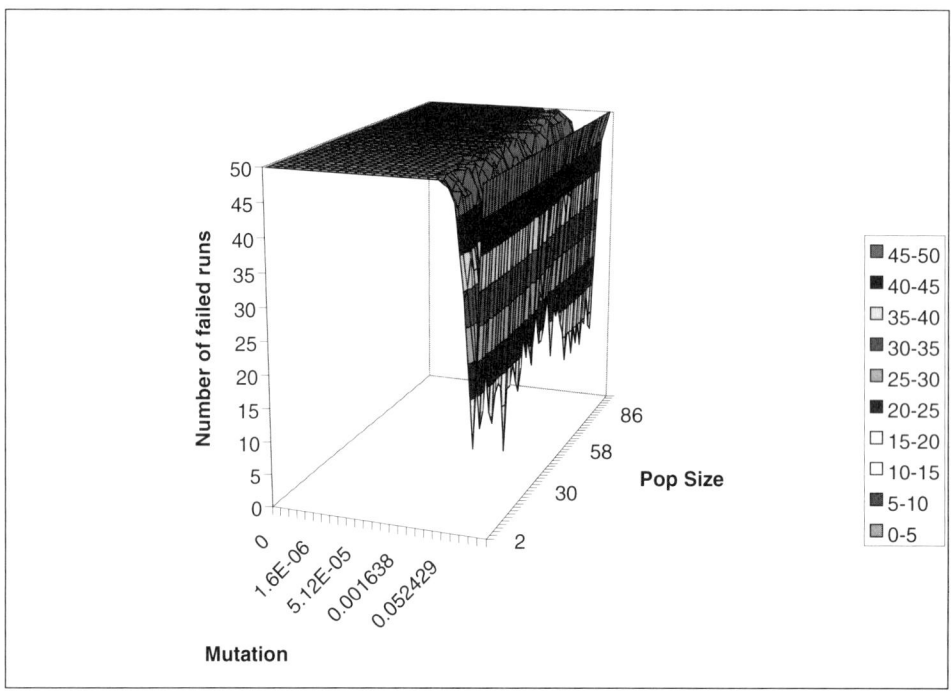

Figure 14. 23 Failure surface for breeder GA on hard problem at 20,000 evaluations limit.

Figure 14.23 shows the failure rate surface for the 20,000 evaluation limit plot showing distinct similarity with Figure 14.20. Here it can clearly be seen again that for the vast majority of the tuning surface, the GA consistently fails to find good solutions. The 'trough of opportunity' is, however, slightly wider at mutation rates between 10% and 41%, and deeper giving a consistently good performance almost 80% of the time over a wide range of population sizes at around 20% mutation.

14.9   The Effects of Skewed Crossover

As was clearly shown earlier and in Oates *et al.* (1998a; 1999), a modified form of two point crossover, referred to as 'Skewed Crossover', can give significantly better results on the B scenario problem with the 'plus used' evaluation function. The results of using this operator in conjunction with a steady-state, three way single tournament GA are shown in Figure 14.24. Here the population size axis is deliberately reversed to show more clearly the peak/ridge feature at 0.04% mutation and the trough at 1.3%. This GA was allowed 5000 evaluations, but being steady-state needed no generational adjustment for different population sizes.

It is important to note the occurrence of the linear feature relating evaluations exploited to population size. The presence of this feature lends further weight to the argument that this feature is caused by increased availability of multi-gene schema with increased population size, rather than any generational feature, as this skewed tournament GA is steady-state. (The effects of diversity depletion with time is the subject of further research by the authors.) It is interesting to note that the mutation rate at which the trough occurs is much less than in all previous cases (either at the 5000 evaluation limit or any other shown), now being at 0.04%. This feature had not previously been seen to be affected by problem complexity and may be a measure of the suitability of the crossover operator. The position of the trough, however, remains unchanged for the hard problem at 1.3%, implying this is a feature of underlying problem complexity and not the ease with which any particular operator can search the landscape. Further results are needed here.

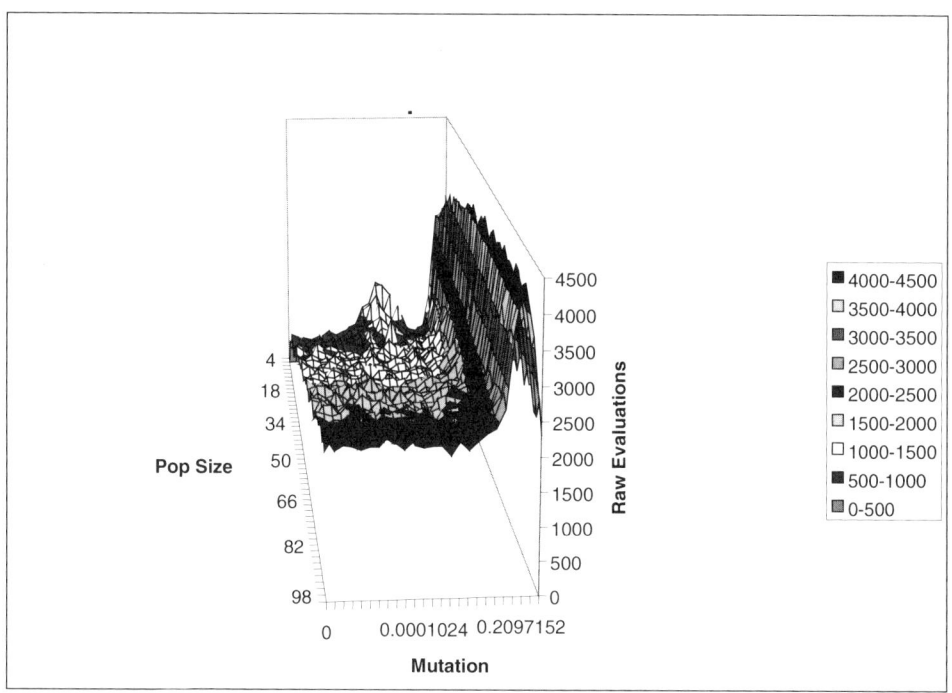

Figure 14.24 Skewed tournament GA on hard problem at 5000 evaluations limit.

Figure 14.25 shows the error surface for the Skewed Tournament GA at 5000 evaluations on the hard problem. Again, the population axis is reversed for easier comparison with Figure 14.17. Although superficially similar, it is important to note two distinct differences in the plots. Firstly, the minimum population size needed to give consistently good performance, without the need for mutation, has dropped considerably from around 200 members in Figure 14.17 to around 30 members here. Secondly, whilst Figure 14.2 showed that the Breeder GA was intolerant of very high levels of mutation (>20%), the Skewed Tournament GA on this problem brings this threshold down to 10%. Clearly, anything approaching random search is ineffective, whilst the lower population threshold implies that as soon as the specialised crossover operator has enough diversity available to it, it has no difficulty in consistently manipulating schemata to find optimal solutions.

Once again, at small population sizes, the narrow trough in Figure 14.25 coincides with the trough in Figure 14.24, occurring between 0.16% and 2% mutation. This region represents optimal tuning for the GA on this problem allowing it to consistently deliver good solutions in a minimum number of evaluations.

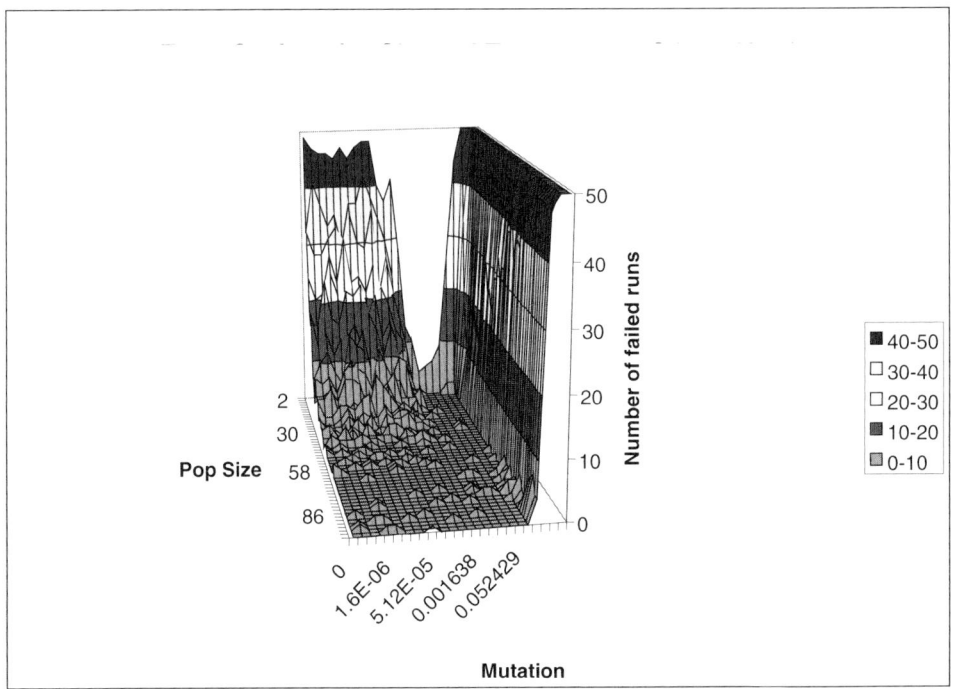

Figure 14.25 Failure Rate Surface for Skewed Tournament GA on hard problem at 5000 evals limit.

The contrast between Figures 14.20 and 14.24 should also be emphasised as they represent different GAs, but on the same hard problem and with the same evaluation limit. The fact that Figure 14.24 looks more like Figure 14.17 shows the significantly improved performance advantage of the skewed crossover operator on this problem.

14.10 Performance with No Mutation

Finally, it is worth examining the relative slopes of the linear features seen at little or no mutation in Figures 14.16, 14.19, 14.21 and 14.24. These are compared in Figure 14.26. As a benchmark, the plot for Breeder GA with 5000 evaluations on the simple problem (SA BDR +A 5K) shows that after an initial super-linear region, the GA is can exploit around 137 evaluations per 10 members of the population. This linear feature has been shown to continue up to population sizes of at least 500 on this problem set (see Oates *et al.*, 1999a).

However, for the Breeder GAs with only uniform crossover on the hard problem (plots SB BDR +U 5K and SB BDR +U 20K) it appears that this slope is much reduced to only 87 evaluations per 10 members of the population, regardless of whether the GA is allowed to run for 5000 or 20,000 evaluations. Clearly, the rate of diversity depletion is greater, suggesting good schema are less common.

In contrast, the plot for the Skewed Tournament GA on the hard problem at 5000 evaluations limit (SB SKT +U 5K) shows a much steeper slope of 177 evaluations per 10 members, with a sub-linear initial phase (however, very small population sizes with three way single tournaments are likely to produce questionable results). As was demonstrated in Oates *et al.* (1999a) with respect to the simpler problem, no significant slope or general performance differences were noted when comparing the generational, elitist Breeder GA with the steady-state, three way single Tournament GA, when both were using uniform crossover. Further, unpublished, results show that skewed two-point crossover is responsible for the improved performance seen here.

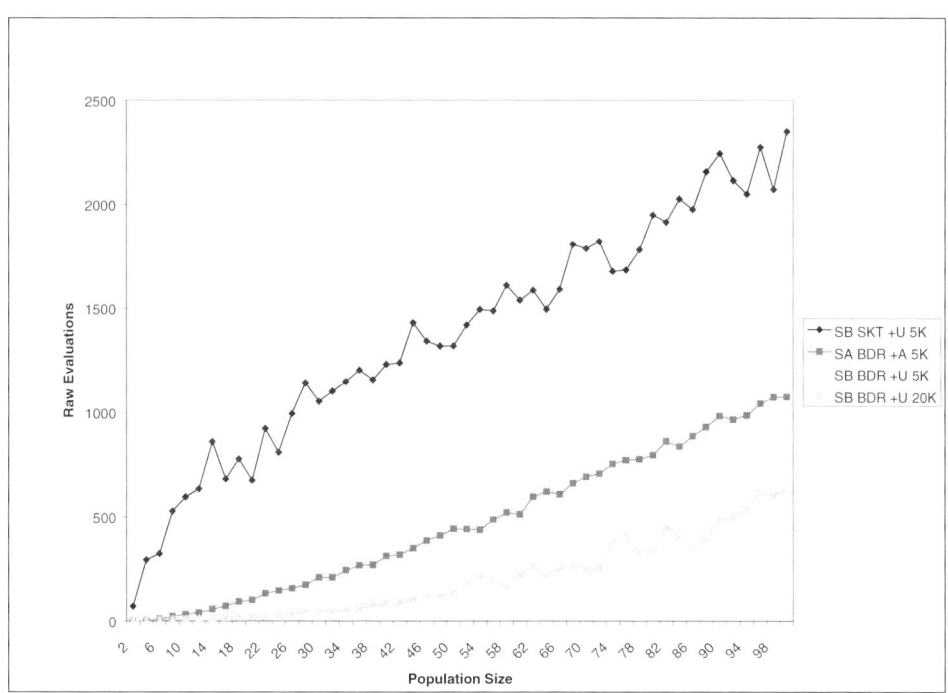

Figure 14.26 Relative slopes of raw evaluations exploited at 0% mutation.

14.11 Conclusions

This chapter has demonstrated the use of some simple evolutionary computational techniques for finding good configurations for distributed information systems. Whilst still at an early stage, current results are encouraging. The choice of specific model used as the evaluation function can clearly have a dramatic effect on the complexity of the search space and hence on how easy it is to find global optima. This in turn can lead to the requirement for more complex optimisers. Basic Hillclimber is shown to be quite ineffective on a more 'realistic' model.

It has been shown that simple averaging of the fitness values of solutions neighbouring the global optimum is not sufficient to indicate or explain critical features of the search space landscape. A richer analysis will often be required, and a candidate technique for visualising non-ordinal multi-dimensional landscapes has been outlined here.

The use of the 'skewed' crossover operator is demonstrated here, proving particularly effective in this case. In the more general case, further work is needed to get a clearer understanding of the applicability of this operator.

The effects of varying levels of mutation on the efficiency of a Genetic Algorithm can be seen to produce a complex performance landscape. In particular, two peaks can be seen in the number of raw evaluations exploited by the GA as mutation is increased (particularly at low population sizes), the first occurring as error rates are falling and the second occurring as error rates rise. Optimal performance is to be found in the intervening trough, where both error rates and number of raw evaluations exploited are low.

There is evidence supporting the hypothesis that the bi-modal performance profile appears to be a generic characteristic of genetic search, here demonstrated over a range of evaluation limits and on two distinct problems and with different crossover operator.

Whilst the mutation rate producing the troughs appear to be independent of population size, (although for low evaluation limits the effect is only visible for small population sizes), the mutation rate producing the peaks and troughs appears to reduce in value with increasing evaluation limit, as more opportunity for recombination is preferred to progress by chance random mutation. There is evidence to suggest that the position of the trough (where the GA can be said to be optimally tuned) is affected by problem complexity, the more difficult the search landscape, the lower the optimal mutation rate. The mutation rate at which the first peak occurs seems to be unaffected by problem complexity.

The mutation rate at which the optimally tuned trough occurs may be independent of operator, no difference being seen here between a good and bad operator. However, the position of the feature is shown to vary with problem complexity. The position of the first peak is shown to be affected by the operator but not by problem complexity.

At 0% and low rates of mutation, there appears to be an extensive region of linear relationship between population size and number of raw evaluations exploitable. This is demonstrated over both Breeder and Tournament (Generational and Steady State) strategies and appears independent of evaluation limit – the latter merely affecting asymptotic cut-off.

Further work is currently underway exploring the performance of other strategies, representations, operators and problem sets to establish whether these phenomena are indeed generic/characteristic profiles of the performance of tuned genetic search.

Acknowledgements

The authors wish to thank British Telecommunications Plc for their support for this work.

15

The Automation of Software Validation using Evolutionary Computation

Brian Jones

15.1 Introduction

The software crisis is usually defined in terms of projects running over budget and over schedule, though an equally important aspect is the poor quality of software measured in terms of its correctness, reliability and performance. The consequences of releasing faulty software into service may be devastating in safety-related applications, telecommunications and other areas. When the USA telecommunications system failed and half of the nation was isolated, lives and property were clearly put at risk. Such potential disasters might be avoided by more careful and thorough validation of the software against specified functions, reliability and performance.

The modern world relies on its telecommunications networks in every facet of life, from the ability to use credit cards at automatic teller machines in any part of the world to obtaining the latest pop song over the Internet, from tele-working from home to tele-shopping from home. The telecommunications networks are a vital part of the infrastructure of our economic, social and cultural lives. The risks of software failure must therefore be balanced against the great benefits of using reliable software to support and control the business of telecommunications. The maturing discipline of software engineering must be applied to produce and validate software in which both the suppliers and the users may have confidence. To this end, a number of standards have been developed specifically to encourage the production of high quality software: some examples are the British Computer

Society (BCS) Standard for Software Component Testing (Storey, 1996) and the generic IEC 61508 (1997) Standard for the Functional Safety of Electrical/Electronic/Programmable Electronic Safety-related Systems. The BCS Standard deals with all types of software and, in addition to the normal approaches to functional and structural testing relevant to the business software of the telecommunications industry, it deals with such methods as Finite State Machine (FSM) testing and cause-effect graphing relevant to the communications software itself. Automation is crucial if the costs of this essential validation are to be controlled. In this respect, genetic algorithms come into their own, since testing may be viewed as a search within the input domain for combinations of inputs that will cover the whole of the software's functionality and structure. GAs are able to test software to a level that neither manual testing nor random testing could achieve.

This chapter describes the essence of software validation and how genetic algorithms may be used to derive a set of tests that will cover some pre-defined attribute of the software's function, structure or performance. A case study describes in detail the use of GAs to ensure that every branch in the software is exercised, and the current use of GAs in testing software and micro-electronic circuits is reviewed.

The creation of correct computer software ranks as one of the most complex tasks of human endeavour, demanding high levels of skill and understanding. Software complexity relates to difficulties in human perception, rather than any problems that the machine experiences in executing the program; complex problems are more difficult for engineers to analyse, result in more complicated designs, and the end product will be more difficult to test. Hence, it is argued that complex software will be more likely to contain faults that are harder to locate by testing or static analysis. Software must not only be written, but also read and understood later by software engineers who were not involved in the original design and implementation, but who are called upon to correct faults and extend functionality (corrective and perfective maintenance respectively).

There are many metrics for complexity (Fenton and Pfleeger, 1997; Zuse, 1991). Metrics range in sophistication from counting lines of code to invoking information theory. Whilst the absolute value of complexity produced by a metric is not important, metrics must:

1. generate a unique value for a program that enables programs to be ranked according to their complexity, and
2. the complexity must increase if lines of code are added or memory requirements are increased or specified execution time decreased.

Measures of complexity fall broadly into two groups: structural and linguistic metrics. Members of the latter group may be identified since linguistic metrics do not change when the lines of code are shuffled. Two of the early and widely used complexity metrics are McCabe's structural metric and Halstead's linguistic metric. Both are calculated easily and relate to cyclomatic complexity and to program length, vocabulary and effort respectively. Both have been criticised and literally dozens of complexity metrics have been proposed (Zuse, 1991). However, the ability to rank programs according to their complexity suggests a ranking of the difficulties of testing and the chance of faults remaining after testing.

The British Computer Society Special Interest Group in Software Testing (BCS SIGIST) produced a standard for testing software units that has subsequently been adopted by the

British Standards Institute (Read, 1995; Storey, 1996). S. H. Read acted as Chair of the committee for the final stages of their deliberations, which included the preparation of a glossary of terms relating to software testing. A *mistake* is a human misunderstanding that leads to the introduction of a *fault* into the software. Faults are synonymous with *bugs* in common parlance, though the use of the term bug is discouraged because bugs happen by chance rather than by mistakes. Faults cause *errors* in the expected results of executing the software; a *failure* is a deviation in some way from the expected behaviour.

Faulty software is different from faulty hardware in that hardware failures tend to be random whereas software failures are systematic. Hardware tends to wear out by a physical process such as the diffusion of impurities in integrated circuits at high operating temperatures; this process may be modelled assuming a Poisson distribution and a mean time to failure predicted. Software does not wear out in this sense; software faults discovered after several years of trouble-free operation have always been present, entering the system through an incorrect design and having been missed through inadequate testing. Such failures are systematic, arising whenever certain circumstances and data combine. Such failure modes are difficult to predict through models.

The approaches to software testing covered by the BCS standard rely on the existence of a specification, so that if the initial state of the component is known for a defined environment, the validity of any outcome from a sequence of inputs can be verified. The standard defines test case design techniques for dynamic execution of the software and metrics for assessing the test coverage and test adequacy. Criteria for testing are decided initially, and the achievement of those criteria is a measure of the quality of testing. The standard advocates a generic approach to testing and identifies a sequence of steps that must be undertaken: test planning; test specification; test execution; test recording; checking for test completion. The standard covers unit or component testing only, and specifically excludes areas such as integration testing, system testing, concurrent/real-time testing, and user acceptance testing. A number of approaches are discussed in detail, including statement and branch coverage, data flow testing, and Linear Code Sequence And Jump (LCSAJ) testing.

Random testing of software is not commonly used, though statistical testing is used to measure the reliability of the software. Statistical testing is defined as testing at random where the inputs are chosen with a profile that models the use-profile of a particular customer. Different people may therefore have different views as to the reliability of a software package which is defined as the probability that the program performs in accordance with the user's expectations for a given period of time. A metric for reliability, R, has been defined in terms of the Mean Time Between Failure, $MTBF$:

$$R = MTBF\ /(1 + MTBF)$$

Whereas reliability relates to normal usage, robustness is a term that expresses the ability of software to recover successfully from misuse or usage not covered by the specification. This is an important issue for safety-related software, and the standard IEC61508 suggests techniques for evaluating probabilities of failure for different safety integrity levels and for systems running continuously and those executing on demand.

IEC 61508 (1997) is a generic standard for any electronic programmable device in a safety related application and has been instantiated for specific applications, e.g. 00-55 for the UK Ministry of Defence and 178B for the Aircraft industry. They specify a raft of techniques for the static and dynamic validation and verification of software. Not surprisingly, the demands are much more rigorous than those suggested by either the BCS standard (Storey, 1996) or the ISO9000 standard. Validation is defined as producing software to satisfy the user's expectations and requirements; verification is checking that the software is a true implementation of the specification, even though the specification may be incomplete or wrong. The IEC61508 standard includes formal, mathematical methods of specification amongst the recommended methods of verification; the software source code may then be verified mathematically. The standard recommends that tests be devised to check that each module and integrated sub-system of modules performs its intended function correctly and does not perform an unintended function. Attention must also be paid to the integration of modules and sub-systems which must interact correctly in terms of both functionality and performance.

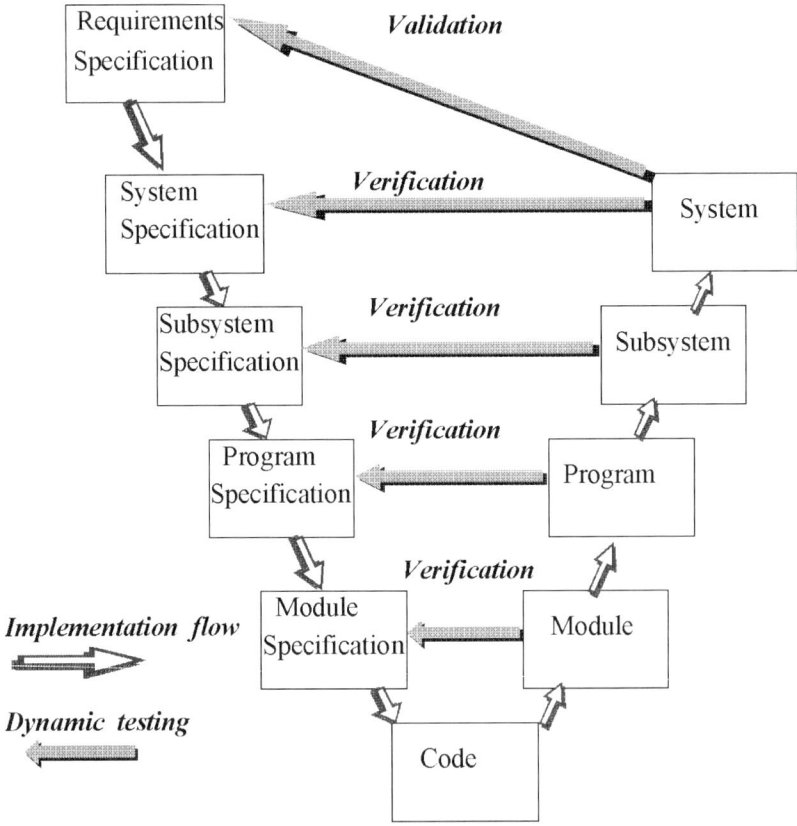

Figure 15.1 The V-model of the software lifecycle, plus validation and verification.

There are many different approaches to deriving test sets and cases for software and the two most common are structural and functional testing. In structural testing, the tests are derived in order to cover some structural attribute such as all branches, and the adequacy of testing is assessed by an appropriate metric such as the percentage of all branches. Such testing is commonly employed at the unit testing level. Functional tests are derived from the software specification and user requirements documents; the appropriate metric is to ensure that every aspect of functionality has been tested. These approaches are complementary since the omissions of one are covered by the other. Structural testing will not reveal the omission of functionality and functional testing may leave much code untested.

As software units are integrated, further testing must check the validity of bringing them together, typically by checking the ranges of variables passed between the units. Further testing covers the building of sub-systems and finally the entire system. The V-model of the software lifecycle clearly relates the role of testing at each stage to the validation and verification of the software (Figure 15.1). Finally, the user is invited to undertake acceptance testing using real data to validate the end product.

The cost of testing the software may amount to half the total development cost especially for safety related projects. Such an enormous commitment demands automation whenever possible. Some aspects of testing such as regression testing have been automated successfully in various Computer Aided Software Testing (CAST) tools (Graham, 1991). The tests are invariably derived manually and entered into a database; Graphical User Interfaces (GUI) may be tested by capturing mouse movements and clicks and replaying them later as a regression test. The motivation for the work described in this chapter is to automate the derivation of test sets by searching the domain of all possible combinations of inputs for an appropriate test set. Genetic algorithms are an ideal and widely-used searching tool for optimisation problems and they have been applied successfully in deriving test sets automatically.

15.2   Software Testing Strategies.

The most important decision facing a software engineer is to identify the point at which testing may cease and the software can be released to the customer. How can sufficient confidence in the correctness of the software be accumulated? This question relates to the issue of test adequacy which is used to identify the quality of the test process and also the point at which the number of tests is sufficient, if not complete. Test adequacy relates to a pre-defined measure of coverage of an attribute of the software. This attribute may relate to coverage of all statements in the software, that is all statements must be executed at least once, or to the functionality, that is each aspect of functionality must be tested. The choice of attribute is arbitrary unless it is prescribed by a standard. Whilst this approach to test adequacy may appear to be unsatisfactory, it is the only practical approach. Furthermore, coverage is rarely complete, and software systems that are not related to safety are frequently released with only 60% of all statements having been executed. A fundamental difficulty of software validation and verification is that there is no clear correlation between a series of successful tests and the correctness of the software.

Metrics play an important role in testing as shown by the emphasis given to them in the BCS Testing Standard (Storey, 1996). They are relatively easy to apply to structural testing where some attribute of the control flow or data flow in the software defines the test adequacy. It is less straightforward to apply metrics to the coverage of functionality. Complexity metrics play a part in deciding how to define test adequacy. As the complexity increases, so the difficulty of design increases and the expectation of a high fault density increases with it. The logical outcome of this is to tighten the definition of test adequacy and to demand functional testing complemented by a form of sub-path testing, rather than simple statement testing. Metrics are essential to any engineering discipline. In this context, software engineering is in its infancy compared to traditional engineering areas and considerable work is still needed to develop effective and to educate software engineers in their use.

In general, software testing aims to demonstrate the correctness of a program. A dynamic test can only reveal the presence of a fault; it is usually impossible to test software exhaustively, and hence it is impossible to prove that a program is completely correct by standard testing methods alone. In contrast, fault-based testing identifies a common fault and probes the software deliberately in order to show its presence. Whereas structural and functional testing aim to give blanket coverage of the software, fault testing targets particular problems that are known to persist. Beizer (1990) gives an analysis of typical faults in software. One such fault lies in the design of a predicate where '>' should have been written as '>='. Such faults are most unlikely to be revealed by standard structural or functional coverage, and testing for this fault would demand a boundary value analysis of the input domain, where the boundaries of the sub-domains are defined either by the branches in the software or the functionality.

Another approach to testing is to generate a finite state machine to model the states that the system may occupy, and the events that cause transitions between the states, and the consequences of the transitions. Test cases comprise the starting state, a series of inputs, the expected outputs and the expected final state. For each expected transition, the starting state is specified along with the event that causes the transition to the next state, the expected action caused by the transition and the expected next state. State transition testing has been investigated by Hierons (1997) and is suited to real-time applications such as telecommunications software.

Cause-effect graphing is another black box testing method that models the behaviour of the software using the specification as the starting point. The cause-effect graph shows the relationship between the conditions and actions in a notation similar to that used in the design of hardware logic circuits. The graph is then re-cast as a decision table where the columns represent rules which comprise all possible conditions and actions. Each column represents a test for which the conditions are set either true or false, and the actions set to be either performed or not. The associated test effectiveness metric in this case is the percentage of feasible cause-effect combinations that are covered.

Most testing aims to check that the software functions correctly in terms of giving the expected output. However, there are other aspects of software execution that must be checked against the specification. Examples of non-functional correctness testing are investigating the temporal properties and maximum memory requirements. Worst-case and best-case execution times are particularly important in real-time applications where data may be lost if their appearance is delayed.

The effort required to test software thoroughly is enormous, particularly when the definition of test adequacy is determined by the demands of validating safety-related software; the IEC61508 standard estimates that as much as 50% of the development effort is taken up by testing. Every time the software is modified, the affected modules must be re-tested (regression testing). Automation is therefore essential not only to contain the cost of verification, but also to ensure that the software has been tested adequately. At present, automation is typically based on capture-replay tools for GUIs, where sequences of cursor movements and mouse button clicks together with the resulting screens are stored and replayed later. This still requires manual derivation of test cases and is most useful in relieving the boredom of regression testing. Ince (1987) concludes his paper on test automation with the sentiment that automatic derivation of test cases is an elusive but attractive goal for software engineering.

Random generation of test sets is a relatively straightforward technique and it will usually achieve a large coverage of the software under normal circumstances. As the software increases in size and complexity, however, the deeper parts of the software will become more and more difficult to reach. Random testing starts to falter under these circumstances and a guided search becomes necessary. Checking that each test yields the expected result is tedious and labour-intensive. Manual checking is not normally feasible because of the effort involved, and two solutions have been proposed. In the first, post-conditions are defined and checked for violation. This is appropriate when the software has been formally specified in a mathematically-based language such as Z or VDM. In the second, a minimum subset of tests that span the test adequacy criterion is calculated and checked manually, though this may be still an enormous task depending on the size of the software and the test adequacy criterion (Ince, 1987; Ince and Hekmatpour, 1986). In the case of testing for single-parameter attributes other than functional correctness (worst case execution time, for example), it is easy to check for a violation of the specified extreme value of the attribute.

15.3 Application of Genetic Algorithms to Testing

The objective of adequate testing is to cover some aspect of the software, be it related to the control flow, the data flow, the transition between states or the functionality. The problem of equivalence partitioning points to the wasted effort of executing different tests that exercise the same aspect of the software. A thousand tests that cause the same flow of control from entry to exit and execute as expected tell the tester nothing new and represent wasted effort. The input domain may be divided into sub-domains according to the particular test objectives such as branch coverage. Figure 15.2 (left) shows a simple, contrived program which has two integer variable inputs, X and Y, each with ranges of (0..10) and sequences of statements $A..E$. Figure 15.2 (right) shows a graphical representation of the 2D input domain of X and Y where the sub-domains of inputs that cause the different sequences to be executed are sets of (X,Y) and are labelled $A...E$. Assuming that the program terminates, the sequence E is executed for all combinations of (X,Y); therefore, E represents the whole domain. The sub-domain A covers all inputs that cause the *while-loop* to be executed at least once; F is the domain of (X,Y) that fail to execute the loop at all. Hence $E = A \cup F$.

The sub-domain that causes the control flow to enter the loop is represented by A followed by one of B, C or D. Hence, $A = B \cup C \cup D$, and since B, C and D are disjoint, we also have $B \cap C \cap D = \emptyset$, where $\emptyset$ represents the empty set. The set C comprises a series of isolated, single values; other sub-domains may overlap or may be disjoint depending on the details of the code. If a test adequacy is defined in terms of coverage of all sub-paths defined as an LCSAJ, the pattern of sub-domains for the program in Figure 15.2 (left) would differ from the domain diagram in Figure 15.2 (right).

```
integer X,Y;

begin
get(X); get(Y);
while X<=8 loop
   A;
   if (X>5) and(Y<6)
      then B;
   elsif (X+Y=5)
      then C;
   else
      D;
   end if;
end loop;
E;
end;
```

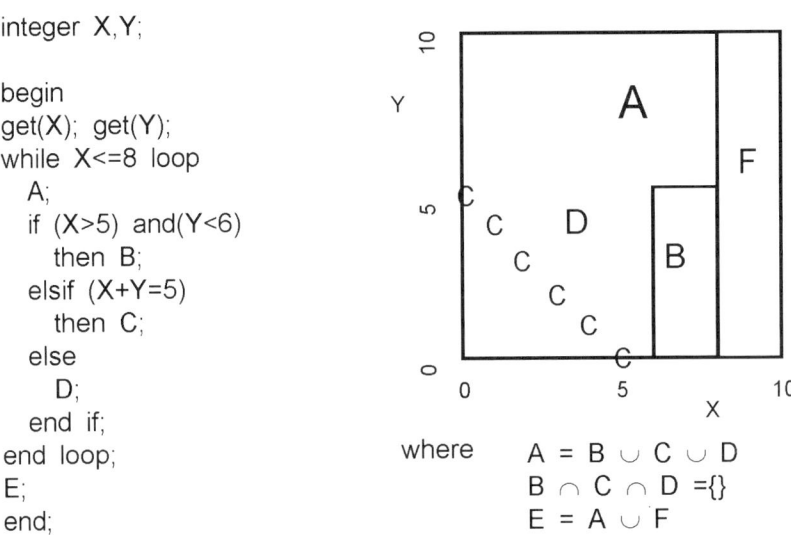

Figure 15.2 The code of the program under test (left). The domain split into sub-domains for branch coverage of the program under test (right).

If test sets are to be derived to give full coverage in Figure 15.2 (right), that is to cover all branches, then the derivation becomes equivalent to searching the domain for members of each sub-domain. Most of the sub-domains in Figure 15.2 (right) may be found easily at random because the cardinality of every sub-domain, apart from sub-domain C is large. The probability of finding a pair (X,Y) that belongs to C is about 1 in 20 for our example. As programs become larger, the domain is split into finer and finer sets of sub-domains because of a filtering effect of nested selections, and random testing becomes less and less effective; it also becomes rapidly more difficult to derive tests manually to cover each sub-domain. All test sets within a sub-domain are equivalent except for those falling near to their boundaries where the probability of revealing faults is increased. The competent programmer hypothesis states that programmers write code that is almost correct, but that some faults occur frequently. These faults often cause the sub-domain boundary to be shifted by a small amount that could only be detected by test sets that fall at or adjacent to the boundary. Boundary value analysis is a stringent method of testing that is effective at revealing common programming mistakes.

Genetic and other evolutionary algorithms have been used to good effect in deriving test sets of high quality by searching the input domain for inputs that fall close to the sub-domain boundary. An important factor in successful application of GAs is the derivation of the fitness function which is based on the definition chosen for test adequacy. If branch coverage is taken as an example, a selection such as '*if* $(A = B)$ *then...else ...end if;*' may occur at some point in the program. Whilst it may be easy to find inputs to ensure that $A \neq B$, it will usually be much more difficult to find inputs to give $A=B$ and to satisfy the demands of boundary value analysis when input values satisfying $A=succ(B)$ and $A=pred(B)$ must be found. One possibility is to define a fitness function based on the reciprocal of the difference between the values of A and B; hence

$$\text{fitness}(X) = (|A(X) - B(X) + \delta|)^{-1}$$

where X is the vector of input values and δ is a small number to prevent numeric overflow. The functions $A(X)$ and $B(X)$ may be complicated and unknown functions of the inputs; this does not present a problem for the GAs, since only the values of $A(X)$ and $B(X)$ need be known at this point in the program. These values are made available by adding instrumentation statements immediately before the *if*-statement to extract the values. This approach works reasonably well for numeric functions $A(X)$ and $B(X)$, though there is a tendency to favour small values of $A(X)$ and $B(X)$ since they are more likely to give higher fitness values. This may not be a problem. However, numerical fitness functions are less suitable when strings and compound data structures such as arrays and records or objects are involved. In these cases, the Hamming distance between values is more appropriate; genetic algorithms using fitnesses based on Hamming distances perform at least as well as those based on numerical fitness functions, and often perform much better (see section 5.4 in Jones *et al.*, 1998). The fitness function is modified to account for constraints such as limits on the ranges of input variables or to avoid special values such as zero.

The fitness function depends on the particular objective of testing and is designed in an attempt to direct the search to a each sub-domain in turn. Not all testing is directed at verifying the logical correctness of code; a considerable amount of effort goes into verifying the software's temporal correctness against the specification of real-time and safety-related applications by seeking the Worst and Best Case Execution Times (WCET and BCET respectively). The input domain is split according to the execution times for each combination of input parameters; in this case, the fitness is simply the execution time.

An issue in applying genetic algorithms to software testing is the frequency of multiple optima, which may arise in testing any attribute such as covering all branches or execution times. In Figure 15.2 (right), the subdomain, C, corresponding to satisfying the predicate $(X+Y = 5)$, is a series of single points. In this case, the points are adjacent, but in general, non-linear predicates give rise to disconnected and often distant points. In attempting to cover this branch, two individual combinations of inputs may approach different optima but have similar fitnesses because they are the same distance from a solution. Whilst the phenotypes of the individuals are similar, the genotypes may be quite different, so that crossover operations may force the children away from a solution. This results in a degraded performance in terms of the number of generations required to satisfy the predicate. Fortunately, most predicates in data processing software are linear; White and Cohen (1980) found in a study of 50 production COBOL programs that 77.1% of the predicates involved one variable, 10.2% two variables, and the remainder were independent of the input variables. Only one predicate was non-linear.

In a similar study of 120 PL/1 programs, Elshoff (1976) discovered that only 2% of expressions had two or more logical operators, and that the arithmetic operators +, −, * and / occurred in 68.7%, 16.2%, 8.9% and 2.8%, respectively, of all predicates. Most arithmetic expressions involved a simple increment of a variable.

Knuth (1971) arrived at similar statistics for FORTRAN programs in which 40% of additions were increments by one and 86% of assignment statements were of the form *A=B*,

$A=B+C$ and $A=B-C$. Although these studies are now dated, they covered a wide range of application areas and there is no reason to suppose that current software will be markedly different. The extra effort needed by genetic algorithms to cope with multiple, isolated optima will not present a substantial problem.

The representation of the input parameters is a key decision for success in automating software testing. Since variables are stored in binary formats in the RAM, a natural and straightforward option is to use this memory-image format for the individual guessed solutions in a traditional genetic algorithm. This is particularly convenient for ordinal, non-numeric and compound data types. The parameters are concatenated to form a single bit string forming an individual chromosome for crossover and mutation operations (Jones et al. 1996). When floating point types are involved, binary representations may cause problems when the exponent part of the bit string is subjected to crossover and mutation, giving rise to wildly varying values. Under these circumstances, an evolutionary strategy as opposed to a genetic algorithm (Bäck, 1996) is more effective. Evolution strategies use numerical representations directly and define crossover as weighted averages between two parents and mutation as a multiplication by a random factor. Whereas crossover is the dominant operator in genetic algorithms, mutation is the dominant operator for evolution strategies.

A further problem with a memory-image binary representation is that small changes in numeric value may cause substantial changes to the binary representation, for example incrementing 255 to 256. Sthamer (1996) investigated the use of grey codes as a substitute with considerable success. The disadvantage is the need to convert between the two binary representations and the actual values of the parameters to pass to the program under test.

15.4   Case Study: The Glamorgan Branch Test System.

A case study to illustrate the application of evolutionary algorithms in test automation will be developed using the example code in Figure 15.2 (left). The aim of testing is to find input pairs of $\{X,Y\}$ that will execute all branches; in the case of the *while-loop*, this would mean that the loop would be bypassed without execution and would be executed once. Most of the branches would be found easily at random, as may be seen from the domain diagram in Figure 15.2 (right). The diagram is drawn to scale and the area of each sub-domain is an indication of the number of combinations $\{X,Y\}$ that would cause a branch to be executed. The chance of finding such combinations at random is the ratio of the sub-domain area to the total area. Following this approach, finding inputs to exercise sub-domains *B* and *C* presents most difficulty. This may be surprising at first, since 25 of the total 100 combinations satisfy the predicate $(X >5$ and $Y < 6)$. The predicate controlling the *while-loop* has a filtering effect such that values of X above 8 are filtered, effectively reducing this to 15. As programs grow in size and complexity, even apparently straightforward, linear predicates combine to produce very small sub-domains. In general, the predicates with the smallest sub-domains are those involving equalities; only six combinations of $\{X,Y\}$ will satisfy $(X+Y = 5)$ and cause sub-domain *C* to be exercised. The power of genetic algorithms is in guiding the search in the domain to those sub-domains that are unlikely to be found at random and are difficult to evaluate by hand. In this case study, we concentrate on satisfying the predicate $(X+Y = 5)$.

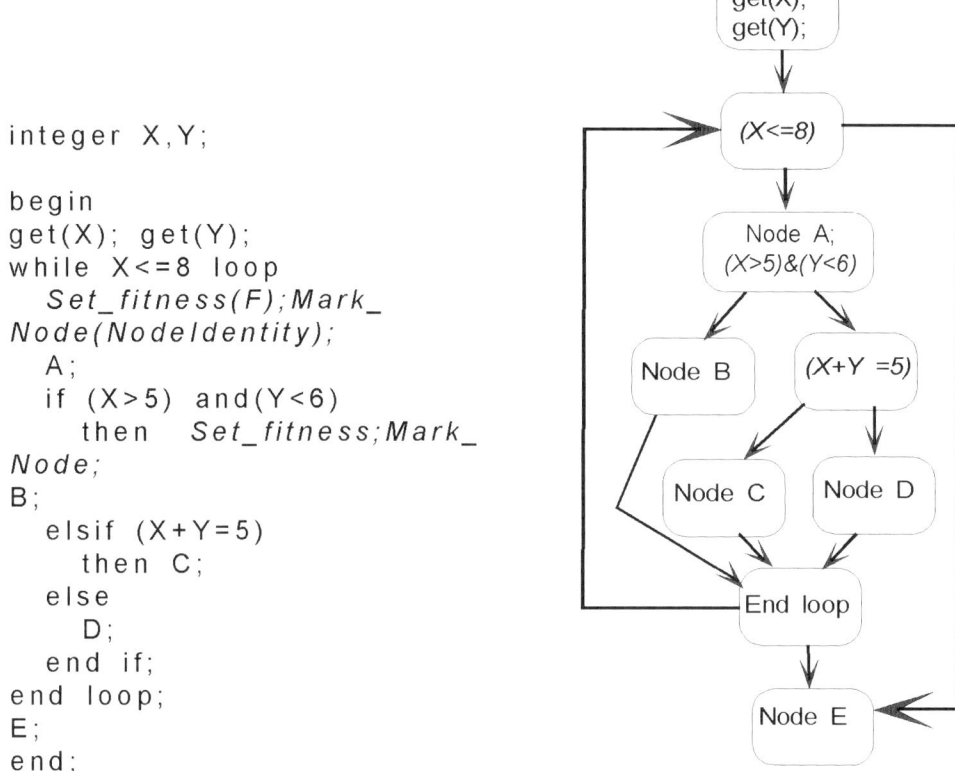

```
integer X,Y;

begin
get(X); get(Y);
while X<=8 loop
  Set_fitness(F);Mark_
Node(NodeIdentity);
  A;
  if (X>5) and(Y<6)
    then    Set_fitness;Mark_
Node;
B;
  elsif (X+Y=5)
    then C;
  else
    D;
  end if;
end loop;
E;
end;
```

Figure 15.3 The instrumented code (left); the control flow graph (right).

The code is instrumented to automate the coverage of all branches. Instrumentation requires the insertion of statements at key points (1) to determine when a branch has been exercised, and (2) to evaluate the fitness of the current evolving individual. The instrumented program is shown in Figure 15.3 (left), where italicised extra lines of code are inserted at the start of each sequence following a decision. The extra code comprises calls to procedures *Mark_Node(NodeIdentity)*, which indicates that the sequence has been exercised and *Set_Fitness(F)*, which sets the fitness to an appropriate value. *Mark_Node* simply sets a flag in the program's control flow graph to indicate which nodes have been executed. *Set_Fitness* adjusts the fitness corresponding to the current input test set according to the following rules. In this case study, the system is searching for *Node D* (Figure 15.3 (right)):

1. suppose the control flow has passed far from it to *Node B*; the fitness is set a small value;

2. suppose the control flow passes to the sibling node (*Node C*); the fitness is set to $abs(X+Y-5+\delta)^{-1}$ where δ is a small value to prevent numeric overflow;

3. suppose the control flow passes through the desired *Node D*; the fitness is set to a large value (δ') and the search will move to the next node to be exercised or terminate if all nodes have been visited.

A possible evolution of a solution over two generations is given in Figure 15.4. The input test sets $\{X,Y\}$ are two four-bit integers in a population of only four individuals which are initialised at random to the values shown in the left hand column for the first generation. The fitnesses calculated according to the above rules are given in column three; all combinations exercise the sibling *Node C* apart from the third individual which exercises *Node B* and the fitness is set to an arbitrarily low value of 0.05. Column four contains the corresponding bit string where the first four bits represent X and the last four Y. Single point crossover occurs between the first two parents and the last two between bit positions 6/7 and 2/3 respectively to give the children's bit strings in column 5. These convert to the test sets $\{X,Y\}$ in columns six and seven, which are applied to the program under test to give the fitnesses in column eight. The selection strategy chooses the second and fourth members of both parents and children. The process is repeated when the third child in the second generation finds a solution.

	PARENT			CHILDREN				
X	Y	Fitness	Bit string	Bit String	X	Y	Fitness	Remark
Generation 1								
5	10	0.10	01011010	01011011	5	11	0.11	Crossover at position 6/7
4	3	0.50	01000011	01000010	4	2	1.00	
6	1	0.05	01100001	01100111	6	7	0.125	Crossover at position 2/3
2	7	0.25	00100111	00100001	2	1	0.50	
Generation 2								
4	2	1.00	01000010	01000011	4	3	0.50	Crossover at position 7/8
4	3	0.50	01000011	01100010	6	2	0.05	Mutate position 3
2	1	0.50	00100001	00100011	2	3	**SUCCESS**	Crossover at position 6/7
2	7	0.25	00100111	00100101	2	5	0.50	

Figure 15.4 Two generations of an example evolution of test sets to cover branch predicate ($X+Y=5$).

15.5 Overview of the Automation of Software Validation using Evolutionary Algorithms.

The automatic validation of software has been a long term goal of software engineering from the early work of Clarke (1976), who used symbolic execution to generate tests, and Miller and Spooner (1976), who used numerical methods for floating point inputs. Since the early work of the Plymouth and Glamorgan University groups which started independently just before 1990, it is no exaggeration to say that there has been an explosion of interest and activity in automating the derivation of test sets using genetic algorithms. Research is now pursued actively within the UK (Strathclyde University, York University/Rolls-Royce), Germany (Daimler-Benz/Glamorgan University and Humboldt University), Austria (Vienna University) and the USA (Stanford University /Daimler-Benz).

The work at Glamorgan is probably the most mature and covers structural testing (Jones *et al.*, 1995, 1996, and 1998), fault-based testing (Jones *et al.*, 1998), functional testing (Jones *et al.*, 1995) and temporal testing (Wegener *et al.*, 1997). The structural testing work centres on branch coverage and the case study described earlier is based on this work. In addition to exercising each branch, all *while-loops* are exercised zero times, once, twice and an arbitrary number of times. For example, in testing a binary search program where an integer variable value is sought in an ordered array of 1000 integers, the number of *while-loop* iterations were controlled to be 1,2,3... The use of genetic algorithms controlled the number of iterations to be 1, 2 and 8 with a factor of over 50 times fewer tests than for random testing (Jones *et al.*, 1998). This example also shows the greater efficiency of basing fitness on the Hamming distance rather than a numerical difference between values of the functions involved in predicates. Full branch coverage was achieved with more than two orders of magnitude fewer tests than random testing in some cases. The greatest improvements were obtained for programs whose domains were split into small sub-domains that had a low probability of being hit at random. This ability to achieve full branch coverage is a major step forward when a typical system is released with only 60% of the statements exercised. The technique should scale to larger programs since the effort using genetic algorithms increases in a sub-linear way (an index of about 0.5) whereas the effort required for random testing increases with an index of about 1.5 (Jones *et al.*, 1998).

Obtaining full branch coverage is not the only issue. The test data should be of high quality; a high quality test may be loosely defined as the probability of the test to reveal faults. Fault-based testing means that tests are derived to reveal common, specific faults. Beizer (1990) classified the faults occurring in software and found that some 25% related to mistakes in the predicate. The Glamorgan group has used genetic algorithms to generate test sets that will reveal such faults. The substitution of ($A>B$) for ($A>=B$) is a typical mistake in a predicate, for example; this is detectable by finding test sets that cause functions A and B to have the same value, and for A to take successor and predecessor values of B. This approach to testing is known as boundary value analysis, since test sets search close to and on the sub-domain boundaries. The quality of the test sets have been evaluated using mutation analysis where single syntactically correct faults are introduced deliberately (Budd, 1981;de Millo *et al.*, 1991). Mutation analysis produces many similar versions, or mutants, of the original software. The aim of mutation testing is to apply the same test sets to both the original and the mutated programs and to obtain different outputs from them; if so, the mutant is said to be killed and the test set is of high quality, since it has revealed the deliberate fault. Some mutations do not generate different outputs; they are said to be equivalent mutants to the original program, and hence cannot be revealed by dynamic testing. A mutation score, *MS*, is defined as:

$$MS = \frac{K}{M - E}$$

where *M* is the total number of mutants, *E* is the number of equivalents and *K* is the number killed.

Jones *et al.* (1998) used genetic algorithms to reveal faults in predicates of the type described above. The fitness function was defined in triplicate to be the reciprocal of the absolute differences between *A* and *B*, *succ(B)*, *pred(B)* successively. The search was continued until the fitness was satisfied, and the test sets applied to both original and mutant programs. It is not guaranteed that these fitness functions can be satisfied, especially for non-linear predicates; for example, for a predicate such as $B*B = 4*A*C$ with integer variables *A*, *B* and *C*, there may be no values satisfying $B*B = 4*A*C + 1$. DeMillo and Offutt (1991) developed a mutation analysis system (MOTHRA) for testing software and derived test sets by evaluating constraints on different paths through the software. They achieved an average mutation score of 0.97. Offutt (1992) suggested that a mutation score of 0.95 indicated a thorough testing regime. Typical mutation scores of 0.97 were obtained with our genetic algorithm system, easily satisfying Offutt's criterion for adequate testing.

Genetic algorithms were being applied to the structural testing of software; the work was pursued concurrently and independently of the Glamorgan group in a European project. The collaborators were based in Toulouse, Plymouth and Athens, and Xanthakis (1992) presented their work at the Fifth International Conference on Software Engineering. The aim of the work was to derive test sets that cover the directional control flow graph of a program, i.e. to ensure that control visited every node in the graph. In this case, the graph comprises nodes representing linear instructions or conditional instructions, arcs, an entry state and an exit state from the program. The coverage metric was defined to be the percentage of nodes visited. The group developed a structural testing prototype known as TAGGER (Testing and Analysis by General Genetic Extraction and Resolution) whose first action was to generate a qualitative control flow graph for the software. Each arc in the graph was labelled with +, – or ? depending on whether increases in the variable at the start of the arc caused the variable at the end of the arc to increase, decrease or change in an indeterminate way. For each node in turn, the relevant predicates on the path from the entry to the node were determined and evaluated for an input test set to give a fitness. They used a conventional genetic algorithm with the addition of a maturation operator, which modified the chromosomes in a way that depended on the fitness. TAGGER was used successfully on a number of small programs which implemented numerical algorithms, but which were not described in detail. TAGGER often achieved a coverage of 100% and outperformed random testing, though no comparative figures were given. Another member of the Plymouth group, Watkins (1995), extended the work to include simulated annealing and tabu search with some success.

Roper (1997), at Strathclyde University, has applied genetic algorithms to the generation of test sets to cover every statement and every branch in a program. Roper's work differs from the two previous approaches in that the fitness is not based on any information about the internal structure of the program, even though the aim was to cover some aspect of the program's structure. The fitness of an input test set was the coverage achieved, for example, the percentage of branches exercised. The areas of the program visited were recorded in a bit string which was compared with the corresponding bit string for other individuals in the population. When the whole population had been evaluated in this way, individuals were selected for survival and subjected to crossover and mutation. The population evolved until a subset of the population achieved the prescribed level of coverage. Encouraging results were achieved on small, contrived programs.

The most common aim of software testing is to validate its functional correctness. There are a number of other attributes that must frequently be verified, and ranked high amongst these in importance is the software performance. Considerable effort has been expended in establishing the BCET and, perhaps more importantly, the WCET, to ensure that the timing constraints specified for the software are satisfied. Performance is clearly of great interest in real-time systems where tasks must be scheduled correctly to achieve the desired effect.

Timing software is more difficult than may appear at first. There are many pitfalls arising from caching effects, queuing, interrupts and so on; Kernighan and van Wyk (1998) attempted to compare the performances of scripting and user-interface languages and concluded that the timing services provided by programs and operating systems are woefully inadequate – their paper is entitled "Timing trials or the trials of timing"! Attempts to time a program using the real-time clock of an IBM-compatible personal computer will run into a number of problems. The tick rate is too coarse leading to an uncertainty in timing an event of more than 55 ms. In principle, this problem could be overcome by timing a large enough number of executions of the program. There are other less tractable problems:

1. processors that use caching give unpredictable timings;
2. execution of a program may be suspended unpredictably by a multi-tasking operating system;
3. the time taken by a processor to execute an operation may depend on the values of the data (for example, the multiplication of two integers).

The first two problems result in different timings between runs of the program so that the input domain cannot be sub-divided in any sensible way. The third problem does allow the domain to be split consistently, but the sub-domains may be large in number and contain only a few input test sets. The task of searching so many sub-domains becomes enormous and one that is tailor-made for genetic algorithms where the fitness for the WCET is simply the execution time (or its reciprocal for the BCET). The first two problems above preclude the use of the real-time clock since the timings (and fitnesses) would be inconsistent. The Glamorgan group, in collaboration with Wegener and Sthamer (originally of Glamorgan) of Daimler-Benz, Berlin, have applied genetic algorithms and evolutionary systems to the timing problem using a package *Quantify* from Rational to measure the number of processor cycles used during a program run (Wegener *et al.*, 1997). *Quantify* instruments the object code directly using the patented method of Object Code Insertion. *Quantify* is intended to identify bottlenecks in software rather than timing program executions and the results from a program run are communicated to the genetic algorithm via a disk file; the result is to slow down the genetic algorithm software. Nevertheless, useful results have been obtained and the technique promises to be useful as a standard for assessing software performance.

Experiments were made on a number of programs of different lengths and having different numbers of parameters. The performance of genetic algorithms was compared with that of random testing. Genetic algorithms always performed at least as well as random testing, in the sense that equally extreme WCET and BCET were found, and with fewer tests and often, more extreme times were found (Wegener *et al.*, 1997).

One of the most difficult decisions to make in searching for the WCET is when to stop; there is no clear and simple criterion for deciding when the extreme execution time has been found. The search may be terminated if

1. the specified timing constraints have been broken,
2. the fitness is static and not improving, or
3. after an arbitrary number of generations.

The last two criteria are unsatisfactory since the input domain may have large sub-domains corresponding to the same execution time, with a small sub-domain associated with a more extreme time. O'Sullivan *et al.* (1998) have used clustering analysis on the population relating to the latest generation to decide when to terminate the search. Individuals that lie closer together than a specified threshold distance form a cluster. Clustering may be based on their distance apart or on their fitness. The behaviour of the clusters as the threshold distance is decreased is displayed as a cluster diagram and suggests whether or not to terminate the search. The search should be terminated in those cases when a single cluster is formed and only breaks into smaller clusters as the threshold distance is decreased to small values. This does not indicate that the global extremum has been found, but rather that that no further improvements are likely. In contrast, those clusters that split quickly as the threshold distance is reduced into a highly structured tree are much more likely to discover more extreme values.

Müller and Wegener (1998) compared the usefulness of static analysis with that of evolutionary systems for determining the WCET and BCET. Static analysis identifies the execution paths and simulates the processor's characteristics without actually executing the program or applying an input test set. Static analysis tends to suggest a pessimistic value for the WCET. Evolutionary systems generate input test sets and execute the program. As the system evolves to find the extreme execution time, the results are clearly optimistic. Müller and Wegener (1998) conclude that static analysis and evolutionary systems are complementary, together providing an upper and lower bound on the WCET and BCET.

Hunt (1995) used genetic algorithms to test software used in the cruise control system for a car. The chromosome included both inputs such as speed set or not set, brake on or off, clutch engaged or not and the output which indicates whether the throttle should be opened or closed. The fitness function was based on the rules defined in the original specification. Hunt found that GAs could be used to search efficiently the space of possible failures, but did not give the significant advantage hoped for. Schultz *et al.* (1993) have also tested control software for an autonomous vehicle with the aim of finding a minimal set of faults that produce a degraded vehicle performance or a maximal set that can be tolerated without significant loss of performance. In this case, the chromosome includes a number of rules that specify certain faults as well as a set of initial conditions. Whereas Hunt's approach relates to functional testing, Schultz's approach assumes that the tester has full access to the structure of the code.

O'Dare and Arslan (1994) have used GAs to generate test patterns for VLSI circuits in searching for those patterns that detect the highest number of faults that remain in the fault list. The test set produced is passed to automatic test equipment for simulation to check the result. They concluded that the GAs were able to produce effective test sets with high percentage coverage of the faults.

Corno *et al.* (1996) have used GAs to generate test patterns automatically for large synchronous, sequential circuits. They achieved encouraging results for fault-coverage and conclude that GAs perform better than simulation or symbolic and topological approaches for large problems, in terms of both fault coverage and CPU time.

The pace of applying genetic algorithms to testing problems is increasing. The Software Testing group at York University is engaged on using genetic algorithms for structural and temporal testing of real-time software (Tracey *et al.*, 1998). A group in Vienna (Puschner and Nossal, 1998) is investigating worst case execution times using GAs. At the time of writing, Voas in the USA is preparing to publish some of his work in this area.

15.6 Future Developments

The demands of ever stricter quality standards for software are putting an increasing pressure on software engineers to develop repeatable processes for software development and for ensuring that the product is of high quality. The frustrations of unreliable software are acute in a culture that has become so dependent on computers, and totally unacceptable in situations where life and property are at stake. Whereas in the past, software testing has been the Cinderella of the software lifecycle because of its tedium and expense, in future, software testing tools will assume a central role. Genetic algorithms have already taken their place in the armoury of some industrial companies for routinely determining the worst case execution time of software. Genetic algorithms have proved their worth in deriving test sets automatically to test different aspects of the software, be they functional, structural, fault-based or temporal. Their application to integration testing is long overdue. Genetic algorithms have shown better performance on some occasions when they are combined with other techniques such as simulated annealing and tabu search, and with deterministic heuristics which in some circumstances may achieve a solution quickly. The integration of GAs with other approaches will be a fruitful line of research.

The key to the success of genetic algorithms in software quality is to incorporate them seamlessly into a CAST tool so that they provide coverage of the software reliably and repeatably with the minimum of human intervention. If this were ever achieved, it would amount to an enormous step forward. Genetic algorithms have come of age in the arena of software engineering, but there is still much to do.

Acknowledgements

I would like to acknowledge the constant support and friendship of Mr. D. E. Eyres; he and I jointly started the work at Glamorgan on the automation of software testing. I thank my research students over the years for their enthusiasm in developing ideas: Harmen Sthamer, Xile Yang, Hans Gerhard Gross, and Stephen Holmes. Mr Joachim Wegener of Daimler-Benz in Berlin has been an invaluable source of advice and has helped to make the projects relevant to the needs of industry. I have enjoyed many fruitful and interesting discussions with Dr Colin Burgess of Bristol University. Professor Darrel Ince of the Open University, first introduced to the challenge of automating software testing through one of his papers and through subsequent discussions.

16

Evolutionary Game Theory Applied to Service Selection and Network Ecologies

Sverrir Olafsson

16.1 Introduction

Today an ever increasing number of social and financial services can be provided over data networks. Individuals and businesses alike initiate and complete a large number of commercial transactions over the Internet or other specially constructed networks. Home banking, home shopping, video on demand, buying and selling of stocks and other financial securities can now be undertaken from almost any part of the world where an access to a local network can be achieved. New services are being offered all the time, and when successful face an immediate competition from a large number of companies and individuals with sufficient knowledge base, financial resources and access to the Internet.

The increasing competition in service provision on the Internet provides individuals with an increasing choice, lower prices and previously unknown opportunities. But this also creates greater difficulties in making the right choices at the right time. The problem is not only that of making the right choice, but also to acquire the necessary information so that educated decisions can be made. Clearly, all network users or agents browsing the network can in principle acquire a complete information on all competing network services. However, acquiring complete information can be excessively expensive, relative to the only marginal improvements it may provide the holder of that information. The problem therefore is to strike the balance between sacrificing information and the penalties one has to pay for making decisions when equipped with only limited information.

Also, the increasing speed with which services are introduced requires an increasing complexity of networks as well as the need for efficient distributed control mechanism. The supply of a multitude of continuously changing and interacting services requires the simultaneous or sequential use of many network resources, possibly distributed over locations large geographical distances apart. That requires reliable algorithms, which identify the required sources and support their efficient utilization. It is in principle possible to burden every network and network user with the communication overheads, which make knowledge of all network resources available. Such an overhead, on the other hand, would severely limit the processing capacity of the network and be very time consuming for the user. In principle, the most appropriate resource can be found for the execution of every task, but the computational expense and the time required for achieving that might well exceed the benefits from using that resource as compared with another, slightly less effective one. This problem is of a very general nature and relates to a number of optimization problems, as well as many management issues where performance has always to be compared with the cost of achieving it.

As a result of these problems soft, biologically based, approaches to resource management have become increasingly popular in recent years (Huberman, 1988). Predominantly these approaches have been based on a neural network approach and various implementations of genetic algorithms. For example, neural networks have been developed for resource allocation (Bousono and Manning, 1995) and for the switching of network traffic (Amin *et al.*, 1994). Genetic algorithms have been applied to resource allocation (Oates and Corne, 1998b), routing or file management in distributed systems (Bilchev and Olafsson, 1998a), just to mention a few. In this paper we take a different approach, based on evolutionary game theory. Game theoretic approach to network utilisation has been considered by a number of authors in recent years (Olafsson, 1995). Common to most of these approaches is the notion of an agent utility, which dictates the dynamics of the allocation of service requirements to network resources. The system performance resulting from this dynamic therefore depends strongly on the selected agent utility function, i.e. with what rationality criteria the agents have been programmed.

A potential problem with agent based approaches is the fact that agents sometimes consider only their own demand for processing requirements. Where each agent is guided only by its own interests, the performance of the whole system can still be very low. In view of this, it seems essential to design systems whose agents are concerned, not only for their own service requirements, but also those of the whole community of agents. This may requires some co-operation between agents, which is aimed at improving the mean efficiency of the whole system. Occasionally, this can be detrimental to the interests of a few agents, but the whole community of agents is likely to benefit from the co-operation. In fact, this problem is very similar to that of the prisoners dilemma (Axelrod, 1984) or the tragedy of the commons (Hardin, 1984), which are representative for many competitive situations in areas as diverse as, politics, commerce, marketing and real biological systems.

The approach taken here is based on evolutionary game theory (Maynard Smith, 1989). Agents distribute their service requirements, i.e. they select strategies on the basis of their personal utility function and their present knowledge of the available services. This knowledge includes present availability, price, quality and possible time delays. Generally, the agents are in the possession of only limited information and may therefore have to test the different service options. The service options are considered to be the strategies available to the agents. On the basis of the utility (fitness value) achieved by applying any set of strategies the agents try to evolve their service selection towards higher utility and lower cost. This process of strategy selection, leading to more efficient resource utilization, resembles the strategic behaviour of an animal community, which on the basis of what it learns evolves its strategies to a more efficient performance. Learning and the updating of strategies based on new information is hence an essential part of the approach shown here.

This approach allows us to monitor the system evolution in terms of game theoretic concepts such as Nash equilibrium. In the evolutionary game theory, the concept of Nash equilibrium has been extended to that of an Evolutionarily Stable Strategy (ESS) (Maynard Smith and Price, 1973) which is a strongly biologically motivated concept. The ESS gives a formalised definition of a strategy, which has the ability to resist the invasion of new strategies (phenotypes). A population or an ecosystem, which plays the evolutionarily stable strategy, can therefore not be invaded by opponents playing different strategies.

It is useful to comment on a few important differences between the evolutionary game theory approach discussed here and other evolutionary models, such as genetic algorithms (Holland, 1975; Goldberg, 1989), evolutionary strategies (Schwefel, 1981) and genetic programming (Koza, 1993). Evolutionary game theory is a local theory, whereas both genetic algorithms and evolutionary strategies are essentially centralized models requiring a simultaneous comparison of the fitness of all individuals existing at the same time and competing in the same environment. Even though evolutionary computing models have some parallel attributes, this mutual comparison of all individuals presents a potentially limiting bottleneck in their implementation. Recenetly there has been considerable work on more distributed implementations of genetic algorithms (Blickle, 1993), but the absence of a theoretical framework has been very noticeable. In evolutionary game theory the mentioned bottleneck can be avoided as each individual's fitness is compared against the aggregated mean fitness of all coexisting individuals (phenotypes). The information required for strategy update is therefore local because no centralized procedure is needed. Evolutionary game theory could thus provide some theoretical basis for analyzing distributed systems.

This chapter is organized as follows. In section 15.2 we give a brief description of genetic algorithms and how by applying schema theory these can be formulated as a dynamic replicator algorithm. In section 15.3 we provide a brief discussion of evolutionary game theory, followed by a section on basic replicator dynamics. Section 15.5 formulates evolutionary game theory as a replicator dynamic, and in section 15.6 we discuss in general terms how evolutionary game theory can be applied to resource allocation on networks. Section 15.7 looks at the links between evolutionary game theory and dynamic models driven by general agent expectations. In section 15.8 we discuss how load distribution on server networks emerges from an expectation dynamic, and give sufficient conditions for this load distribution being stable. In this section we also discuss two examples, and give the conditions for stable load distributions in both cases. We end with general conclusions and comments on the outlook for this approach to service selection on networks.

16.2 Connecting Genetic Algorithms and Evolutionary Game Theory

In spite of implementation differences genetic algorithms and evolutionary game theory share some very fundamental analogies, based essentially on the principles of a replicator dynamic. These similarities are in fact so fundamental that both models can be viewed as two different dynamic implementations of a general replicator dynamic. Of course this is not surprising as both models have their roots in two biological domains, one considering the molecular basis of evolution whereas the other studies the competition between species (phenotypes) in an ecological context. In this section we will discuss these similarities and demonstrate how both algorithms lead to a mean fitness increasing dynamics.

Genetic Algorithms (GA) (Holland, 1975; Goldberg, 1989) draw upon principles from the theory of natural selection and were initially introduced as a possible model of real evolution. It was only later realized that genetic algorithms could successfully be applied to a number of different optimisation problems (De Jong, 1975). The basis for the application of GA to optimisation is the fact that they define a search strategy that is biased towards search points of high fitness, i.e. points, which tend to improve the value of the cost function. GAs have similarities with simulated annealing (Geman and Geman, 1984) in that they employ random search strategies. The main difference between the two methods, however, is the fact that GAs have intrinsic parallel features which render them powerful in solving a number of important optimisation problems. GAs have turned out to be particularly valuable in dealing with problems where the external circumstances cannot be modeled precisely.

Within the framework of GAs and evolutionary models in general, evolution can broadly be viewed as a two-step process of random structure modification followed by selection. Random mutations or cross over processes generate new structures, or modify existing genes, whose phenotypic manifestations are then tested against the requirements of the environment. However, in the language of optimisation each randomly generated structure offers a solution to the defined cost function. Depending on how good that solution is, when compared with other already known solutions, it is accepted or rejected by the selection process. GAs distinguish themselves from other optimization algorithms in that at any moment in time they maintain a collection of generated structures, or points in search space, rather that simply one point. All these points or structures are competing solutions to the optimization problem. It is the implementation of the selection algorithm that decides which structures to keep and which ones to get rid of.

The fundamental entities in most implementations of GA are strings of variable or fixed length N. In the case of fixed length strings the state space can be presented as the set:

$$\Omega_N = \{0,1\}^N$$

of all binary strings of length N. Within a given problem framework each element in this space can be associated with a given fitness; hence there is a fitness function:

$$f : \Omega_N \to \Re$$

where $\Re$ is the set of real numbers. The objective of a GA is therefore to find the optimal value of this fitness function over the space of binary strings:

$$f_m = \max_{s \in \Omega_N} f(s)$$

The collection of all strings kept at any moment in time defines the present population of strings. All members of this population compete to be accepted as a solution to the given problem.

Generally, it is not known which regions in Ω_N are likely to provide good or even acceptable solutions to a given problem. In the absence of any such knowledge the best one can do is to generate randomly an initial population of strings. Assume that the initial population size is M then the initial population at time $t=0$ is presented by a $N \times M$ matrix:

$$S(0) = (s_1(0), ..., s_M(0))$$

In this notation each position $s_i(0)$ presents an N component binary string:

$$s_i(0) = (s_{i,1}(0), ..., s_{i,N}(0))$$

From this initial population subsequent populations are generated by the use of mutation and/or crossover followed by a selection. Formally we write for this process:

$$S(0) \rightarrow S(1) \rightarrow ... \rightarrow S(T) \rightarrow ... \qquad (16.1)$$

The competitive pressure in the evolutionary process is based on the fact that the population size is usually kept constant at each generation. Of the generally large number of offspring produced by each generation, only a few are selected to survive and to produce their own offspring. Assume that at some time t we have the population $S(t)$ still of size M. After mutations and crossovers, within this population, offspring are generated and the population increases in size. From the new enlarged population only M are allowed to survive to produce new offspring. The selection of survivors to be kept for further breeding generally depends, in a stochastic manner, on the fitness of the candidates. Each member of the population, say $s_i(t)$, has some fitness $f_i(t)$ and his survival into the next generation depends upon this fitness value. Essentially the chance of survival to the next generation is given by the expression

$$p_i = \frac{f_i}{\sum_j f_j} \qquad (16.2)$$

Therefore, the fitter an individual is the more likely it is to survive to produce offsprings by a mutation or a crossover with other fit individuals, which were also selected for reproduction.

16.2.1 Schema Theory

Consider a binary string where some positions are fixed, whereas the remaining positions can take arbitrary values, i.e. either 0 or 1. To follow a general convention we take the symbol * to represent 0 or 1. An example of a string in this notation is $s = \{10*1**01\}$. A schema H is defined as a subset of strings in the space of strings Ω_N. The structure of the schema is provided by the fact that certain positions in the strings that belong to H are fixed. The remaining positions, indicated by $*$ are variable. An example of a set of four bit strings belonging to the same schemata $H = (*11*0)$ is given by the set:

$$S = \{(01100), (01110), (11100), (11110)\} \tag{16.3}$$

$H = \{*11*0\}$ is therefore a compact notation for the set S.

As the strings belonging to given schema are presented by three symbols $\{0,1,*\}$ there are in total 3^N different possible schemata. This is in fact much larger than 2^N the number of possible strings. Therefore, any given string will belong to 2^N different schemata.

To develop a dynamic based on preferential selection of high fitness let $m(H,t)$ be the number of strings, which at time t belong to schema H. Remember, $S(t)$ presents the whole set of strings in existence at time t. Therefore the number of strings $m(H,t)$ has to be in the intersection $H \cap S(t)$. Each set of strings can be presented in terms of a smaller set of schemata classes. Formally, one can decompose any set of strings into a set of disjoint schemata classes:

$$S = \bigcup_i H_i$$

Now, let $f(H,t)$ be the average fitness of strings in $H \cap S(t)$ and $\langle f(t) \rangle$ the average fitness of all strings in the population $S(t)$ at time t. It can be shown (Goldberg, 1989) that the number of strings still belonging to schema H at time $t + 1$ is given by:

$$m(H,t+1) = m(H,t) + \Delta m = \frac{f(H,t)}{\langle f(t) \rangle} m(H,t) \tag{16.4}$$

As we are mainly interested in the change in the number of strings belonging to the schema H, we rewrite this expression as:

$$\Delta m(H,t) = \left(\frac{f(H,t) - \langle f(t) \rangle}{\langle f(t) \rangle} \right) m(H,t) \tag{16.5}$$

From this it is obvious that as long as the fitness of strings in H is larger than the mean fitness of all strings the change is positive, i.e. $\Delta m(H,t) > 0$. Otherwise, the population of strings belonging to H is reduced.

There is, however, a small caveat to this argument as the genetic operations can remove some of the strings from the schema H. It is possible to provide some analysis on the type of operations, which leave the schema structure H invariant. For that purpose one defines two new quantities. The order of a schema, $o(H)$ is simply the number of fixed position on the string. The defining length of a schema $\delta(H)$ is the distance between the first and the last fixed positions on the schema. It can be shown (Goldberg, 1989) that a schema is invariant with respect to a cross over when the cross over position is chosen outside the strings defining length. If P_c denotes the probability with which the crossover operator is applied, then:

$$p_{s,c} = 1 - \frac{P_c \delta(H)}{N-1} \qquad (16.6)$$

is the probability that the string survives the crossover operation. Similarly, if P_m is the probability of a mutation then:

$$p_{s,m} = (1 - P_m)^{o(H)}$$

is the probability that a string will still belong to the schema H after the effects of a mutation. From this we find that the expected number of strings belonging to the schema H after the application of crossover and mutation is given by:

$$m_{c,m}(H,t) = p_{s,c} p_{s,m} m(H,t) < m(H,t) \qquad (16.7)$$

For small values of P_m we have $(1 - P_m)^{o(H)} = 1 - o(H)P_m$ and therefore:

$$p_{s,c} p_{s,m} = \left(1 - \frac{P_c \delta(H)}{N-1}\right)(1 - o(H)p_m)$$

Multiplying out the bracket and ignoring the term containing $P_c P_m$ we have:

$$p_{s,c} p_{s,m} = 1 - \frac{P_c \delta(H)}{N-1} - o(H) p_m \qquad (16.8)$$

Equation 16.5 has to be modified slightly to accommodate for the effects of these probabilities. One finds:

$$m(H, t+1) \geq \frac{f(H,t)}{\langle f(t) \rangle}\left(1 - \frac{P_c \delta(H)}{N-1} - o(H)P_m\right)m(H,t) \qquad (16.9)$$

If $M(t)$ is the total population at time t then we can look at:

$$p(H,t) = \frac{m(H,t)}{M(t)}$$

as the probability that a randomly selected string from the total population belongs to the schema H. In terms of this probability we rewrite equation 16.9 as:

$$p(H,t+1) \geq \frac{f(H,t)}{\langle f(t) \rangle}\left(1 - \frac{P_c \delta(H)}{N-1} - o(H)P_m\right)p(H,t) \qquad (16.10)$$

The interpretation of this equation is as follows. The expression $p(H,t)$ gives the probability that a string selected at time t belongs to the schema H. The expression in the large bracket is the probability that the children of this selected string will also belong to the schema H. The probability that these children belong to H is further biased by:

$$\frac{f(H,t)}{\langle f(t) \rangle}$$

which measures the ratio of their fitness to the mean fitness of the whole population of strings. Finally, the change in probability of a string belonging to schemata H from one generation to another can be written as

$$\Delta p(H,t) \geq \frac{p(H,t)}{\langle f(H,t) \rangle}\left(f(H,t)\left(1 - \frac{p_c \delta(H)}{N-1} - o(H)p_m\right) - \langle f(H,t) \rangle\right) \qquad (16.11)$$

If the probability that a string leaves a schema as a result of crossover or mutation can be neglected the above expression simply reads:

$$\Delta p(H,t) = p(H,t)\left(\frac{f(H,t) - \langle f(H,t) \rangle}{\langle f(H,t) \rangle}\right) \qquad (16.12)$$

Equations 16.11 and 16.12 demonstrate the basic principle of a replicator dynamic which is to be interpreted in the following way. The fitness of each schema is to be compared with the mean fitness of the whole community of strings. If the schema fitness turns out to be larger than the mean fitness of all strings then the number of strings belonging to that schema class is increased, i.e. it is rewarded by a higher number of offspring. Structurally the equations 16.11 and 16.12 are very similar to the dynamic replicator equations considered in evolutionary game theory as will be explained in the next section.

16.3 Evolutionary Game Theory

The application of game theory to evolution goes back to the works of Lewontin (1960) and later Slobodkin and Rapoport (1974). These authors used game theory mainly to analyse the struggle of a species against nature. In the game, a species seeks to construct strategies, which maximise its fitness (probability of surviving and producing offspring) under given environmental conditions. Later, Maynard Smith and Price (1973) extended the theory and applied it to animal conflict situations as they occur in territorial claims or competition for mates or dominance rights. Central to their approach to evolutionary game theory is the analysis of pairwise contests between opponents. In every contest each opponent can apply only a limited number of strategies or different combinations of strategies.

An essential feature of evolutionary game theory is its dynamic nature. The probability distribution on the set of available strategies evolves as a result of how successful the strategies are. Maynard Smith and Price (1973) introduced the concept of evolutionarily stable strategy as an uninvadable strategy. They demonstrated that, in a number of cases, the strategy distribution eventually evolved towards an evolutionarily stable strategy. This concept has been immensely popular amongst evolutionary biologists. It has also been successful at explaining some evolutionarily stable conditions such as the sex ratio in many species (Hamilton, 1967). More recently evolutionary game theory has also received some interest from game theorists and economists (Weibull, 1995).

The application of conventional game theory to economics and decision theory has not been as successful as some of the field's initiators had hoped for. It was indeed noted by von Neumann that some of game theory's most serious limitations were due to its lacking of dynamics. Clearly, some economists and game theorists have, therefore, watched with interest the emergence of the dynamic evolutionary game theory. Also, game theory has been applied to resource allocation and routing on communication and data networks. In view of the rapid changes that are taking place in the provision of networked services in recent years it appears that dynamic modeling is more appropriate for dynamically changing service environment. In this environment, users have to make decisions as to what services are most suitable to satisfy their needs.

This is why it was suggested in Olafsson (1996) that evolutionary game theory may be well suited for the modeling of modern service provision. But also evolutionary game theory is not entirely without problems. In spite of its dynamic nature the concept of equilibrium plays a fundamental role. Even though evolutionarily stable strategy has a well defined meaning in ecological systems (Maynard Smith, 1989) its meaning in competitive markets and service provision systems is far from clear. Also, what decision-making processes in human, market, social or service selection scenarios are responsible for the strategy distribution moving towards an evolutionarily stable strategy? Or, in more basic terms, under what general conditions can one expect market competitors and/or service selectors to play the evolutionarily stable strategy?

When this criticism is considered, one has to be aware that the concept of equilibrium (evolutionarily stable or not) is, even in evolutionary theory a difficult one as evolution demonstrates a process of continuous change. It does nevertheless supply the evolutionary biologist with valuable tools for the study of animal to animal or animal to environment conflict situations. Even though the system trajectories are generally far more difficult to analyse than are the equilibria, they contain important information on the dynamic of the learning process. Some studies on the learning process leading to evolutionarily stable strategies have been undertaken in recent years (Zeeman, 1981).

The user service selection process is naturally modeled as a learning process. With an increasing speed with which products and services are designed and then brought to market, accompanied by an ever shorter life expectancy, the continuous consumer learning process becomes an ever more important issue. This dynamic and volatile environment calls for a dynamic and adaptive modeling. In Olafsson (1995a; 1996) we have argued that these scenarios are appropriately modeled by evolutionary game theory, where both the transitory learning dynamics and the resulting equilibria may be of considerable interest.

There is one final point worth making. Whereas conventional game theory has a clear notion of a player which when facing other players applies some given strategy, evolutionary game theory puts the main emphasis on the strategy itself. There is no clear definition of a 'player'. The player concept is replaced by a large population of players, which can select from a finite number of strategies. At any time, when the whole population of players is in a certain state of a 'mixed strategy', fitness can be associated with this mixed strategy, and therefore the whole population, as well as to individual strategies. A player, if one wants to hold on to the concept, becomes in principle only an anonymous member of the population, which applies any of the strategies with some probability.

16.4 Replicator Dynamics

In this section we give a brief description of the concepts, problems and ideas behind the application of evolutionary game theory to the analysis of strategy evolution in ecological systems, as developed by Maynard Smith and Price (1973). As in most evolutionary systems the model discussed here describes an evolutionary system driven by increased fitness expectations. Generally fitness is not simply some kind of a deterministic function but a probabilistic linear or non-linear combination of some options available to an individual. We demonstrate in this section that evolutionary game theory is simply one instance of an application of competitive replicator dynamics, where the fitness of each individual is compared against the mean fitness of the community it belongs to. A dynamic of this kind is generally (not always) mean fitness increasing, i.e. the mean fitness of the community tends to increase under the evolutionary dynamics.

We start with a set of possible strategies presented by a set $S = \{s_1, s_2, ..., s_N\}$. We assume that individuals have to choose from this set of possible strategies. Assume that the fitness of strategy s_i is given by f_i and that the vector $X = (x_1, x_2, ..., x_N)$ gives the distribution of the population over the different strategies, i.e. x_i is the fraction, which selects strategy s_i. Generally, the fitness depends upon this distribution, $f_i = f_i(X)$. The mean fitness of all individuals is given by:

$$\langle f(X) \rangle = \sum_{i=1}^{N} x_i f_i \tag{16.13}$$

Each individual's fitness $f_i = f_i(X)$ will be measured against this mean fitness. The difference:

$$\Delta_i = f_i - \langle f \rangle$$

gives the fitness individuals selecting s_i have in excess of the mean fitness of all individuals. Only if $\Delta_i > 0$ will selecting s_i be more rewarding than following the selection mixture as presented by the vector $X = (x_1, x_2, ..., x_N)$. On the other hand, if $\Delta_i < 0$ then the pure strategy s_i is weaker than the mixed strategy $X = (x_1, x_2, ..., x_N)$. In evolutionary terms Δ_i can be viewed as the growth rate, positive or negative, of the ith strategy or phenotype. The selection dynamic itself can then be written as:

$$\dot{x}_i = x_i \left(f_i - \langle f \rangle \right) \qquad (16.14)$$

Restricting the dynamic to the unit simplex, i.e. setting:

$$\sum_{i=1}^{N} x_i = 1$$

sustains a competitive pressure such that an increase in the selection of one option decreases the fraction of individuals selecting some other option. This follows from the fact that the dynamic (equation 16.14) is norm preserving. It is straightforward to establish that the equilibrium states for the dynamic (14) have to satisfy one or the other of the two conditions:

$$x_i = 0; f_i \neq \langle f \rangle \qquad (16.15)$$

$$x_i \neq 0; f_i = \langle f \rangle \qquad (16.16)$$

In other words, in equilibrium all contributing strategies have the same fitness, which again are the same as the system's mean fitness. In these circumstances there is no incentive to change strategy.

Note that the set of equations (16.14) has a very similar structure to the population equations derived from the schema theory in GA. The only difference is that in the GA framework the fitness function for a schema H has to be discounted by their probability of surviving the crossover and/or the mutation process. Also, in the case of GA the right hand side of the replicator equation is divided by the mean fitness of the whole population. As long as the mean fitness is scaled in such a way that it does not become negative, both sets of equations, i.e. with or without the division by the mean fitness, have the same dynamic and attractor properties (Olafsson, 1995a).

16.5 From Replicator Dynamics to Evolutionary Game Theory

In this section we will briefly describe how evolutionary game theory can be placed in a replicator dynamic context. Let $S = \{s_1, s_2, ..., s_N\}$ be the set of strategies available to all individuals and let $X = \{x_1, x_2, ..., x_N\}$ define the present probability distribution over the strategy set. The first step to introduce game theoretic principles into the picture is to define a utility matrix whose elements:

$$U = \{U_{i,j}; i, j = 1, ..., N\} \qquad (16.17)$$

specify the payoffs, or gains, expected from a contest between individuals. Their precise meaning is as follows: U_{ij}, $i, j = 1, ..., N$ is the payoff to an individual applying strategy s_i against an individual using strategy s_j. Note that this definition does not imply that the gain matrix is symmetric as in general, $U_{ij} \neq U_{ji}$ for $i \neq j$. Essentially, the introduction of the utility matrix allows for the mixing of strategies.

Assuming that the population as a whole selects the strategies as described by the distribution $X = \{x_1, x_2, ..., x_N\}$, then the fitness of an individual selecting strategy s_i is given by the expression:

$$f_i = \sum_{j=1}^{N} U_{i,j} x_j \tag{16.18}$$

where the weighting of each utility component is given by the corresponding probability. Therefore, the mean fitness of the whole population is given by the quadratic form:

$$\langle f \rangle = \sum_{i=1}^{N} x_i f_i = \sum_{i,j=1}^{N} x_i U_{i,j} x_j \tag{16.19}$$

Generally, the replicator dynamic tends to increase the mean fitness of the community, but there are some exceptions to that, see Olafsson (1996). However, in the special case when the utility matrix is symmetric, i.e. $U_{ij} = U_{ji}$ for all i and j, then the quadratic form (equation 16.19) is a Lyapunov function for the replicator dynamics, i.e.

$$\frac{d}{dt} \langle f \rangle = \frac{d}{dt} (xUx^t) \geq 0 \tag{16.20}$$

It is essentially this fact that makes the evolutionary game theory a useful tool for tackling optimisation problems.

16.6 Evolutionary Games and Network Ecologies

We consider a network of servers, which hold databases and other service provision facilities accessible to a large number of users. Generally each user will only have some limited information about the services available on the network. This information can be achieved through direct network exploration or through the evaluation of bids received from network agents. Figure 16.1 shows an example of such a network of services.

We consider a collection of users (agents) seeking to allocate their service requirements to service providers subject to limited information. We assume that the system consists of N distributed servers. The way in which the users distribute their tasks on the servers is given by the fractional load vector $X = (x_1, ..., x_N)$, where $x_i = x(s_i)$ is the load allocated to server s_i. If a user opts for server s_i, when the total load distribution is given by X, then the utility received by that action has to be measured against the mean utility associated with the total load distribution. If the utility from selecting s_i is given by $f_i(X)$ and the utility associated with the total distribution X is given by $f_X(X)$, then the further selection of server s_i is given by a replicator dynamics of the type of equation 16.14.

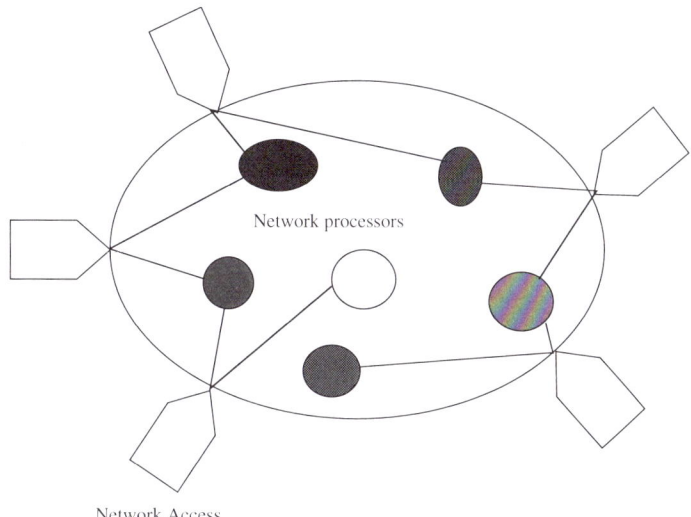

Figure 16.1 A schematic presentation of a network of services (network processors) that can be accessed at various points by user agents. The agents distribute their service requirements on the network processors on the basis of their perceived utility.

The way in which the user receives information on the availability and the usage condition of different servers is in fact not important for the implementation of the evolutionary dynamics. Users can ask for bids from service providers and then compare them before making a decision. However, they can just as well actively explore available network services and compare their findings. Similarly, it is possible to work on the basis of secondary findings, like information from other users, or simply make a decision on the basis of past experiences, i.e. the user would run some kind of an expectation based dynamics.

There are various quantities and qualities of a service the user will be particularly concerned about. These include, price, quality and waiting time. It is possible that the user receives information about the workload x_i on a particular server s_i. This information can be supplemented, for example, by the expected waiting time $t_i = t(x_i)$ and the likely quality of service $q_i = q(x_i)$, both as functions of the load. The information I_i regarding the state of one service provider s_i is then presented by a quantity of the form:

$$I_i = \{x_i, t(x_i), q(x_i), ...\} \qquad (16.21)$$

For each contacted server, or each server on whom some information is available, a similar quantity can be written down. Assuming that M servers have been contacted, the information available to the user can be written as:

$$I = (I_1, ..., I_M) \qquad (16.22)$$

A user could well have clearly defined criteria, which would enable him to make a straightforward server selection. Generally, the situation is not that simple as the selection is based on a number of different criteria which may be so complex that the user is unable to understand the implications of his choice. This is where the replicator model can serve as a useful tool for the decision making process.

To make a comparison between the different service providers the user will be interested in the anti-symmetric criteria function:

$$\Lambda_{i,j} = I_i - I_j = \{x_i - x_j, t(x_i) - t(x_j), q(x_i) - q(x_j), ...\} \tag{16.23}$$

which presents a hyper-surface in multidimensional criteria space. Generally different users will have different criteria functions. For example, for one user the quality might be more important than the price whereas some users might be more concerned about the price of the service.

The information contained in equation 16.23 can be used in many different ways. With appropriate time and quality functions it can be used to define the user utility for individual service providers. Instead of using equation 16.23 directly we can construct a function of $\Lambda_{i,j}$ as a candidate for a utility function. One example is:

$$U(\Lambda_{i,j}) = U_{i,j} = \frac{1}{\sqrt{2\pi\sigma_{i,j}^2}} \int_{-\infty}^{\Lambda_{i,j}} \exp\left(-\frac{\Lambda_{i,j}'^2}{2\sigma_{i,j}^2}\right) d\Lambda_{i,j}' \tag{16.24}$$

This accumulative distribution has the following interpretation: when making a choice between the two servers i and j $U_{i,j}$ presents the probability that i is selected and therefore $U_{j,i} = 1 - U_{i,j}$ is the probability that the user selects j. The variance $\sigma_{i,j}$ presents the uncertainty involved when comparing the utility from using the two servers. In the extreme case when $\sigma_{i,j} \to 0$ the decision function becomes deterministic, i.e.

$$U(\Lambda_{i,j}) = \begin{cases} 1 & \text{i.e. select } s_i \text{ when } I_i > I_j \\ 0 & \text{i.e. select } s_j \text{ when } I_j > I_i \end{cases} \tag{16.25}$$

In the next section we will develop the necessary concepts that enable us to cast the decision making process into a framework appropriate for the evolutionary game theory.

16.7 Expectation Dynamics as an Evolutionary Game

Let $X = (x_1, ..., x_N)$ present the anticipated load distribution on the contacted servers and let $U_{i,j}$ be some arbitrary utility function dependent on the anti-symmetric criteria function (equation 16.23). Note that equation 16.24 only presented one instance of utility function. In the following, the structure of the utility function is arbitrary. We interpret the expression

$$E\{U_i, X\} = \sum_{j=1}^{N} U_{i,j} x_j \qquad (16.26)$$

as the expected utility from using server s_i. Therefore, the expected mean utility to all users subscribing to the servers according to $X = (x_1, x_2, ..., x_N)$ is given by:

$$\langle E(U, X) \rangle = \sum_{i,j=1}^{N} x_i U_{i,j} x_j \qquad (16.27)$$

The dynamic we set up for the service selection is based on the notion that each user has some understanding of the expected mean utility from the usage of services from a given category. When considering a particular service it will be compared against this expected mean utility. From section 16.4, the replication rate for the considered service is given by:

$$\lambda_i = E\{U_i, X\} - \langle E\{U, X\} \rangle \qquad (16.28)$$

We view this expression as the expected growth rate of service s_i. From this we derive the explicit form for the service selection dynamic as:

$$\dot{x}_i = x_i \left(E\{U_i, X\} - \langle E\{U, X\} \rangle \right) \qquad (16.29)$$

Here we have not made any specific assumptions about $U_{i,j}$. It is simply some matrix quantity, which codes the information available to the user on the basis of which he makes a service selection. Note that the growth rate (equation 16.28) depends upon the distribution $X = (x_1, x_2, ..., x_N)$, generally in a non-linear manner. A service can initially be attractive but then, as its general usage increases to and above some critical, service-dependent value its growth rate may well turn negative. Similarly, a service may have the feature of becoming ever more popular the more it gets chosen by the general user. The dynamic of many different utility functions has been studied in Olafsson (1996). Before we continue with the general development, we consider one particularly simple example.

16.7.1 A Simple Example

Consider a network server with capacity c_i and fractional load distribution x_i. We define the expected user utility for this server as $E_i(c_i, X) = c_i - x_i$ and the mean utility as

$$\langle E(C, X) \rangle = \sum_{k=1}^{N} x_k (c_k - x_k) \qquad (16.30)$$

The replicator dynamics for the server system is given by:

$$\dot{x}_i = x_i \left(c_i - x_i - \sum_{k=1}^{N} x_k (c_k - x_k) \right) \quad (16.31)$$

Either one of the two following conditions is sufficient for the dynamics to reach an equilibrium:

$$x_i = 0 \text{ and } E_i(c_i, X) < \langle E(C, X) \rangle \quad (16.32)$$

$$x_i > 0 \text{ and } E_i(c_i, X) = \langle E(C, X) \rangle \quad (16.33)$$

It follows that if $x_i, x_j > 0$, then:

$$E_i(c_i, X) = E_j(c_j, X) = \langle E(C, X) \rangle$$

Equation 16.33 implies the following relationship between server load and capacity:

$$c_i - c_j = x_i - x_j \quad (16.34)$$

$$c_i - x_i = \sum_{k=1}^{N} x_k (c_k - x_k) \quad (16.35)$$

As the fractional load distributions lie within the unit simplex these conditions can only be satisfied if the server capacities are scaled to satisfy the same normalization condition:

$$\sum_{k=1}^{N} c_k = 1$$

The Jacobian for the dynamic of equation 16.31 evaluated at $x_i = c_i$; $\forall i$, is:

$$\Omega|_{X=C} = \begin{pmatrix} c_1^2 - c_1 & c_1 c_2 & & c_1 c_N \\ c_2 c_1 & c_2^2 - c_2 & & \\ & & & \\ c_N c_1 & & & c_N^2 - c_N \end{pmatrix} \quad (16.36)$$

As each of the diagonal elements in Ω is negative, $\Omega_{i,i} = c_i^2 - c_i < 0; \forall i$ the condition $x_i = c_i$ provides a stable load distribution. As stability is an important issue for the dynamic implementation of service allocation, we will analyse it in some detail in the following section.

16.8   Stability of Load Distribution

Consider the case of two different load distributions $X = (x_1,..., x_N)$ and $Y = (y_1,..., y_N)$ on a server system. The question one can ask is which distribution supplies the users with the higher utility value? To be able to answer that question, we need to generalize the definitions from the previous section in the following way. Let:

$$E_X\{U,Y\} = XUY^t \qquad (16.37)$$

be the expected utility to users complying to the X load distribution when the system is presently loaded according to Y. Then the quantity:

$$M(X,Y) = E_X\{U,X\} - E_Y\{U,X\} \qquad (16.38)$$

gives the efficiency of the X-distribution when compared with the Y-distribution. The following statement can be proved (Olafsson, 1996): The load distribution X is preferred to the distribution Y if one of the following conditions is satisfied:

$$M(X,Y) \geq 0 \qquad (16.39)$$

$$M(X,Y) = 0 \text{ and } \nabla_X M(Y,X) \cdot (Y - X) > 0 \qquad (16.40)$$

We now give necessary and sufficient stability conditions when utility depends upon the load distribution. Let $U_{i,j}; i, j = 1,2,..., N$ be the utility matrix for the system from which the expected utility can be constructed, and assume that the load distribution dynamic is given by a replicator dynamic of the type of equation 16.29. Then the resulting dynamic system is stable in an equilibrium distribution X_0 if the real parts of the eigenvalues of

$$\Omega_{i,j}(X_0) = \Delta_{i,j}(X_0) + x_{0,j}\left\{\sum_{k=1}^{N}\frac{\partial U_{i,k}}{\partial x_j} - \sum_{k,l=1}^{N} x_k\left(\frac{\partial U_{k,l}}{\partial x_j}\right)x_l\right\}_{X=X_0} \qquad (16.41)$$

with:

$$\Delta_{i,j}(X) = x_i\left(U_{i,j} - \sum_{k=1}^{N}(U_{j,k} + U_{k,j})x_k\right) \qquad (16.42)$$

are negative (Olafsson, 1996). Notice that if the utility does not depend upon the load distribution, i.e. if:

$$\frac{\partial U_{i,j}}{\partial x_k} = 0; \forall i, j, k$$

then equation 16.41 takes on the much simpler form, $\Omega = \Delta$. For a more thorough discussion of the stability properties of the expectation dynamic, see Olafsson (1996).

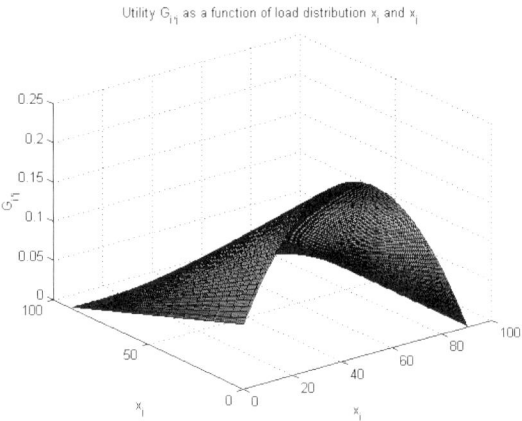

Figure 16.2 The utility function (equation 16.43). The surface is parameterized by the two variables x_i and x_j.

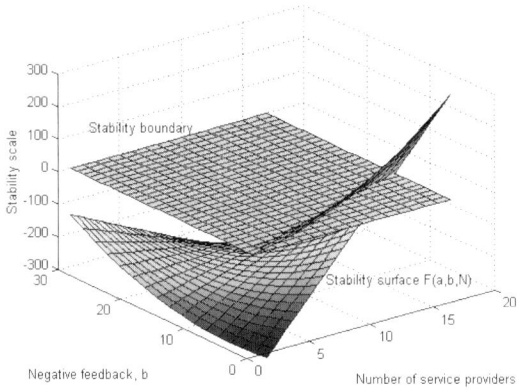

Figure 16.3 The figure displays the conditions for stable load distributions. The part of the curved surface below the horizontal stability boundary consists of those parameter combinations which lead to stable load distribution.

16.8.1 Example

We consider an example, which is of some relevance for a number of applications. In this case the utility a user receives from the usage of a service increases with its general usage but only up to a certain critical value. After that value is exceeded, the utility from the service is reduced. One way to model a utility function of this type is by:

$$G_{i,j} = (ax_i - bx_i^2)(1 - x_j) \qquad (16.43)$$

The utility as a function of two x variables is presented in Figure 16.2.

From the form of the utility function it is clear that the utility to selector of strategy s_i increases with x_i up to the critical value $x_{i,c} = a/2b$ beyond which the benefits from using the service are reduced. Olafsson (1997) proves that the emerging load distribution is stable provided that the following condition is satisfied, where N is the number of servers:

$$F(a,b,N) = aN^2 - 2(a+b)N + 3b < 0 \qquad (16.44)$$

The surface graph, Figure 16.3, gives the condition for stability for fixed a and running values for N and b. The parts of the curved surface which lie below the flat plain provide parameter combinations which lead to a stable load distribution.

16.8.2 Example

We consider a network of five different servers that can be accessed by a large number of users. Consider the case of a fixed utility matrix given by:

$$H_0 = \frac{1}{5} \begin{pmatrix} -1 & 7 & 3 & 5 & 4 \\ 3 & 2 & 5 & 3 & 2 \\ 4 & 3 & -1 & 2 & 3 \\ 4 & 2 & 3 & 1 & 4 \\ 2 & 4 & 3 & 3 & 1 \end{pmatrix} \qquad (16.45)$$

In this case the users do not negotiate the execution of their requirements, but simply allocate them to service providers as dictated by the fixed utility (equation 16.45). Running the replicator dynamics with this utility results in the load distribution of Figure 16.4.

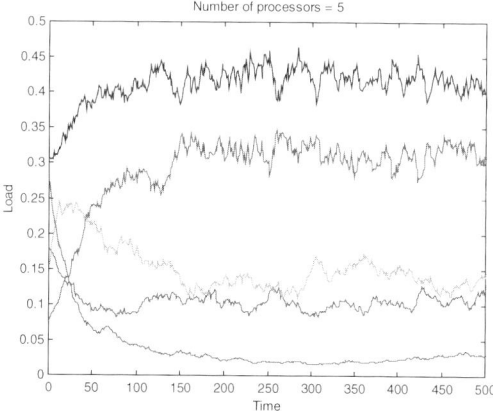

Figure 16.4 Load distribution based on the fixed utility matrix (). The large variance in the load distribution implies poor resource utilization.

The jiggery load curves result from the system noise, which is used to model the limited information available to the agents. The noise which is normally distributed is added to the utility as follows, $H_0 \rightarrow H_0 + \alpha\eta$, with $\alpha = 0.5, \eta \in N(0,1)$. The load configuration in Figure 16.4 is stable but it does not provide an efficient usage. The mean usage of the five servers over the considered time is $x_m = (0.41, 0.29, 0.11, 0.15, 0.04)$ with a standard deviation $std(x_m) = 0.15$.

To improve user utility an exploration of available servers and a negotiation with service providers is enabled. This activity is coded for by supplementing the utility matrix H_0 with an evolving term G, which dynamically adjusts utility to the evolving load distribution. We consider the case where the evolving utility matrix is given by the expression:

$$G_{i,j} = \frac{1}{1 + \exp(-\beta(e_j x_j - e_i x_i))} \quad (16.46)$$

When running the replicator dynamics with G alone, the resulting equilibrium distribution satisfies the following condition:

$$e_j x_j - e_i x_i = 0 \quad (16.47)$$

as in this case all servers are estimated to provide the same utiliy. The components e_i quantify the inverse demand for the service s_i. We evaluate the derivatives of the utility:

$$\frac{\partial G_{i,j}}{\partial x_k} = G_{i,j}(1 - G_{i,j})[e_i \delta_{i,k} - e_j \delta_{j,k}] \quad (16.48)$$

from which we derive the special cases:

$$\frac{\partial G_{i,j}}{\partial x_k} = \begin{cases} 0, & \text{if } k \neq i \ \& \ k \neq j \\ G_{i,j}(1 - G_{i,j})e_j, & \text{if } k = j \\ -G_{i,j}(1 - G_{i,j})e_i, & \text{if } k = 1 \end{cases} \quad (16.49)$$

It is clear that the utility from using server i increases when x_j increases by fixed x_i. Similarly, the utility from using i decreases with increasing x_i by fixed x_j. The changing attractions are proportional to the inverse demands e_i.

After some lengthy calculation, we find for the stability matrix:

$$\Omega_{i,j} = x_i(G_{i,j} - \delta_{i,j}) + x_i \left(e_i \delta_{i,j} \sum_k G'_{i,k} x_k - G'_{i,j} e_j x_j \right) - x_i \sum_l (x_j G'_{j,l} e_j x_l - x_l G'_{l,j} e_j x_j)$$

$$(16.50)$$

where $G'_{i,j} = G_{i,j}(1 - G_{i,j})$. From this we find that the condition:

$$e_i \sum_k G'_{i,k} x_k (1 - 2x_i) < 1 ; \forall i \qquad (16.51)$$

is sufficient for the equilibrium being a stable distribution.

It is possible to use the equilibrium condition and the interpretation of e_i to construct a utility matrix, which improves the user utility received from H_0 alone. We will briefly describe how this can be done, but refer to Olafsson (1997a) for details. If H_0 alone determines the load distribution we end up with the load on the kth component being proportional to:

$$x_k \sim \sum_i H_{k,i}$$

Because of this, we make the following assumption for the service demand components:

$$e_k = \sum_j H_{k,j}$$

Now simulating the system with the utility matrix $U = H_0 + G$ leads to different results. Figure 16.5 shows the distribution resulting from H_0 in the time interval $[0,500]$ and for the modified utility matrix $U = H_0 + G$ for the time interval $[501,1000]$.

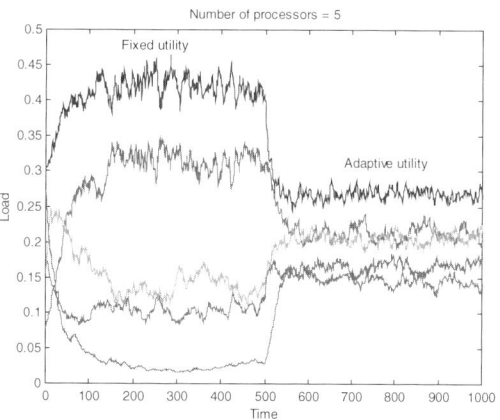

Figure 16.5 Load distribution with a fixed utility leading to a large standard deviation, $t = 0$ to 500, followed by a load distribution with adaptive utility leading to more efficient utilization, $t = 501$ to 1000.

The mean load distribution in the time interval $[501,1000]$ turns out to be $x_m = (0.27, 0.22, 0.15, 0.20, 0.16)$. This leads to considerably lower variance in the load distribution, a standard deviation of $std(x_m) = 0.049$ as opposed to 0.15, when the distribution is controlled by H_0 alone.

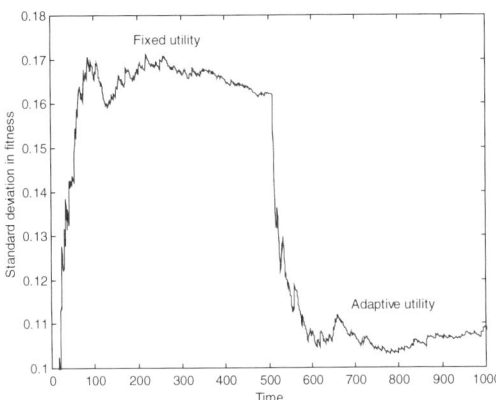

Figure 16.6 The standard deviation in the mean fitness for a fixed utility function, t = 0 to 500 and for an adaptive utility function, t = 501 to 1000.

Also, by introducing an adaptive utility function, the standard deviation in the mean fitness is substantially reduced. This is shown in Figure 16.6.

16.9 Conclusion

In this work we discussed the application of evolutionary game theory to the distribution of service requirements on network servers. Before we introduced evolutionary game theory and how it can be applied to the management of load distribution on a network of servers, we briefly discussed the dynamic similarities between schema theory of genetic algorithms and evolutionary game theory. We demonstrated how both models can be formulated as simple dynamic replicator models. However, in this work, only evolutionary game theory is considered for concrete applications.

It is generally accepted that the increasing complexity of network services requires more efficient distributed control algorithms. In this work, we suggest that competitive replicator dynamics can be used to improve service allocation on heterogeneous server networks. The service allocation dynamic is based on the expected user utility from using any of many competing service providers. We constructrf a network game played by users and network agents. The users negotiate the execution of their requirements by the appropriate service provider. The essential feature of the evolutionary game dynamics is that the service distribution is driven by the user expectation of utility enhancement.

Each user selects a server on the basis of expected utility. Various possible servers are considered and their associated utility functions compared. The expected utility from using any server is compared with the expected mean utility from receiving the required service. Servers associated with user utility above this mean value are more likely to be selected than those associated with below average utility. The service expectation dynamics therefore provides an example of replicator dynamic which increases the mean utility of all active users.

The implemented replicator dynamic is highly non-linear as it depends upon continuously evolving load distribution on the server network. It therefore provides an example of an ongoing optimization process in a continuously changing environment. Conditions for stable load distributions are derived and these associated with high user utility. We discussed the concept of evolutionarily stable strategies, and suggested that these can be identified with resource allocation processes which satisfy the needs of most users and therefore define a stable load distribution. This load distribution remains stable only as long as the utility function does not change.

We discussed three examples of a service distribution process driven by the replicator dynamics of an evolutionary game. We derived necessary conditions for stable load distribution, and displayed them graphically by introducing the concept of stability boundary. It is also demonstrated that a more efficient utilization of network resources can be achieved when the utility function is dependent on the evolving load distribution. This scenario is contrasted with the less efficient one of a static utility function. It was also shown that adaptive utility function leads to a load distribution, which has lower standard deviation than the scenario where the utility function is kept unchanged.

17

Intelligent Flow Control under a Game Theoretic Framework

Huimin Chen and Yanda Li

17.1 Introduction

In recent years, flow control and network management in high speed networks have drawn much attention by researchers in the telecommunications field. In large-scale network environments with high complexity, decentralized control and decision making are required. Many research papers concerned with multidisciplinary approaches to modeling, controlling and managing networks from a computational intelligence point of view appear in the key telecommunication journals every year. However, only a few of them concentrate on the similarity between resource allocation in a network environment and the market mechanism in economic theory. By taking an economic (market based and game theoretic) approach in the flow control of a communication network, we seek solutions where the intelligence and decision making is distributed, and is thus scalable, and the objective of a more efficient and fair utilization of shared resources results from the induced market dynamics.

Unlike other decentralized approaches, our main focus is on using computational intelligence to model the distributed decision agents and to study the dynamic behavior under game-theoretic framework in allocating network resources. By viewing a network as a collection of resources which users are selfishly competing for, our research aims at finding efficient, decentralized algorithms, leading to network architectures which provide explicit Quality of Service (QoS) guarantees, the crucial issue in high speed multimedia networks.

There are several parallel research projects from different institutions dealing with this complicated issue. COMET at Columbia Unversity (http://comet.columbia.edu/research/)

calls the decentralized decision making process *Networking Games*. Using network game approaches, they intend to solve the crucial issue of pricing from the ground-up of the network from the engineering point of view, rather than having, as in existing communication networks, a necessarily arbitrary price structure imposed *post facto*. By taking a market-based approach to distributed systems, the Michigan MARX Project (http://ai.eecs.umich.edu/MARX/) seeks solutions to enable adaptive allocation of resources in large-scale information systems. A very useful collection of links and various resources on network economics and related matters is maintained at Berkeley by H.R. Varian (http://www.sims.berkeley.edu/resources/infoecon/Networks.html). However, there are few published works which attempt to solve call admission control as well as traffic management in telecommunication networks from a game-theoretic point of view by using computational intelligence. In this chapter, we organize our work into two parts. The first part is dedicated to the Connection Admission Control (CAC) process in ATM networks. By taking dynamic CAC as a resource sharing problem, we use a cooperative game model with the product form of a certain combination of user preferences to optimize the call block probability while maintaining the negotiated QoS at the call setup stage. After deriving the product form of the user preference function, we use a genetic algorithm to optimize this objective function to maintain fair shares during the traffic transmission phase. The second part of our work models non-cooperative user behavior in the resource allocation process under a generalized auction framework. We propose an optimal auction rule that leads to equilibrium in a repeated dynamic game. We also use neural networks to model the user strategy in bidding, and briefly discuss the formation of common knowledge in repeated games and the allocation efficiency. The common trait for these two parts is that we use some computational intelligence techniques to solve the complicated modeling and optimization problems in resource allocation.

17.2 A Connection Admission Control Scheme Based on Game Theoretic Model in ATM Networks

17.2.1 Preliminaries of Congestion Control in ATM Networks

Asynchronous Transfer Mode (ATM) has gradually become the standard in Broadband Integrated Services Digital Networks (B-ISDN). It aims to integrate all of the digital communication services, such as voice, image, video and data transfer, into a single integrated network. A fixed-length packet in an ATM network is called a *cell*. The network takes advantage of statistical multiplexing to improve the link utilization, while ensuring the requirements of different types of Quality of Service (QoS) from various traffic sources. The demand for intelligent flow control to prevent possible network congestion has become a major issue for network management. The congestion control schemes of the ATM network can be classified into two types: preventive and reactive control. The first approach is mainly applied to Constant Bit Rate (CBR) and Variable Bit Rate (VBR) services, while the second is applied to Available Bit Rate (ABR) services. Preventive congestion control includes Connection Admission Control (CAC) and traffic enforcement. CAC decides whether to accept or reject a user request for connection establishment. If a connection is established, the bandwidth (and possibly the cell buffers at each switching node) for the incoming service along this connection has to be explicitly allocated during the call holding period. The number of connections accepted at any time has an upper bound due to finite network resources, which mainly depends upon statistical characteristics of traffic sources provided by the end users, QoS constraints and the pricing scheme added on each connection. Traffic enforcement regulates the traffic load in each connection by shaping or smoothing the incoming traffic according to the user declared traffic descriptors.

There exist a lot of CAC schemes for bandwidth allocation. Early attempts often used a static approach to model the incoming traffic: during the call setup phase, the end user provides two sets of parameters, one indicating the QoS requirements (e.g. cell loss rate, maximal tolerable delay) and the other containing all the statistical descriptors of the traffic. If the network has enough bandwidth and buffers to transmit the new traffic source, then it will accept this user request and allocate the required resources. Otherwise, the call will be blocked or rejected. However, because of its reliance on the traffic descriptors provided by the end user, the above method has the following problems (Hsu and Walrand, 1996):

- How to choose statistical parameters for different traffic sources has not yet been agreed upon.
- Usually, at the call setup phase, the end user is not able to determine the traffic descriptors accurately.
- Once the incoming traffic is multiplexed with other cell streams, its statistical characteristics will change, and the user declared parameters may not have the same accuracy for all traversed links during transmission.

To overcome these problems, one simple solution is to reduce the parameters of the traffic descriptors from the end user (e.g. peak cell rate and mean cell rate), and the network makes a conservative resource allocation by assuming the 'worst' traffic pattern the user will provide. Details of various traffic models for 'worst' case service patterns can be found in Perros and Elsayed (1996). Although quite simple, this approach often leads to inefficient bandwidth utilization. Thus, it is necessary to develop a scheme using a dynamic allocation mechanism which can estimate the required bandwidth of each source by monitoring user-specified QoS parameters during cell transmission. A review of dynamic allocation strategies for ATM networks can be found in Chong *et al.* (1995). However, the diversity of traffic characteristics and the variety of QoS constraints make the optimization effort of dynamic bandwidth allocation a complicated problem. Various analytical methods and queuing models have been used, but most of them are highly computationally expensive (e.g. even the simple heterogeneous on/off queuing model in Lee and Mark (1995) has no analytical solution in closed form). Recently, heuristic approaches based on equivalent bandwidth have been presented using neural networks and fuzzy set theory. For example, Hiramatsu (1991) proposed a CAC and a traffic enforcement framework using a neural network; Chang and Cheng (1994) proposed a method of traffic regulation based on a fuzzy neural network; Ndousse (1994) implemented a modified leaky bucket to smooth the traffic by using fuzzy set theory; in Zhang and Li (1995), the authors derived a CAC scheme based on a fuzzy neural network, and integrated it with a genetic algorithm to optimize the network parameters. All of these schemes can be seen as a trade-off between the accuracy of traffic modeling and the implementation complexity. A major drawback is the lack of fairness among different classes of service, especially when different pricing schemes are introduced to the individual user. The recent development of the differentiated service Internet model recalls the interest of fair shares among different services, and the CAC scheme also requires improvement to fit this need.

In the following, we first briefly describe the measurement-based method to monitor the equivalent bandwidth of traffic sources; then we propose a cooperative game model to derive the measure of fair shares in resource allocation. Under the game theoretic framework, the fairness criterion of resource allocation can be derived through maximizing the product form of the user preference function (also called the *utility function*). Given the analytical form of the utility function based on the bandwidth-delay-product, we use a modified genetic algorithm to solve the optimization problem and derive the available bandwidth for each type of traffic. The CAC criterion is modified to maintain the fairness of accepting/rejecting the incoming calls. Simulation results show that our method can achieve a desired fairness for different traffic types and also maintain good link utilization.

17.2.2 Cooperative Game Model for Dynamic Bandwidth Allocation

The equivalent bandwidth model of traffic sources

We first briefly introduce the method to estimate the equivalent bandwidth of multiplexed traffic based on traffic measurement. Due to the diversity of the traffic arrival process, when heterogeneous traffic sources are multiplexed, the bandwidth allocated for each single traffic source has an equivalent representation, assuming that no multiplexing is induced. In terms of equivalent bandwidth, the estimation is based on large quantities of traffic multiplexing. Denote the traffic arrival process as $\{X_1, X_2, \ldots\}$ and the service rate as S. In a given traffic measurement interval, there are W such discrete arrivals, and we assume W is large enough. With regard to its asymptotic distribution, we have:

$$\hat{X}_1 = \sum_{k=1}^{W} X_k, \quad \hat{X}_2 = \sum_{k=W+1}^{2W} X_k, \ldots \quad (17.1)$$

Assume the arrival process is stationary; the new process in equation 17.1 asymptotically has identical independent distribution. According to large deviation theory, the equivalent bandwidth b can be calculated by the following equation (Chen, 1996):

$$\hat{\lambda}_n^W(\theta) = \frac{1}{W} \log \frac{W}{n} \sum_{i=1}^{n/W} e^{\theta \hat{X}_i} \quad (17.2)$$

$$\hat{\delta}_n^W = \sup_{\theta > 0} \{\hat{\lambda}_n^W(\theta) \le S\theta\} \quad (17.3)$$

$$b = \lim_{n \to \infty} \frac{\hat{\lambda}_n^W(\hat{\delta}_n^W)}{\hat{\delta}_n^W} \quad (17.4)$$

Details about the equivalent bandwidth can be found in Lee and Mark (1995), and measurement-based admission control procedures in Jamin et al. (1996). It is necessary to point out that the equivalent bandwidth of the individual traffic represents the required resource capacity that the network manager has to allocate for the established connection in order to meet the QoS requirement. The accepted traffic sources can also be regulated via end-to-end traffic enforcement (Chen, 1996). The above provides a brief illustration of dynamic bandwidth allocation, which is also the preliminary in developing a CAC strategy that we will discuss next.

The fair share criterion based on the cooperative game model

Early research on dynamic resource allocation mainly concentrated on improving link utilization under certain QoS constraints. Less attention has been paid to the fair shares issue among different service types when many incoming calls are competing for network connections and some calls must be blocked. Recently, more work has emphasized the problem of fairness among end users, as well as different types of traffic streams using network resources. For example, in Bolla et al. (1997), the authors try to balance the call block probability among the incoming calls of different types. In Jamin et al. (1998), the authors combine the CAC strategy with a usage-based pricing scheme to achieve a certain degree of fairness. In general, these works propose different objective functions that have to be optimized when admission control and the models of the objective function can be classified into two classes. One formulates the optimization problem taking the Grades of Service (GoS) as constraints of the objective function. The other approach assigns each class of traffic a certain weight or a reward parameter, and the distribution of the GoS is adjusted in resource allocation by changing the value of the weights. The implementation of the above approaches is by no means easy due to the lack of an efficient algorithm to optimize the objective function on-the-fly. Thus, it is difficult to satisfy the needs of a real-time CAC decision using the above schemes. However, from the user-preference point of view, the problem of fair shares among different traffic types is suitable for modeling as multi-player cooperative games. Early studies such as Sairamesh et al. (1995) and Dziong and Mason (1996) discussed the call block probability in resource allocation under the game theoretic model. In Dziong and Mason (1996) the Raiffa, Nash and modified Thomson solutions were compared with regard to their characteristics and the CAC boundary to apply the solution, but only a few simple traffic models are used in simulation to achieve nearly equal call blocking probabilities among the different traffic types.

In the following, we first introduce the cooperative game model, and then define the utility function based on the form of bandwidth-delay-product of the incoming traffic. Using this model, we propose a CAC scheme to achieve a certain trade-off of call block probabilities among different traffic types. Finally, we try to optimize the objective function used in the cooperative game model by applying a genetic algorithm. To simplify the notation, suppose there are two types of incoming traffic competing for the network resources. All possible bandwidth allocation strategies between the different types of traffic form a strategy set, and the outcome of the game is evaluated with two players' utilities namely $u = \{u_1, u_2\}$, $u_i \in R$. Denote U as the set of all possible outcomes of the game between the two players, and assume that U is convex and compact. Note that this assumption is introduced in order to derive the unique optimal solution under the cooperative game model. However, in solving the optimization problem using a genetic algorithm, we do not need this property of the set U.

The outcome of the cooperative game with two players has the following properties: the increase of one player's utility will decrease that of the other; each player's cooperative utility is no less than the non-cooperative outcome. Suppose that when two types of traffic sources demand network bandwidth, the network manager uses a centralized decision rule to ensure the fairness of resource sharing, such as certain coordination between the users transmitting different types of traffic. This procedure is called a cooperative game. For convenience, assuming that the elements in U have been normalized to [0, 1], we define the player's preference as his *utility function*, given below:

$$v_1 = u_1 + \beta(1 - u_2) \tag{17.5}$$

$$v_2 = u_2 + \beta(1 - u_1) \tag{17.6}$$

where β is a weight factor. In the cooperative game model with two players, by solving the optimization problem:

$$\max_{u \in U} \{v_1 \cdot v_2\} \tag{17.7}$$

the two players' bandwidth utilization may achieve a 'relatively' fair share at some operation point with respect to β. When $\beta = 0$, the solution of equation 17.7 is called the Nash point; when $\beta = 1$, the solution is called the Raiffa point; and when $\beta = -1$, the solution is called the modified Thomson point. In the multi-player cooperative game model, player j's utility function has the form

$$v_j = u_j + |\beta(N-1)| - \sum_{j \neq i} u_i \tag{17.8}$$

The economic meanings of each solution corresponding to equation 17.7 can be found in Shubik (1982). When using the cooperative game model, choosing an appropriate utility function is the key issue in ensuring fairness among the various services. Except for the diversity of the bandwidth requirements of the traffic sources, the statistical characteristics of the traffic and utilization of the resources along the end-to-end links should also be considered. In deriving a utility function of an appropriate form, we may want to include the above factors, and also impose a weight factor that controls the influence on each player's satisfaction at different GoS levels. Denote C to be the original data length of the traffic source; denote C' to be the average length after compensating the data retransmission due to cell error or cell loss. We use the empirical formula $C' = C \cdot [1+\alpha(L)]$ to model the traffic amount, where L is the cell loss rate at one switching node and $\alpha(L)$ is a function of L called the *influence factor*, due to cell retransmission. Assume that B is the available bandwidth to be allocated to the user, and T is the mean end-to-end cell delay, then the average cell transmission time can be written as follows:

$$D(L,B) \approx \frac{C'}{B} + k\left(\frac{C'}{B} + 2T\right)\frac{nC'L}{1-nC'L} \quad (17.9)$$

In equation 17.9, k is a coefficient modeling the sliding window for traffic shaping, and can be chosen as half of the window size under a window-based traffic shaping mechanism. Notice that k may also be a variable in connection with the burst curve of the arrival traffic. The upper bound of k should not exceed the capacity of the *leaky bucket* under leaky bucket flow control. n is the number of nodes and/or switches that the traffic traverses from source to destination. In equation 17.9 we potentially assume that the cell loss rate at each node is approximately the same. Using the result derived above, the utility function based on the bandwidth-delay-product is defined as follows

$$C(L,B) = f(L)\left[C' + k(C' + 2BT)\frac{nC'L}{1-nC'L}\right] \quad (17.10)$$

In equation 17.10, $f(L)$ is a weight function that decreases with the increase in the cell loss rate, depending on different QoS requirements. From equation 17.10, we can see that the number of traffic sources and the bandwidth-delay-product represent the availability of network resources to a certain type of traffic during the call setup phase. $f(L)$ is considered as the nominal QoS given that the connection is established for the incoming traffic. Hence we consider $C(L,B)$ as the utility function of various types of traffic sources. Note that the utility function for each type of traffic is given as:

$$u_i = C_i(L_i, B_i), \quad i = 1,2,...,N \quad (17.11)$$

then the outcome of the cooperative game given in equation 17.7 ensures the fair share of bandwidth among different traffic sources in a quantitatively simplified measurement. Note that utility functions of other forms may also be introduced for differentiated or best effort services following a similar derivation; the genetic algorithm does not require the objective function to have a specific analytical form. Note also that in our approach, the cooperative game is not played by the end users but the network manager, who sets certain operation points of the CAC boundary according to the fair shares criterion.

The CAC strategy with fair shares among various traffic sources

After introducing the cooperative game model, we propose the CAC strategy, which includes the following stages: when the ith class of traffic arrives, the network estimates its equivalent bandwidth $\tilde{b}_i$ using equations 17.1–17.4 proposed in section 17.2.2. If the statistical parameters of the incoming traffic are available, the equivalent bandwidth can also be calculated by means of queuing analysis through various numerical approaches. At the same time, the network manager uses a genetic algorithm to search for the optimal solution of the optimization problem of equation 17.7, thus achieving the fair shares bandwidth B_i available for this class of service. In case the bandwidth allocated to each traffic is per-flow guaranteed (e.g. static allocation), the network end can run genetic search off-line or at an earlier stage; while a network using dynamic bandwidth allocation, B_i is adjusted online via genetic optimization to achieve the desired optimal point. Assume the bandwidth for the *i*th class of traffic sources is $\bar{B}_i$, as a result, the network decides to accept or reject the incoming traffic depending on whether $\tilde{B}_i - B_i$ is greater than b_i. In fact, when the *i*th and *j*th classes of traffic compete for the network resources, the network will accept the class at a higher priority whose allocated bandwidth has not exceeded its maximal bandwidth in fair share. In the next subsection, we will give the detailed procedure using a GA to solve the optimization problem of equation 17.7 and we will show that by properly choosing the weight factors of the utility function in equations 17.5 and 17.6, the desired fairness among various classes of service can be achieved.

17.2.3 Call Admission Control with Genetic Algorithm

The basic principles of genetic algorithm
Genetic Algorithms (GAs) are stochastic optimization methods that usually require the objective function to have the form given below:

$$\min\{f(C) | C \in IB^N\} \qquad (17.12)$$

Usually, we have:

$$\forall C \in IB^N = \{0,1\}^N, \ 0 < f(C') < \infty$$

We can see that equation 17.7 satisfies the above requirement. Chapter 1 has introduced the basic operation of a genetic algorithm. In the following, we concentrate on the results of applying the technique in our application.

Computer Simulation Results
In our work we use the GA proposed in Zhang *et al.* (1994; 1995), which uses modified mutation and crossover operators. The parameters coded into strings for GA optimization are the bandwidth available for each type of traffic source. When there are n_i sources accepted by the network with service type *i*, the cell loss rate of this type of service at the switch can be estimated by using equivalent bandwidth (Chen, 1996). Normalizing equations 17.10 and 17.11 and substituting them into equation 17.7, we can get the fitness value for GA optimization. The simulation results show that by using an improved GA approach, the global optimum (strictly speaking, it is sub-optimum or near optimum) is achieved with around only 20 iterations in GA evolution.

The simulation scenario is chosen as follows. We assume there are two types of traffic source. One (traffic I) has a peak bit rate of 64kbps, a mean bit rate of 22kbps and an average burst length of 100 cells. The source-destination route traverses four switching nodes and the mean cell delay is 4ms. The QoS of traffic I requires a cell loss rate less than 10^{-4}. The other (traffic II) has a peak bit rate of 10Mbps, a mean bit rate of 1Mbps and an average burst length of 300 cells. The source-destination route traverses two switching nodes and the average delay is 0.28 ms. The traffic source has the QoS constraint of a cell loss rate less than 10^{-8}. Assume that the link capacity of the network is 155.5 Mbps; the weight function in equation 17.10 takes the empirical formula as below:

$$f(L) = e^{-10(L-QoS)}, \quad \alpha(L) = 1.005L \qquad (17.13)$$

The length of one string representing the normalized bandwidth in the GA is 15 bits. In computer simulation, the CAC boundary given different traffic scenarios (data length of traffic sources) as well as different buffer sizes is calculated via the measurement-based approach discussed in earlier. Concerning the fair shares between traffic I and II, the maximal number of connections for each type of traffic during a heavy load period is derived using GA optimization, and is listed in Table 17.1.

Table 17.1 The CAC boundaries with different weight factors of two types of traffic sources.

Weight factor β	Buffer size (in cells)	Data length of traffic I	Data length of traffic II	The number of traffic I	The number of traffic II
Nash 0	10^3	10^5	10^7	4046	3
Raiffa 1	10^3	10^5	10^7	4748	1
Thomson -1	10^3	10^5	10^7	2430	8
-0.25	10^3	10^5	10^7	3460	4
-0.75	10^3	10^5	10^7	2729	7
0.25	10^3	10^5	10^7	4328	2
0.75	10^3	10^5	10^7	4625	1
Nash	10^3	10^5	10^8	2628	7
Raiffa	10^3	10^5	10^8	3296	3
Thomson	10^3	10^5	10^8	1368	13
Nash	10^3	10^3	10^8	1966	12
Raiffa	10^3	10^3	10^8	4338	4
Thomson	10^3	10^3	10^8	76	17
Nash	10^2	10^5	10^7	2495	3
Raiffa	10^2	10^5	10^7	3868	1
Thomson	10^2	10^5	10^7	1245	5

From Table 17.1, we can see that the traffic sources with a longer data length (indicating a longer connection holding period) will get a relatively larger bandwidth. It is clear that traffic II needs a much larger bandwidth than traffic I. If fair share is not considered between the two types of traffic sources, traffic II is more likely to be blocked than traffic I during the heavy load period when both types of incoming traffic compete for network connection. The solution of the cooperative game has such properties that it guarantees certain reserved bandwidth for both of the traffic sources, despite the diversity of their required bandwidth and the number of hops in their routes. The greater the amount of traffic, the more the reserved bandwidth for the usage of this traffic type at the network operation point. Table 17.1 also reveals that the CAC boundary will become smaller when the buffer size decreases, provided other configurations unchanged. However, the solution of the game theoretic model will not exceed the boundary without the constraint of fair shares in a CAC decision.

The selection of the weight factor β of the utility function has a significant influence on the fair share among different types of traffic sources. The network manager may choose an appropriate β to make the trade-off of call block probabilities among different traffic sources. In Table 17.1, we can see that the available bandwidth for traffic I increases as β increases from -1 to 1. When $\beta = 1$, the utility function of each type of traffic considers its own gain as well as the loss of other types of traffic for the amount of bandwidth available. When $\beta = -1$, the net gains of the players' cooperation in sharing the bandwidth are considered. When $\beta = 0$, each player only cares about individual profits through cooperation. Thus, the above three conditions have a clear economic meaning representing certain fair share criteria in a CAC decision. Through various simulation studies, we find that using a Nash solution as an additional constraint to the CAC decision may be a good candidate in making a balance of call block probabilities among different traffic types. As $\beta = 0$ and the data length of traffic varies, the upper bound of the bandwidth available for traffic I is derived from 28% to 87%, while for traffic II it is from 46% to 78%, and the link utilization remains relatively high. In short, the simulation results show that a certain trade-off between call block distribution among different traffic sources and the network efficiency can be achieved by setting an appropriate operation point at the CAC boundary, which is an optimization problem under the cooperative game framework.

17.2.4 Summary of Connection Admission Control using a Cooperative Game

In this section, the CAC mechanisms based on dynamic bandwidth allocation are briefly discussed. The major issue of fair shares of the bandwidth among different traffic sources is analyzed under a cooperative game model. The utility function concerning the bandwidth-delay-product of each type of traffic is proposed, and the final optimization problem is solved using an improved genetic algorithm. To speed up our CAC scheme, the optimization problem to achieve fair shares can be calculated offline. In online tuning, we can set the initial population to include the previous solution for GA optimization. The proposed CAC scheme is still a centralized mechanism to regulate the traffic flows from end users, and we potentially assume that the network decision-maker has enough information (traffic statistics and QoS requirements) on the incoming traffic to force the link utilization to the desired operation point. The cooperative game model also has a clear economic meaning to interpret certain types of solution as the negotiation result among the users to improve all of their profits in using the available network bandwidth. When such information to derive the utility function of each type of traffic is unavailable, a distributed mechanism for resource allocation is required. We will model the decentralized resource allocation process as a generalized auction game in the next section.

17.3 Optimal Auction Design for Resource Sharing

17.3.1 Resource Allocation as an Auction Design Problem

In the previous section, we have studied the connection admission control problem under a cooperative game framework. Under that framework, we assume that the network manager can explicitly control all of the user traffic during call setup and traffic transmission. However, for the traffic using best effort transmission, the cooperative approach may fail to set the operation point along the CAC boundary, due to the inherent nature of non-cooperative users in sharing the network resources. To characterize user behavior in a non-cooperative environment, a generalized auction model will be discussed in the following.

From the non-cooperative game point of view, communication networks are often characterized by what the economists call *externalities*. That means the value of a network to a user depends upon the other users. The positive externalities are that a communication network is more valuable if more people are connected. The negative externalities are that users who cannot or will not coordinate their actions sufficiently to achieve the most desirable allocation of network resources may lead to network congestion. To avoid negative externalities, prices play a key role as allocation control signals. The telephone system and current Internet represent two extremes of the relationship between the resource allocation and pricing. The resources allocated to a telephone call are fixed, and prices are based on the predictable demand at any given time. In the case of the Internet, the current practice of pricing by the maximal capacity of a user's connection decouples the pricing from resource allocation. In the emergence of multi-service networks (e.g. ATM), neither of these approaches are viable; the former because of the wide range of applications will make demand more difficult to predict; and the latter, because once the entry fee is paid, there is no incentive to limit usage, since increasing consumption benefits individual users, whereas limiting it to a sustainable level brings benefits which are shared by all. This makes it vulnerable to the well-known 'tragedy of the commons'. Thus, there is a need to develop new approaches to price network resources as an alternative to flow control. In a distributed pricing environment, the network manager should charge each network user a price dynamically responsive to unpredictable demand. Auctions have long been considered as an efficient tool for resource allocation in a decentralized control system. To analyze the potential performance of different kinds of auction, we follow Vickrey (1961) and study the auctions as non-cooperative games with imperfect information. We consider the use of the auction model as a decentralized mechanism for efficiently allocating resources to the users of the network (called *bidders*). In our approach, allocations are for arbitrary shares of the total available quantity of network resources, as in Lazar and Semret (1997), against the auction of identical items proposed in Vickrey (1962). Following the analytical model in Milgrom (1981), we find that the auctioneer's best policy is a generalized Vickrey auction under certain regularity assumptions. However, any buyer who does not bid his highest possible price for a unit resource might be assigned a nonzero probability that he will get any resource at all when the manager performs optimal policy, which is different from the PSP auction rule derived in Lazar and Semret (1997). The discussion of auction design in section 3 can be generalized to other scheduling problems in distributed computing systems, while the allocation rule remains the same.

In the following discussion, we assume that the resource to be allocated to the potential bidders is a quantity of link capacity, and does not assume any specific mapping to a QoS requirement. Rather, the users are defined as having an explicit valuation (i.e. utility function) toward quantities of resource, and each user is free in choosing his bidding strategy to maximize his own valuation. The evolutionary behavior of the users' knowledge of their opponents' bidding strategies is also studied under the repeated auction game.

17.3.2 Basic Definitions and Assumptions for Optimal Auction Design

Following Milgrom (1981), we use similar notation to derive the problem of optimal auction design with incomplete information on the bidders. Given the resource of quantity C, and a set of players $N = \{1, 2, \ldots, n\}$ submitting bids, i.e. declaring their desired share of the total resource and the prices they are willing to pay for it, an auction mechanism is the rule used by the auctioneer to allocate resources to the players based on their bids. We will use i and j to represent typical bidders in N. The auctioneer's problem derives from the fact that he does not know how much the bidders are willing to pay for the resource capacities they need. That is, for each bidder i, there is some value t_i which is bidder i's estimate of the unit resource capacity he wants to bid for. We shall assume that the auctioneer (e.g. network manager) uncertainty about the value estimate of bidder i can be described by a continuous probability distribution over a finite interval. Specifically, we let a_i represent the lowest possible value that bidder i might assign to a unit resource; we let d_i represent the highest possible value that i might assign to a unit resource. Let $f_i:[a_i,d_i] \to R^+$ be the probability density function for i's value estimate t_i, assuming that $0 < a_i < d_i < +\infty$, $\forall t_i \in [a_i,d_i]$, $f_i(t_i) > 0$. Correspondingly, let $F_i:[a_i,d_i] \to [0,1]$ denote the cumulative distribution function so that:

$$F_i(t_i) = \int_{a_i}^{t_i} f_i(s_i) ds_i$$

Thus, $F_i(t_i)$ is the auctioneer's assessment of the probability that bidder i has a value estimate no greater than t_i.

We will let T denote the set of all possible combinations of a bidder's value estimates:

$$T = [a_1,d_1] \times [a_2,d_2] \times \cdots \times [a_n,d_n]$$

For any bidder i, we let T_{-i} denote the set of all possible combination of value estimates which might be held by bidders other than i, so that:

$$T_{-i} = \underset{j \in N, j \neq i}{\times} [a_j, d_j]$$

We further assume that the value estimates of the n bidders are stochastically independent random variables. Thus, the joint density function on T for the vector $t = (t_1, t_2, \ldots, t_n)$ of individual value estimates is:

$$f(t) = \prod_{j \in N} f_j(t_j)$$

Of course, bidder i considers his own value estimate of unit resource to be a known quantity, not a random variable. However, we assume that bidder i assesses the probability distributions for other bidders' value estimates in the same way as the auctioneer does. That is, both the resource seller and bidder i assess the joint density function on T_{-i} for the vector:

$$t_{-i} = (t_1, \cdots, t_{i-1}, \cdots, t_{i+1}, \cdots t_n)$$

of values for all bidders other than i to be:

$$f_{-i}(t_{-i}) = \prod_{j \in N, j \neq i} f_j(t_j)$$

We denote the auctioneer's personal value estimate for unit resource, if he were to keep it unsold, to be t_0. We assume that the auctioneer has no private information about the resource, so that t_0 is known to all the bidders. Suppose that the resource is infinitesimally divisible and the bidders' budget limit in one-shot as well as repeated bidding game is $B = \{b_1, b_2, \cdots, b_n\}$, so that bidder i can always request for b_i / t_i resource capacity, given that his value estimate is t_i.

17.3.3 Feasible Auction Mechanisms

Given the density function f_i and each bidder's evaluation estimate t_i, the network manger's problem is to select an auction mechanism to maximize his own expected utility. We must now develop the notation to describe the auction mechanisms that the manager might select. To begin, we will restrict our attention to a special class of auction mechanism: the *direct revelation* mechanism (Milgrom, 1981). In the direct revelation mechanism, the bidders simultaneously and confidentially announce their value estimates to the manager; then the manager determines who gets the (partial) resource according to their request, and how much each bidder should pay for unit resource capacity, as some function of the announced value estimates $t = (t_1, t_2, \cdots, t_n)$ and required capacity $r = (r_1, \cdots, r_n), r_i = b_i / t_i$ of total resource. Thus, a direct revelation mechanism is described by a pair of outcome functions (p, x) of the form $p: T \rightarrow R^n$ and $x: T \rightarrow R^n$ such that, if t is the vector of the announced value estimates of unit capacity, then $p_i(t)$ is the probability that bidder i gets r_i of the total resource and $x_i(t)$ is the expected amount of money that bidder i must pay for unit resource. Notice that in this case, we assume that the bidders' budget limit B is deterministic, and for the case that all bidders bid for a capacity no less than the total resource, the auction problem reduces to a traditional one item auction.

We shall assume throughout this section that the manager and the bidders are risk neutral and have additive separable utility functions for resource capacity being sold or consumed. Thus, if bidder i knows that his value estimate of unit capacity is t_i, then his expected utility from the auction mechanism described by (p, x) is:

$$U_i(p,x,t_i) = \frac{b_i}{t_i} \int_{T_{-i}} [t_i p_i(t) - x_i(t)] f_{-i}(t_{-i}) dt_{-i} \qquad (17.14)$$

where $dt_{-i} = dt_1 \cdots dt_{i-1} dt_{i+1} \cdots dt_n$.

Similarly, the expected utility for the network manager from this auction mechanism is:

$$U_0(p,x) = \sum_{j \in N_T} \int \frac{b_j}{t_j} [t_0(1-p_j(t)) + x_j(t)] f(t) dt \qquad (17.15)$$

where $dt = dt_1 \cdots dt_n$.

Not every pair of functions (p, x) represents a feasible auction mechanism. There are three types of constraint which must be imposed on (p, x). First, since there are only C unit capacities to be allocated, the function p should satisfy the following probability conditions:

$$\forall i \in N, t \in T, 0 \le p_i(t) \le 1 \quad \text{and} \quad \sum_{j \in N} p_j(t) \frac{b_j}{t_j} \le C \qquad (17.16)$$

Secondly, we assume that the manager cannot force a bidder to participate in an auction which offers him less expected utility than he could get on his own. If the bidder did not enter the auction, he would not pay any money, and his utility payoff would be zero. Thus, to guarantee that the bidder will participate in the auction, the following *individual-rationality* conditions must be satisfied:

$$\forall i \in N, \forall t_i \in [a_i, d_i], U_i(p,x,t_i) \ge 0 \qquad (17.17)$$

Thirdly, we assume that the manager cannot prevent any bidder from lying about his individual value estimate of unit capacity if the bidder expects to gain from lying. Thus, the direct revelation mechanism can be implemented only if no bidder ever expects to gain from lying. That is, honest responses must form a Nash equilibrium in the auction game. To guarantee that no bidder has any incentive to lie about his value estimate, the following incentive-compatibility conditions must be satisfied:

$$\forall i \in N, \forall s_i, t_i \in [a_i, d_i],$$
$$U_i(p,x,t_i) \ge \frac{b_i}{s_i} \int_{T_{-i}} [t_i p_i(t_{-i}, s_i) - x_i(t_{-i}, s_i)] f_{-i}(t_{-i}) dt_{-i} \qquad (17.18)$$

given t_i is the true value estimate of bidder i. We call (p, x) a *feasible* auction mechanism if and only if equations 17.16–17.18 are all satisfied.

Given an auction mechanism (p, x), we define:

$$Q_i(p,t_i) = \int_{T_{-i}} p_i(t) f_{-i}(t_{-i}) dt_{-i} \qquad (17.19)$$

to be the conditional probability for any bidder i that he will get r_i resource capacity from the auction mechanism (p, x), given that his value estimate of unit resource capacity is t_i.

Our first result is a simplified characterization of the feasible auction mechanism.

Lemma 1
(p, x) is a feasible auction mechanism if and only if the following conditions hold:

$$\forall i \in N, \forall s_i, t_i \in [a_i, d_i], \text{ if } s_i \leq t_i, \text{ then } Q_i(p, s_i)/s_i \leq Q_i(p, t_i)/t_i \qquad (17.20)$$

$$\forall i \in N, \forall t_i \in [a_i, d_i], U_i(p, x, t_i) = U_i(p, x, a_i) + b_i \int_{a_i}^{t_i} \frac{Q_i(p, s_i)}{s_i} ds_i \qquad (17.21)$$

$$\forall i \in N, U_i(p, x, a_i) \geq 0 \qquad (17.22)$$

and together with equation 17.16.

The proof is given in the appendix. Note that in a resource auction, there also exist many non-direct revelation approaches. For example, the Progressive Second Price (PSP) auction proposed in Milgrom (1981) is an auction rule with a 2-dimensional message space, and the author shows that this auction has a Nash equilibrium, hence the design objective is met at equilibrium. In this case, a PSP auction is feasible as $a_i = d_i$ in equation 17.20 when complete information of the valuations among all bidders is known to all participants (often called *common knowledge*). However, Lemma 1 may not apply to more general cases of a divisible auction unless one can guess the right mapping of direct revelation to desired mechanism and build it into the allocation rule from the start, as in Wurman (1997)
.

17.3.4 Optimal Auction Design for Allocation of Arbitrarily Divisible Resource

In the following, we will show the optimal condition for the auction design problem in the set of feasible mechanisms defined previously. With a simple regularity assumption, we can compute the optimal auction mechanism explicitly in an efficient way. This is important for the implementation of an auction game in a decentralized network control environment.

Lemma 2
Suppose that $p: T \rightarrow R^n$ satisfies that $\forall i \in N, t \in T, p$ maximizes

$$\int_T \left\{ \sum_{i \in N} \frac{b_i}{t_i} \left[t_i - t_0 - \frac{1 - F_i(t_i)}{f_i(t_i)} \right] p_i(t) \right\} f(t) dt \qquad (17.23)$$

subject to the constraints of equations 17.16 and 17.20. Assume also that

$$x_i(t) = t_i \left[p_i(t) - \int_{a_i}^{t_i} \frac{p_i(t_{-i}, s_i)}{s_i} ds_i \right] \qquad (17.24)$$

Then (p, x) represents an optimal auction.

With a simple regularity assumption, we can compute an optimal auction mechanism directly from Lemma 2. We say that the auction problem with a non-empty feasible mechanism is *regular* if for any $i \in N$, the function:

$$c_i(t_i) = t_i - \frac{1 - F_i(t_i)}{f_i(t_i)}$$

is strictly increasing in t_i. Now consider that the manager keeps the total resource if:

$$t_0 > \max_{j \in N} c_j(t_j)$$

otherwise he first gives r_i capacity of the resource to bidder i who has the highest $c_i(t_i)$ with a probability $p_i(t) = t_i / d_i$ if $C - r_i \geq 0$. If $C - r_i < 0$, the manager only allocates capacity of an amount C to that bidder with the same probability. If:

$$c_i(t_i) = c_j(t_j) = \max_{k \in N} c_k(t_k)$$

then the manager may break the tie by considering the bidder requiring smaller quantity of resource first, or by some other allocation rule. Ties will only happen with zero probability in the regular case. Given that there is still a certain amount of resource left after the first allocation, the manager can choose the bidder with the second highest $c_i(t_i)$ from among the remaining bidders according to the probability constraints (17.20), and perform allocation recursively until the maximal value of $c_i(t_i)$ is less than t_0 or there is no capacity left at all. Let $M = \{1, 2, \cdots, m\}$ be the re-ordered index of bidders satisfying that $\forall i, j \in M, i < j$; we have $c_i(t_i) \geq c_j(t_j) \geq t_0$. Thus, the specific auction rule given above can be defined as follows:

$$Er_i = \left\{ r_i \wedge \left[C - \sum_{j=1}^{i-1} r_j p_j(t) \right]^+ \right\} p_i(t) \qquad (17.25)$$

where Er_i means the expected amount of resource that bidder i will get in the sense of repeated games of this auction. The probability assignment has the form $\forall i \in N$:

$$p_i(t_{-i}, s_i) = s_i / d_i \qquad (17.26)$$

if bidder i is chosen by the auctioneer to have non-zero probability to win (i.e. $i \in M$).

Theorem
Given the auction problem with regularity assumption defined above, the auction rule (p, x) satisfying equations 17.24–17.26 represents an optimal auction mechanism.

Several remarks are worth noting. In the auction considered above, any bidder needs to pay if he has a non-zero probability to win a certain amount of resource capacity. The money (whether relating to real money or 'funny money' based on quotas) he pays to the auctioneer is the minimal price he needs to have the same probability to win the required capacities as that assigned when he bids at his value estimate given his opponents' profile. For identical bidders (their a_i, d_i, f_i, b_i are the same), the manager's rule is like generalized Vickrey auction (Vickrey, 1961), but there still exists economic inefficiency in the optimal auction mechanism due to incomplete information (lack of common knowledge) among the bidders and the auctioneer. Any bidder will have some risk of not getting any resources at all, even when there exists some resource capacity larger than he needs and his opponents bid lower than his value estimate of unit capacity. Note also that the network manager will bid like a resource buyer by marking his reserved price to $c_i^{-1}(t_0)$ for unit capacity. For example, suppose there are 10 bidders with a budget of 1000 each to buy 10 unit resources in an auction game. To model the incomplete information among the bidders, we let $a_i = 100$, $d_i = 1000$, $t_0 = 100$, $f_i = 1/900$. By straightforward calculation, we get the manager's reserved price of 550, which is much higher than his own value estimate. This price does not change when the numbers of bidders and each bidder's budget limit varies. Implementing the auction rule given in equations 17.24–17.26 used by the manager in a network environment is just like that in Lazar and Semret (1997), with minor modifications. An extended version of a PSP auction with a simulation study in a complicated network environment can be found in Lazar and Semret (1998). Our approach can also be applied to the decentralized allocation of network resources with the assumption of the bidders' probability density functions and their budget limits as *common knowledge*. The auction game can be repeatedly played by the same users or some kind of software agents. The users and the manager can set an *a priori* probability density function and tune the function in the repeated games. If the manager has a channel to send a message indicating the users' demand (e.g. average price of the unit capacity from previous bidding), the users will tune their bidding seals smartly so that the whole game will converge to the desired equilibrium.

17.3.5 Simulation Study of Individual Learning to Form Common Knowledge

An issue of obvious concern is how the convergence of common knowledge scales with the number of bidders, as well as each bidder's bidding strategy in the repeated game using the proposed auction rule. It should be noted that, in the repeated auction game, both the bidder and the auctioneer have an incentive to modify their valuation of the probability density function of their opponent's bidding profile, according to the history information. The dynamic behavior of the auction game will be studied experimentally in the following. We use TREX (implemented by Java – see Semret (1996)) as the prototype to simulate an interactive auction game using software agents. Unlike the scenarios of interconnected networks studied in Lazar and Semret (1998a), our simulations are limited to a single route with a divisible resource capacity, and add *learning strategies* to the bidders as well as the auctioneer to formulate certain system dynamics (e.g. convergence of probability density to the true value estimation of the unit capacity). Detailed descriptions are in Chen (1998).

In computer simulation, the resource capacity is normalized to 1, and five bidders with a budget constraint of 0.5 each enter the repeated auction game with incomplete information on their opponents' true value estimations. Assume the bidders are named from 1 to 5 and their true value estimations of unit capacity are 0.6, 0.7, 0.8, 0.9 and 1.0, respectively. Initially, the bidders and the auctioneer assume their opponents' true value estimations to be uniformly distributed in [0, 10]. In every round of the auction game, the auctioneer performs the optimal auction rule derived in section 3.4 to allocate resources to the bidders. At the end of each round, the bidders and the auctioneer use learning strategies to tune their model of the opponents' Probability Density Functions (PDFs) of true value estimation. We expect that the PDFs of each bidder's valuation (as well as the auctioneer's valuation of each bidder) will converge to the true value estimate in the repeated auction games.

The learning strategy for the bidders and the auctioneer is very important to the formation (and perhaps the evolution) of common knowledge. There exist various bidding strategies optimizing the expected utility of each bidder in a non-cooperative manner. For simplicity, we only simulate two strategies: a Parzen window and a neural network approach to model the learning behavior of each participant in a repeated auction. To modify the opponents' PDFs, each bidder needs to collect the history information of the opponents' bids, and maximizes his own utility by setting the best sealed bid for the next round. In Parzen window learning, the PDFs are updated by using the new information of the opponents' bidding seals after each round of the auction. In the neural network approach, each bidder uses a feedforward neural network to predict his opponents' bids according to the bids collected historically. Backpropagation training is used to tune the parameters of the neural network after 10 rounds of the auction. The PDFs are updated according to the empirical risk minimization rule. Extensive discussion on the simulation study of an optimal repeated auction can be found in Chen (1998). After 500 rounds of the auction, we pick bidder 5 to see his modified PDFs from bidders 1 to 4. Other bidders use the same type of learning strategies and yield similar dynamic behaviors. The simulation result using the Parzen window approach is given in Figure 17.1, and the result using the neural network approach is shown in Figure 17.2.

Similar results are derived for bidders 1 to 4 and the auctioneer, and the plots are omitted. We can see that the peaks of the PDFs from bidders 1 to 4 are centered approximately at the bidders' true value estimates of unit capacity after enough rounds of the repeated auction game. The neural network approach yields better estimation of true PDFs than the Parzen window in the same number of rounds. We expect that a good learning strategy may speed up the formation of common knowledge in the repeated auction game. In the extreme case, all the bidders and the auctioneer know their opponents' true value estimates of the unit capacity, therefore, the auction game reduces to a generalized Vickrey auction with complete information. As a limit of the repeated game, the best strategy for the bidder is simply to bid at his truthful value estimate with maximal possible capacity according to budget constraints. The allocation is economically efficient in that the bidder with the highest value estimate of unit capacity gets his resources first, and the network manager allocates the resources obeying exactly the order of the value estimates in sequence from the sealed bids.

Intelligent Flow Control under a Game Theoretic Framework

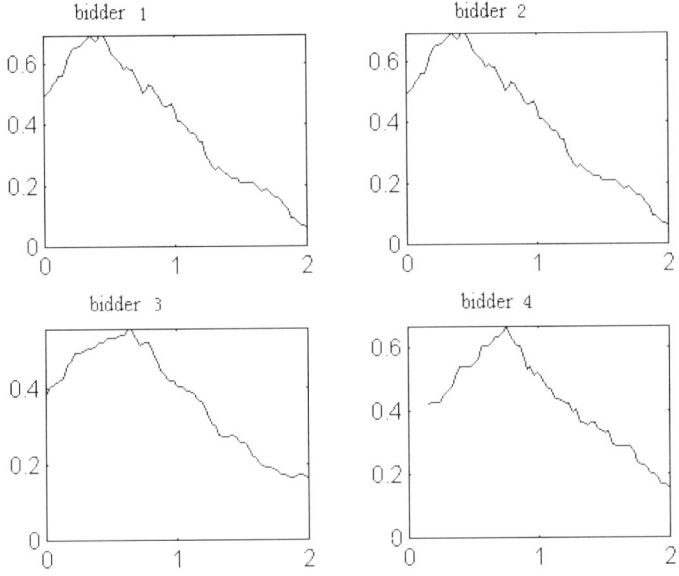

Figure 17.1 The PDFs estimation from bidder 5 using the Parzen window.

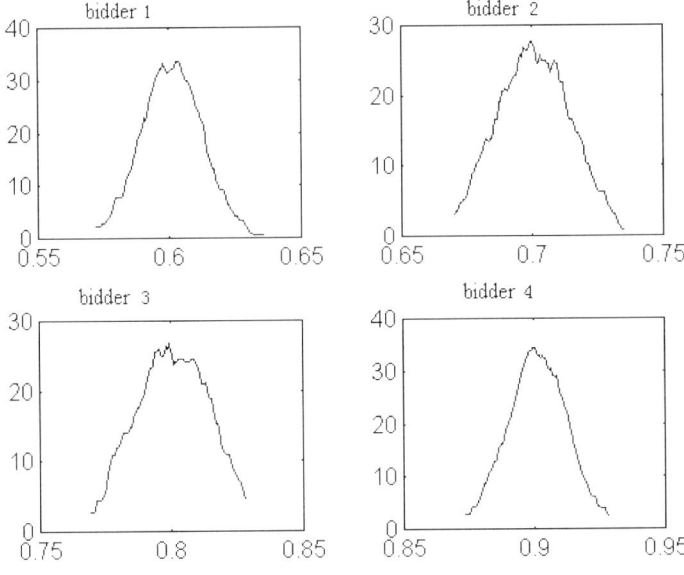

Figure 17.2 The PDFs estimation from bidder 5 using the neural network approach.

17.4 Concluding Remarks

In this chapter, we propose a game theoretic framework to analyze flow control and resource allocation in high speed networks. We organize our work in two parts. The first part deals with the connection admission control problem in ATM networks. By taking dynamic CAC from a different point of view, we propose a cooperative game model that considers a certain combination of user preferences to optimize the call block probability while maintaining the negotiated QoS. After deriving the product form of the utility function of different traffic types as user preferences, we use a genetic algorithm to optimize this objective function to maintain fair shares during the traffic transmission along each connection. The second part of the work tries to model the non-cooperative user behavior in resource allocation under a generalized auction framework. We propose an optimal auction rule that leads to equilibrium in a repeated dynamic game. We use the Parzen window and neural networks to model the user strategies in tuning the PDFs of the opponents, and briefly discuss the formation of common knowledge in the repeated game, as well as the allocation efficiency. Contrary to our CAC scheme as a centralized network control mechanism, the auction game to allocate network resources is suitable for the decentralized control environment. The common trait for these two parts is that we use some computational intelligence techniques to solve the complicated modeling and optimization problem in resource allocation. Our future work will be to study the interactions of bidders' strategies with non-spot entrance and leaving, and to investigate the conditions to maintain a stable auction game with maximal allocation efficiency. Applications of our approach to the economic systems are also under investigation.

17.5 Appendix

17.5.1 Proof of Lemma 1:

Note that:

$$\frac{b_i}{s_i}\int_{T_{-i}}[t_i p_i(t_{-i},s_i)-x_i(t_{-t},s_i)]f_{-i}(t_{-i})dt_{-i} = U_i(p,x,s_i)+\frac{b_i}{s_i}(t_i-s_i)Q_i(p,s_i)$$

Thus, the incentive compatibility (equation 17.18) is equivalent to:

$$\forall i \in N, \forall s_i, t_i \in [a_i, d_i], U_i(p,x,t_i) \geq U_i(p,x,s_i)+\frac{b_i}{s_i}(t_i-s_i)Q_i(p,s_i) \quad (17.27)$$

Using equation 17.27 twice with the role of t_i and s_i switched, we get

$$\frac{b_i(t_i-s_i)}{s_i}Q_i(p,s_i) \leq U_i(p,x,t_i)-U_i(p,x,s_i) \leq \frac{b_i(t_i-s_i)}{t_i}Q_i(p,t_i)$$

Then equation 17.20 follows when $s_i \leq t_i$.

The inequalities can be rewritten as

$$\frac{Q_i(p,s_i)}{s_i}b_i\delta \leq U_i(p,x,s_i+\delta) - U_i(p,x,s_i) \leq \frac{Q_i(p,s_i+\delta)}{s_i+\delta}b_i\delta$$

for any $\delta > 0$. Since $Q_i(p,t_i)/t_i$ is increasing in t_i, it is Riemann integrable. Thus, we get

$$U_i(p,x,t_i) - U_i(p,x,a_i) = b_i \int_{a_i}^{t_i} \frac{Q_i(p,s_i)}{s_i} ds_i$$

Now we must show that conditions in Lemma 1 also imply equations 17.17 and 17.18.

Suppose $s_i \leq t_i$, then equation 17.20 implies that

$$U_i(p,x,t_i) = U_i(p,x,s_i) + b_i \int_{s_i}^{t_i} \frac{Q_i(p,u_i)}{u_i} du_i$$

$$\geq U_i(p,x,s_i) + b_i \int_{s_i}^{t_i} \frac{Q_i(p,s_i)}{s_i} du_i = U_i(p,x,s_i) + \frac{b_i}{s_i}(t_i - s_i)Q_i(p,s_i)$$

Similarly, if $s_i > t_i$, we can get the same inequality form as equation 17.18, and incentive-compatibility obeys. Finally, equation 17.17 directly follows from equations 17.21 and 17.22. □

17.5.2 Proof of Lemma 2:

Recalling the auctioneer's utility function defined in the previous section, we may rewrite it as

$$U_0(p,x) = \int_T \sum_{j \in N} \frac{b_j}{t_j} t_0 f(t) dt + \sum_{j \in N_T} \int \frac{b_j}{t_j}(t_j - t_0) p_j(t) f(t) dt + \sum_{j \in N_T} \int \frac{b_j}{t_j}[x_j(t) - p_j(t)t_j] f(t) dt$$

Using Lemma 1, we know that for any feasible (p, x),

$$\sum_{j \in N_T} \int \frac{b_j}{t_j}[x_j(t) - p_j(t)t_j] f(t) dt - \sum_{j \in N} \int_{a_j}^{d_j} U_j(p,x,t_j) f_j(t_j) dt_j$$

$$= -\sum_{j \in N} \int_{a_j}^{d_j} [U_j(p,x,a_j) + b_j \int_{a_j}^{t_j} \frac{Q_j(p,s_j)}{s_j} ds_j] f_j(t_j) dt_j$$

$$= -\sum_{j \in N} U_j(p, x, a_j) - b_j \int_{a_j}^{d_j} \int_{s_j}^{d_j} f_j(t_j) \frac{Q_j(p, s_j)}{s_j} dt_j ds_j$$

$$= -\sum_{j \in N} U_j(p, x, a_j) - b_j \int_{a_j}^{d_j} (1 - F_j(s_j)) \frac{Q_j(p, s_j)}{s_j} ds_j$$

$$= -\sum_{j \in N} U_j(p, x, a_j) - \int_T (1 - F_j(t_j)) \frac{b_j}{t_j} p_j(t) f_{-j}(t_{-j}) dt \quad (17.28)$$

Using equation 17.28, the auctioneer's problem is to maximize

$$U_0(p, x) = \sum_{j \in N_T} \int_{t_j} \frac{b_j}{t_j} t_0 f(t) dt - \sum_{j \in N} U_j(p, x, a_j) + \sum_{j \in N_T} \int \left\{ \frac{b_j}{t_j} \left[t_j - t_0 - \frac{1 - F_j(t_j)}{f_j(t_j)} \right] p_j(t) \right\} f(t) dt$$

(17.29)

subject to the constraints of equations 17.20–17.22 and 17.16. Note that 17.21 and 17.22 can be rewritten as:

$$\forall i \in N, \forall t_i \in [a_i, d_i],$$

$$U_i(p, x, a_i) = \int_{T_{-i}} [b_i p_i(t) - b_i \int_{a_i}^{t_i} \frac{p_i(t_{-i}, s_i)}{s_i} ds_i - \frac{b_i}{t_i} x_i(t)] f_{-i}(t_{-i}) dt_{-i} \geq 0 \quad (17.30)$$

If the manager chooses x according to equation 17.24, then he satisfies both equations 17.21 and 17.22, and he gets $\forall i \in N$, $U_i(p, x, a_i) = 0$, which is the best possible value for this term in equation 17.29. Furthermore, the first term on the right side of equation 17.29 is independent of (p, x). So the manager's objective function can be simplified to equation 17.24, and equations 17.16 and 17.20 are the only constraints left to be satisfied. This completes the proof of the lemma. □

17.5.3 Proof of Theorem

It can be shown that only those bidders belonging to M have a non-zero probability to get the desired resources they bid. Using regularity assumption, $\forall t_j' > t_j$, we have $c_j(t_j') > c_j(t_j)$. Hence for any bidder j, if his true value estimate of unit capacity is t_j, then the probability that he gets the desired resource by bidding the unit price t_j is greater than that using any bidding strategy with a unit price less than t_j. Furthermore, given the allocation sequences using the auction rule (equation 17.26), if bidder j changes his bid for unit capacity from t_j to s_j while his position in allocation sequences remains unchanged, then his expected utility remains the same as its truthful bidding. Thus the incentive compatibility constraint (equation 17.20) is satisfied. It can be shown that equation 17.25 implies equation 17.16 and the auction is *feasible*. Below we will show that the proposed auction maximize the expected utility of the auctioneer. Assume there exists another auction mechanism (p', x') that yields higher expected utility for the auctioneer, we have:

$$\int_T \left\{ \sum_{j \in N} \frac{b_j}{t_j} [c_j(t_j) - t_0] p_j(t) \right\} f(t) dt > \int_T \left\{ \sum_{j \in N} \frac{b_j}{t_j} [c_j(t_j) - t_0] p_j(t) \right\} f(t) dt \qquad (17.31)$$

Obviously, in the bidding vector t, there exists a certain bidder k with non-zero probability to get his bidding capacity, whose bid for unit capacity should satisfy

$$\int_{T_{-k}} p'_k(t_{-k}, t_k) f_{-k}(t_{-k}) dt_{-k} > \int_{T_{-k}} p_k(t_{-k}, t_k) f_{-k}(t_{-k}) dt_{-k} \qquad (17.32)$$

That means $\exists t_k \in [a_k, d_k]$ such that $Q_k(p', t_k) > Q_k(p', t_k)$. Without loss of generality, we assume the sets M' and M in two different auction mechanisms are the same (otherwise, M' in the auction mechanism (p', x') should be a subset of M), and bidder k maximizes his own utility in both of the auction mechanisms. Notice that the probability that bidder k receives the desired resource capacity is no greater than 1: using incentive compatibility, we have

$$\frac{p'_k(t_{-k}, t_k)}{t_k} \leq \frac{p'_k(t_{-k}, d_k)}{d_k} \leq \frac{1}{d_k} \qquad (17.33)$$

From equation 17.26, equation 17.33 implies:

$$\frac{p_k(t_{-k}, t_k)}{t_k} = \frac{1}{d_k} < \frac{p'_k(t_{-k}, t_k)}{t_k} \qquad (17.34)$$

Consider that bidder k will have a non-zero probability to get an r_k resource capacity in both of the two auction mechanisms, regardless of the sequences bidder k is ordered in M or M', the probability constraint of a *feasible* auction rule should satisfy both equations 17.33 and 17.34. Hence reaching a contradiction. □

18

Global Search Techniques for Problems in Mobile Communications

Bhaskar Krishnamachari and Stephen B. Wicker

18.1 Introduction

In the last two decades, wireless communication systems such as cordless phones, paging systems, wireless data networks, satellite-based and cellular mobile systems have been steadily increasing in both popular importance and technological sophistication. The first-generation wireless systems were developed in the late 1970s and 1980s and were based on analog technology (such as the Advance Mobile Phone Service (AMPS) by AT&T and Nordic Mobile Telephone (NMT) by Ericsson). As demand increased and digital technology matured in the 1980s and 1990s the second-generation digital wireless systems were designed (such as Global System for Mobile Communications (GSM) in Europe and Digital AMPS in North America). These systems, currently in use, offer higher system capacity and improved quality of service. Third-generation systems, referred to as Personal Communication Systems (PCS), are currently under development and expected to be deployed at the beginning of the 21st century (Li and Qiu, 1995; Gibson, 1996).

The utilization of wireless communications has shown growth rates of 20–50% per year in various parts of the world. As the benefits of digital technology are realized, it is expected that there will be demand for the transmission of high-bandwidth, high-quality multimedia information content over these systems. The PCS goal is to provide to every user the ability to exchange such information securely with anyone, anytime, anywhere in the world, using a unique Personal Telecommunication Number (PTN). To meet these challenging expectations, intensive research has been undertaken in recent years to develop a sophisticated PCS with increased network capacity and performance. One of the trends in this research has been the growing incorporation of Artificial Intelligence (AI) techniques into such systems (Muller et al., 1993).

Like all engineering endeavors, the subject of mobile communications also brings with it a whole host of complex design issues. A number of issues concerning resource allocation, design, planning, estimation and decision in mobile communications can be formulated as combinatorial optimization problems. Many of these problems are NP-hard (Garey and Johnson, 1979), characterized by search spaces that increase exponentially with the size of the input. They are therefore intractable to solution using analytical approaches or simple deterministic algorithms that cannot terminate in polynomial time. Heuristic and stochastic optimization procedures offer more appropriate alternatives.

At about the same time as the advent of the first generation wireless systems, three robust and general global search techniques for NP-hard combinatorial optimization were invented – Genetic Algorithms (GA) (Holland, 1975), Simulated Annealing (SA) (Kirkpatrick et al., 1983), and Tabu Search (TS) (Glover, 1986). These techniques have proved to be very successful and their application to a large number of fields such as industrial production, management, financial services, game theory, telecommunications, graph theory, biological modeling and VLSI has been increasing steadily from the mid 1980s through the 1990s. Their application to the field of mobile telecommunications is still in its infancy, with most of the 20-odd papers that have been written on the subject having been published in the last five years. This chapter provides a fairly comprehensive survey of this still-nascent literature on the subject.

The twin goals of this chapter are (a) to familiarize mobile communication engineers with global search techniques and their potential, and (b) to provide those interested in these techniques with an idea of the nature and scope of their application to mobile communications. Accordingly, the rest of the chapter is organized as follows: the three global search techniques are described in section 18.2; section 18.3 provides a survey of the recent papers where these techniques have been applied to optimization for mobile communications; concluding comments are presented in section 18.4.

18.2   Global Search Techniques

First, we note that there is some confusion regarding nomenclature in the literature. These techniques are sometimes referred to as 'local' search techniques (e.g. Aarts and Lenstra (1997)) because they proceed by searching neighborhoods in the search space. In this chapter, we distinguish between local search procedures (such as steepest descent algorithms) that are susceptible to being trapped in local optima, and global search procedures (such as genetic algorithms, simulated annealing and tabu search) that are capable of escaping such minima and providing globally optimal solutions.

An optimization problem seeks to find the optimal solution $x^* = \min\{f(x) | x \in S\}$, where f is the cost function (also known as the objective function) with domain S – the set of possible solutions. A *combinatorial* optimization problem is defined as an optimization problem where S has discrete members. When the problem complexity is low, either an exhaustive search of the space, or deterministic algorithms (such as linear and nonlinear programming (Luenberger, 1973), dynamic programming (Sacco, 1987), branch and bound algorithms and polyhedral cutting plane approaches (Nemhauser and Wolsey, 1988) may be used to obtain the solution. For more difficult problems, heuristic and stochastic search techniques must be employed to find the optimal point in a large solution space.

A subset of S, designated $N(x)$, may be associated with each point $x \in S$. $N(x)$ is referred to as the *neighborhood* of x. Most search techniques of optimization operate by starting with some point x and exploring its neighborhood for solutions. Local search techniques such as steepest descent algorithms explore a neighborhood accepting each successive point as a solution only if it has a lower cost than the current solution. This causes entrapment in the point with the lowest cost function in the neighborhood – a local optimum. Global search techniques provide an escape from such traps by providing the ability to selectively accept successive points, even if they have a higher cost than the current solution. We examine, in turn, genetic algorithms, simulated annealing and tabu search. These algorithms are very simple and easy to implement (perhaps the chief reason for their popularity), and are very general and robust in their simplest form – assuming nothing about the structure of the space being searched. To improve their efficiency when applied to specific problems, they may be modified suitably or even hybridized with heuristics-based local search techniques.

18.2.1 Genetic Algorithms

These algorithms derive their inspiration from the natural process of biological evolution. Solutions are encoded (often in binary) into strings or *chromosomes*. The algorithm operates on a population of these chromosomes, which evolve to the required solution through operations of fitness-based selection, reproduction with crossover and mutation that are fundamentally similar to their natural analogs. Using Markov chain modeling, it has been shown that GAs are guaranteed to converge asymptotically to the global optimum if an elitist strategy is used where the best chromosome at each generation is always maintained in the population (Rudolph, 1994). More detailed descriptions of these algorithms, their implementation and applications can be found in Chapter 1 of this volume, and in Goldberg (1989), Bäck et al. (1997) and Mitchell (1997).

Genetic algorithms have been quite successful in a variety of applications and enjoy massive popularity, at least amongst members of the evolutionary computation community. Among the papers surveyed in this chapter, applications of genetic algorithms are by far the most numerous. The name *evolutionary telecommunications* has recently been suggested for this growing field.

18.2.2 Simulated Annealing

Here we build somewhat on the broad description of simulated annealing given in Chapter 1. The algorithm derives its inspiration from the thermodynamic process by which solids are heated and cooled gradually (annealed) to a crystalline state with minimum energy. Simulated annealing operates on a single point (not a population of points as in GA), and at each step a point x' in $N(x)$ is generated from the current point x. If the point has a lower cost function than x it is accepted unconditionally, but even if it has a higher cost it is accepted probabilistically using the Metropolis criterion described below. This acceptance probability is proportional to the temperature T of the annealing process, which is lowered gradually as the algorithm proceeds.

The Metropolis criterion is as follows: for $x' \in N(x)$, the probability that x' is selected is

$$P_{x \to x'} = \min\left(1, \exp\left[-\frac{f(x') - f(x)}{T}\right]\right) \qquad (18.1)$$

When T is high initially, there is a greater probability of making *uphill* moves, which allows the search to fully explore the space. It has been shown, by modeling SA as Markov processes (Aarts and Korst, 1989), that simulated annealing will converge asymptotically to the global optimum under two conditions:

Homogeneous condition: if the temperature is lowered in any way to 0, the length of the homogeneous Markov sequence (formed by the accepted points) at each temperature is increased to infinite length.

Inhomogeneous condition: if irrespective of the length of these isothermal Markov chains, the cooling schedule is chosen such that T approaches 0 at a logarithmically slow rate.

Since in practice neither of these is possible in finite implementations, polynomial time approximations are used. The choice of cooling schedule and the length of the Markov chains at each temperature affects the quality of the results and the rate of convergence. The definition of the neighborhood function (which is usually based on some heuristic understanding of the problem at hand) determines how new points are visited. The SA is ended if an acceptable solution is found or a designated final temperature is reached.

Simulated annealing also has quite a dedicated following and has been very successful in a broad range of NP-hard optimization problems. Quite a few papers surveyed in this chapter show the application of SA to mobile communications.

17.2.3 Tabu Search

Again, we add some additional detail here to the description of this algorithm given in Chapter 1. Tabu search is based on the premise that intelligent problem solving requires incorporation of adaptive memory (Glover, 1989; 1989a; Glover and Laguna, 1997). Unlike GAs and SA, tabu search is not purely a family of stochastic search techniques; it is a non-random metaheuristic algorithm for combinatorial optimization, with several variants which introduce stochastic elements to the search. Like the other two techniques it also provides means for escaping local minima.

In TS, a finite list of forbidden moves called the tabu list T is maintained. At any given iteration, if the current solution is x, its neighborhood $N(x)$ is searched aggressively to yield the point x' which is the best neighbor such that it is not on the tabu list. Note that is not required that $f(x') \leq f(x)$, only that $f(x') = \min(f(x^+) | x^+ \in N(x) - T)$. As each new solution x' is generated, it is added to the tabu list and the oldest member of the tabu list is removed. Thus the tabu list prevents cycling by disallowing repetition of moves within a finite number of steps (determined by the size of the list). This, along with the acceptance of higher cost moves, prevents entrapment in local minima. It may also be desirable to include in the tabu list attributes of moves rather than the points themselves. Each entry in the list may thus stand for a whole set of points sharing the attribute. In this case, it is possible to allow certain solutions to be acceptable even if they are in the tabu list by using what are called *aspiration criteria*. For example, one such criterion is satisfied if the point has a cost that is lower than the current lowest cost evaluation. If a neighborhood is exhausted, or if the generated solutions are not acceptable, it is possible to incorporate into the search the ability to jump to a different part of the search space (this is referred to as *diversification*). One may also include the ability to focus the search on solutions which share certain desirable characteristic (*intensification*) by performing some sort of pattern recognition on the points that have shown low function evaluations in the past.

Tabu search is a meta-heuristic technique, and it must be adapted to the problem at hand for it to be efficient. The choice of moves that generate the neighborhood of a point is problem-specific. Different implementations can be generated by varying the definition and structure of the tabu list (for example by deciding how tabu attributes are determined), the aspiration criteria, intensification and diversification procedures, etc. To speed up the search, faster ways to determine the best neighbor are required.

Like the other two global search techniques, TS has also been applied successfully to a large number of NP-hard optimization problems and has been shown to compare favorably with GA and SA (Aarts and Lenstra, 1997). A search of the literature, however, has not revealed extensive application of TS in mobile communications though there is certainly no *a priori* reason why it cannot be applied to the same problems as GA and SA.

18.3 Applications to Mobile Communications

In this section, we review papers describing the applications of these global search techniques to optimization problems in mobile communications. The number of problems in mobile communications that have been formulated and recognized as hard combinatorial optimization problems and processed with these algorithms is as yet quite small. Thus, while the papers discussed here by no means cover the entire range of possible applications of these techniques to mobile communications, they do represent nearly all such attempts to date, and should provide a good overview of the subject.

It will be noted that most of the papers surveyed here are concerned with the application of GAs. This is not the result of any inherent superiority in terms of efficiency or ease of programming in using GAs as opposed to SA and TS, but perhaps an indication of its general popularity among researchers interested in optimization for wireless telecommunications.

Several papers discussing global optimization for system-level resource allocation, design and planning in cellular mobile systems are described. Papers covering the use of these techniques for lower-level problems such as optimal multi-user detection in CDMA technology, design of frames in TDMA schemes, blind channel identification and equalization are also described.

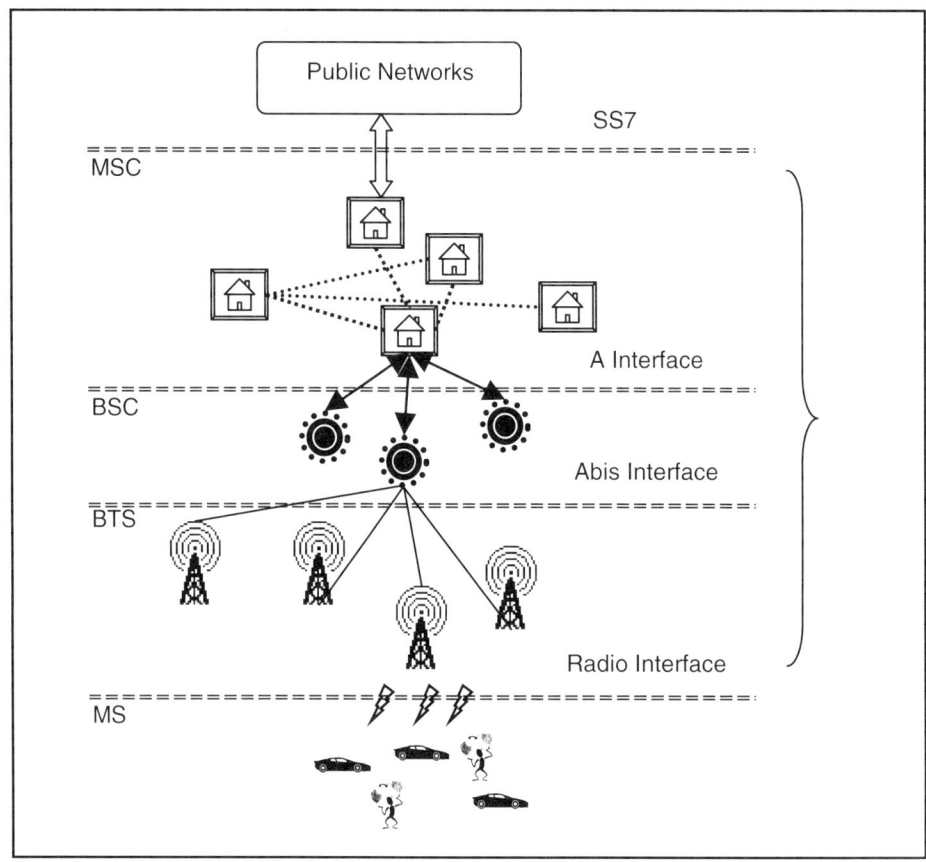

Figure 18.1 GSM network architecture.

18.3.1 Design of Fixed Network Topology

In cellular communication networks, the mobile units are connected to the public networks, such as PSTN, ISDN and other data networks, through a network hierarchy determined by the system architecture. Figure 18.1 shows such an architecture for the European GSM standard (Rappaport, 1996). The Mobile Stations (MS) in each cell communicate over the radio interface with the Base Transceiver Stations (BTS). BTSs connect to Base Station Controllers (BSC) via microwave links or dedicated leased lines on what is called the *Abis interface*. Each BSC may control several hundred BTSs, and some low-level functions such as mobile handoffs are made by the BSC. The BSCs in turn are connected to the Mobile Switching Centers (MSC) via the *A interface*. The MSC controls traffic among all the BSCs and there is a sub-network between MSCs at this level that includes units such as the Home Location Register (HLR), Visitor Location Register (VLR), the Authentication Center

(AuC), and the Point of Interconnect (POI) to the public networks. Thus, there is a large part of the communication hierarchy in a mobile cellular system which is a fixed wired network.

The cost of the topology of a fixed network depends upon several factors, including the cost of the nodes, the cost of links, link flow and capacity, and constraints such as the maximum number of links for each node. The problem of designing minimum cost network topologies is closely related to the minimum spanning problem in graph theory and is often very complicated; an exhaustive enumeration of topologies to obtain the optimal arrangement becomes infeasible for even moderately sized networks. Network topology design is perhaps the most studied application of global search techniques to telecommunications (for example, Celli et al. (1995), Costamagna et al. (1995), Ko et al. (1997a), Pierre and Elgibaoui (1997)).

The design of the fixed portion of the GSM network and the closely related Digital Cellular System 1800 (DCS1800) is performed in Shahbaz (1995) using a GA-based Genetic Optimizer for Topological Network Design (GOTND). The cost function to be minimized is defined as follows:

$$f = \sum_{\forall n} C_n^{NODE} + \sum_{\forall p} C_p^{POI} + \sum_{\forall l \in L^{BTS \to BSC}, L^{BSC \to MSC}, L^{MSC \to MSC}} C_l^{LINK}$$

$$\text{subject to}: F_l \leq C_l, \quad \forall l \quad (18.2)$$

$$\text{and} \quad \xi \leq 0.001$$

where C_n^{NODE} is the cost of all nodes of type n, C_p^{POI} the cost of the pth POI, and C_l^{LINK} the cost of the lth link which is one of types:

$$L^{BTS \to BSC}, L^{BSC \to MSC}, \text{ or } L^{MSC \to MSC}$$

F_l, C_l represent the flow and capacity of each link; and ξ is the call blocking probability.

In this chapter each candidate solution is represented by a set of seven chromosomes. The first four chromosomes represent the x and y coordinates for the BSCs and MSCs. The last three chromosomes describe the existence of links between BTSs and BSCs, between BSCs and MSCs and between MSCs themselves. The standard fitness based selection mechanism is used. Instead of the generic crossover and mutation operations, however, the GOTND employs 17 problem-specific operators. These include "Move one/some BSCs randomly", "move one/some MSCs randomly", "move MSC and BSC coupled randomly", "move MSC to BSC", "push BSC toward all connected BTS(s)", "push BSC(s) toward connected MSC", etc. among others. Different weighted combinations of these operators are tested to determine those that yield a fast convergence rate. Thus the operators that are more successful are used more often than less successful operators during the optimization.

In Shahbaz (1995) the GOTND is applied to an example scenario with 100 BTSs, 4 BSCs and one MSC that is also the POI to the public telecommunication networks. The cost data for the links is assumed to be piecewise linear and was obtained from German Telecom. The results showed that a cost reduction of 19% over the best network in the first iteration could be achieved using the GOTND.

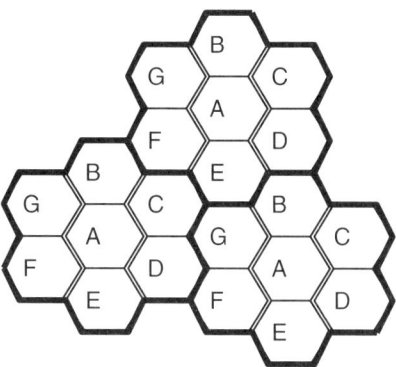

Figure 18.2 Frequency re-use in seven-cell clusters.

18.3.2 The Channel Allocation Problem

One of the most basic limitations of mobile radio communication systems is the restricted spectral bandwidth. Fundamental to the cellular concept is the idea of *frequency reuse* by which the same frequencies/channels can be reused in different geographical locations (Gibson, 1996; Rappaport, 1996). Each such location is referred to as a *cell* (usually hexagonal in shape), and is allocated a set of channels according to the expected demand in that cell. The entire spectrum is allocated to a *cluster* of cells arranged in shapes that allow for uniform reuse patterns. The geometry of the hexagonal cells restricts the number of cells per sector N to the discrete values

$$N = i^2 + ij + j^2 \tag{18.3}$$

where i and j are integers. Thus clusters can only accommodate 1, 3, 4, 7 ... cells. Figure 18.2 shows an illustration of the frequency reuse in a 7-cell cluster system.

The allocation of the limited number of channels to each cell in a mobile communication system is a much-studied problem (Katzela and Nahshineh, 1996). The channels must be allocated in such a way as to satisfy:

Co-channel interference constraints: These constraints are due to radio interference between cells that have the same channels,

Co-cell (or *adjacent channel*) *constraints*: These constraints arise because of the radio interference between adjacent channels in the same cell that are not separated by some minimum spectral distance

Cell demand requirements: These indicate how many channels are required in each cell due to their unique traffic patterns.

It has been shown that the Channel Allocation Problem (CAP) is equivalent to the generalized coloring problem in graph theory, a problem that is known to be NP-complete (Hale, 1980). There are essentially two kinds of allocation schemes – Fixed Channel Allocation (FCA) and Dynamic Channel Allocation (DCA). In FCA the channels are permanently allocated to each cell, while in DCA the channels are allocated dynamically upon request. DCA is desirable, but under heavy traffic load conditions, FCA outperforms most known DCA schemes (Raymond, 1994). Since heavy traffic conditions are expected in future generations of cellular networks, efficient FCA schemes become more important. All of the papers surveyed that apply global search techniques to CAP are FCA schemes (Funabiki and Takefuji, 1992; Duque-Anton et al., 1993; Jaimes-Romero et al., 1996; Lai and Coghill, 1996; Min-Jeong, 1996; Ngo and Li, 1998).

Let us assume that there are C channels to be assigned to N cells. Typically, the interference constraints are modeled by an N-by-N compatibility matrix D. The diagonal elements of this matrix, $D_{x,x}$, represent the co-cell constraint – the number of frequency bands by which adjacent channels assigned to cell i must be separated. The non-diagonal elements $D_{x,y}$ represent the number of frequency bands by which channels assigned to cells x and y must differ. If the compatibility matrix is binary, then these constraints are expressed more simply – if the same channel cannot be reused by cells x and y then $D_{x,y} = 1$, and if it can be reused then $D_{x,y} = 0$.

The traffic requirements for the cells are modeled by a demand vector T of length N that represents the number of required channels in each of the cells. The assignment to be generated is denoted by an $N \times C$ binary matrix A which is chosen such that

$$A_{x,k} = \begin{cases} 1 & \text{if cell } x \text{ is assigned channel } c_k \\ 0 & \text{otherwise} \end{cases} \quad (18.4)$$

In general, the cost due to the violation of interference constraints in this problem is given as:

$$f' = f_{co-cell} + f_{cochannel} = \alpha_1 \sum_{x}^{N} \sum_{i \neq j}^{C_x} \sum_{j}^{C_y} \Phi(i,j) + \alpha_2 \sum_{x \neq y}^{N} \sum_{y}^{N} \sum_{i}^{C_x} \sum_{j}^{C_y} \Psi(x_i, y_j) \quad (18.5)$$

where:

$$\Phi(i,j) = \begin{cases} 0 & \text{iff } |c_i - c_j| < D(x,x) \\ 1 & \text{otherwise} \end{cases}$$

is a measure of the co-cell constraint satisfaction, and

$$\Psi(x_i, y_j) = \begin{cases} 0 & \text{iff } |x_i - y_i| < D(x,y) \\ 1 & \text{otherwise} \end{cases}$$

is a measure of the co-channel constraint satisfaction, x_i, y_j, the assigned frequencies for the ith and j^{th} channels of cells x and y respectively, C_x the number of channels in the xth cell, and α_1, α_2 are constants to weigh the relative importance of the two constraints. The cost due to the violation of traffic demand requirements can be modeled explicitly as an error term:

$$f_{traffic} = \sum_x^N \left(T_x - \sum_k A_{x,k} \right)^2 \qquad (18.6)$$

The cost function to be minimized can be expressed as:

$$f = f' + f_{traffic} \qquad (18.7)$$

If the traffic demand requirements are incorporated implicitly by only considering those assignments which satisfy them, then the cost function only consists of the interference-constraint violation term:

$$f = f' \qquad (18.8)$$

subject to:

$$\sum_k A_{x,k} = T_x, \forall x$$

Some of the papers that describe the application of global search techniques to this problem are described below.

In Lai and Coghill (1996), a genetic algorithm is used to determine an optimal channel assignment. In the encoding chosen in this paper for the GA, chromosomes whose total length is the sum of all channels required for each cell represent each possible allocation. The traffic demand is therefore incorporated into the representation. A typical chromosome consists of a linear arrangement of the channels for each cell listed in order. The standard mutation operator is chosen, while a slightly modified Partially Matched Crossover (PMX) operator is designed that performs a crossover while resolving (correcting) any channel constraint-violations that may arise from the crossover to improve the performance of the algorithm. The algorithm was tested on a homogeneous cellular network of 49 cells where only three channels are available, and on data taken from an actual inhomogeneous cellular network consisting of 25 cells and 73 channels (known to be sufficient). In both cases the algorithm was able to generate conflict-free allocations. In the case of the inhomogeneous network example, the setting of a higher value of α_2 (i.e. emphasizing co-channel constraint satisfaction over co-cell constraint satisfaction) was empirically observed to speed up convergence.

Genetic algorithms are also used in Jaimes-Romero et al. (1996) for frequency planning. A binary compatibility matrix is used to model the co-channel constraints. The paper presents the performance of two versions of genetic algorithms – the simple GA with standard operators and binary representation, and a hybrid GA where an integer representation is used for the channels. For the simple GA, the demand requirements are explicitly incorporated into the cost function as in equation 18.7. For the hybrid GA a representation which has the demand vector built into it is chosen. Thus a modified fitness function is chosen consisting only of the co-channel constraint satisfaction term. The hybrid GA uses uniform crossover instead of 1-point crossover (referring to the number of loci where recombination takes place). Furthermore, a steepest descent (local search) algorithm is incorporated at the end of each generation to perform a more thorough search of the solution space. The neighborhood for this local search is defined by selecting a different channel for each cell in conflict in a particular assignment, and choosing the best such allocation.

The two algorithms were tested on cellular network scenarios with uniform and non-uniform traffic loads for two kinds of cellular systems – the standard planar system (with 12 channels and 21 cells), and a linear system (12 channels and 9 cells laid out side-by-side). It was observed empirically that the hybrid GA performs better than the simple GA, which, for a fixed upper limit on number of generations (around 200), was unable to resolve conflicts in all but the simplest cases.

Yet another attempt at using GA for fixed channel assignment is described in Ngo and Li (1998). The authors describe a modified genetic-fix algorithm that utilizes an encoding technique called the *minimum-separation encoding scheme*. The number of 1's in each row of the binary assignment matrix A corresponds to the number of channels allocated to the corresponding cell. To satisfy the demand requirement this would normally be constant. Each chromosome is a binary string that represents the matrix A through a concatenation of its rows. The genetic-fix algorithm defines mutation and crossover operators for a binary chromosome in such a way that the number of ones (the 'weight' of the vector) is always preserved. The binary encoding is further compressed by taking into account the co-cell constraint that no two channels in the xth cell can be closer than $d_{min} = D_{x,x}$. This would imply that each 1 in each row of the A matrix must be followed by $(d_{min} - 1)$ zeros. The minimum-separation encoding scheme takes advantage of this fact by eliminating $(d_{min}-1)$ zeros following the 1. Thus, the bit string <1000100100> is represented by <1011> when $d_{min} = 3$. This compression reduces the search space further. The implicit encoding of the co-cell constraints and demand requirements in this manner means that the cost function to be minimized only contains the co-channel constraint violation cost.

The genetic fix algorithm was tested on a suite of five problems with varying numbers of cells (4, 21 and 25), and numbers of channels (ranging from 11 to 309), along with different compatibility matrices and demand vectors. The results of this algorithm were compared with those from a paper (Funabiki and Takefuji, 1992) that applied a neural network to the first four of these problems using the frequency of convergence (the ratio of the number of successful results to the total number of runs) as a criterion. The genetic-fix algorithm was found to outperform the neural network in all cases. The fifth problem was for a 21-cell system where the algorithm was able to find a conflict free assignment with fewer channels than previously reported in the literature.

Simulated annealing is applied to the channel allocation problem in Duque-Anton *et al.* (1993). In the cost function used by the authors, only co-channel interference is considered, and the traffic demand term is explicitly specified. The initial temperature T_o for the annealing was chosen by starting with $T=0$, and increasing it until the ratio χ of accepted to proposed transitions was between 0.7 and 0.9. The cooling schedule is chosen such that

$$T' = T \cdot \exp\left(-\frac{\lambda T}{\sigma}\right) \qquad (18.9)$$

where λ is a parameter with value between 0 and 1, and σ is the standard deviation of the cost at level T. The transitions used to define the neighborhood of an acceptable solution include the basic 'flip-flop' (add one channel to a cell, and remove one – preserving the number of channels assigned), and the dense-packing move (find all nearest co-channel cells, pick the channel that is most used in these cells and switch it on at the current cell; randomly select and turn off one of the other channels currently in use in this cell). To further enhance the ability of the annealing algorithm to evade local minima, the authors allow multiple flip-flops in one move (this permits greater movement in the solution space). The paper further discusses how additional constraints imposed on this basic allocation scheme, such as soft interference constraints, hard and soft pre-allocated channels, and minimized spectrum demand, may be incorporated into the cost function.

The simulated annealing algorithm is tested on a 7-cell cluster scheme for an inhomogeneous network with 239 cells, 38 channels, some pre-allocated channels and varying demand. Simple (only flip-flop moves) and sophisticated (flip-flop, dense-packing and multiple flip-flops) neighborhood schemes are compared in terms of convergence percentage, final cost and computing time. As expected, the sophisticated schemes perform significantly better, demonstrating the importance of having a good neighborhood definition.

Simulated annealing and tabu search are compared for the CAP in Min-Jeong (1996). The cost function to be minimized is given as

$$f'' = \sum_{j \in C} \max_{x \in N}\{A_{x,j}\} + \frac{1}{\psi} f \qquad (18.10)$$

where f is as defined in equation 18.8 with the traffic demand defined implicitly. The first term in the expression attempts to minimize the bandwidth allocated, and ψ is a small positive number that gives a larger weight to the second term that represents the interference constraints.

For the SA, the initial temperature is chosen as in Duque-Anton et al. (1993) by using the acceptance ratio χ. The cooling schedule is taken from Aarts and Korst (1989):

$$T' = \frac{T}{1 + \frac{T \ln(1+\delta)}{3\sigma}} \qquad (18.11)$$

where δ is a *distance parameter* chosen to be 0.5 in this paper. For the TS algorithm described in the paper, two tabu lists are maintained – one for the cell (of size 7), one for the channels (size 1). An aspiration criterion is used to allow even tabu moves if they are found to yield lower cost than the current best cost. The dense packing and basic flip-flop moves (described before) are used with equal weights to determine the neighborhood for both SA and TS algorithms.

Applying the algorithms to examples of various sizes (9 to 49 cells with traffic demand between 1 and 10), the authors observe that simulated annealing obtains fairly good solutions, while the Tabu Search is extremely ineffective. It took longer than 24 hours on a SUN SPARC 10 workstation for problem instances with 16 cells or more, and on the problem instance of 9 cells it did not match the performance of the SA algorithm in terms of the minimum number of channels used. It is, however, entirely possible that the dismal performance of the TS in this experiment may have been due to an inefficient implementation of the algorithm.

The speculation regarding the poor results of TS in Min-Jeong (1996) is given some credence by the results in Hao et al. (1998) where the use of tabu search for frequency assignment is investigated more thoroughly. The formulation of the constraints and the expression for the cost function used in this paper are identical to those given in equation 18.4. To minimize the number of frequencies (NF) used in a channel assignment, the tabu search is carried out at an initially high value of NF. If a conflict-free assignment is found then the value of NF is decreased by 1, and this is repeated until one can no longer find such a conflict-free assignment.

The co-cell constraints are incorporated into a candidate assignment x given as follows:

$$x = \langle c_{1,1} ... c_{1,T_1} ... c_{x,1} ... c_{x,T_x} ... c_{N,1} ... c_{N,T_N} \rangle \qquad (18.12)$$

such that $\forall x \in \{1...N\}, \forall i, j \in \{1...T_x\}, |c_{x,i} - c_{x,j}| \geq D_{x,x}$, where $c_{x,j}$ is the jth channel allocated to the xth cell. This assignment satisfies the adjacent channel interference constraint by definition.

A neighbor x' of a solution point x can be obtained by changing the value of any channel that has been allocated in x such a way that the new value always satisfies the co-cell constraint. Further, a candidate list of new neighbors V^* is specified as follows:

$$V^* = \{ x' \in N(x) \mid x' \text{ and } x \text{ differ at the value of conflicting frequency} \} \qquad (18.13)$$

The size of this list varies during the search, and helps the search to concentrate on influential moves while reducing the size of the neighborhood. Since only a part of the frequency assignment is changed during each move, the speed of evaluation is reduced by differentially updating the cost as each move is made. The size of the tabu list, k, is made to be proportional to the length of the candidate list at each iteration, $k = \alpha |V^*|$ and thus varies as the algorithm proceeds. Again, the simple *aspiration* criterion is chosen where the tabu status of a move is cancelled if it leads to a solution better than the best solution encountered so far.

The TS algorithm thus designed was applied on a suite of real-sized problems taken from data provided by the French National Research Center for Telecommunications and compared with a local search algorithm (steepest descent) and the best known results obtained with other methods including simulated annealing. For each TS run of 100,000 iterations, the steepest descent algorithm was run 20 times for 5000 iterations. The local search algorithm repeatedly yielded sub-optimal results while the TS was able to find near-optimal solutions quite speedily for most of the instances selected. In comparing their tabu search algorithm with the simulated annealing results from Ortega *et al.* (1995), the authors showed that the TS algorithm was capable of matching, even outperforming SA in locating the minimal number of frequencies for channel allocation. The TS algorithm was also observed to be faster than the SA. Based upon further experimental evidence, the authors credit the performance of the algorithm to the usage of the conflict-based candidate list strategy and the incorporation of co-cell constraints into the candidate solution points (neither of which was done in Min-Jeong (1996), where TS did not fare well).

18.3.3 Optimal Base Station Location

One of the most challenging design problems in the planning and deployment of a mobile radio communication network is deciding on the optimal locations for base stations. Currently the location of base stations is often done in an *ad hoc* manner, via manual inspection of maps showing propagation properties of the area to be serviced. Given a list of potential sites in a service area where base stations may be located, the goal is to use knowledge of the radio propagation characteristics of the area to select sites in such a way as to minimize their number while maximizing coverage in the area. This would reduce network cost and complexity and be a great improvement upon the current methodology for setting up micro-cellular networks. The problem is very similar to the Minimum Dominating Set (MDS) problem from graph theory, which is known to be NP-hard (Garey and Johnson, 1979).

To perform such an optimization, the cost function necessarily involves the radio propagation characteristics of the area. This may be determined using sophisticated ray-tracing software (which could be a very time-consuming process), or by using empirical propagation models for path loss. Another issue in the design of the cost function is the trade-off between coverage and number of base stations. The more base stations there are, the greater coverage there is, but there is also correspondingly greater radio interference and network cost.

Genetic algorithms are applied to this task in Calegari *et al.* (1997). It is assumed that a list of N possible locations is known beforehand, such that if there were base stations at all those locations, 100% radio coverage is guaranteed. A binary encoding of the problem is chosen as follows: each chromosome consists of N bits, with a one at each bit if there is a base station at the location corresponding to that bit, and a zero otherwise. The simplest GA is first evaluated for the fitness function chosen as follows:

$$f = \frac{R^\alpha}{n} \tag{18.14}$$

where R is the radio coverage (the percentage of locations covered by the base stations selected), and n is the number of base stations that are present. α is a parameter that is tuned to favor coverage with respect to the number of transmitters and is assigned a value of 2 in this paper.

Experimental tests of this algorithm were performed on a 70 × 70 km region in France for $N = 150$, with the radio coverage from each possible location being computed by a radio wave propagation simulation tool. A solution with 52 base stations covering 80.04% was found, but since the algorithm was found to be very slow it was parallelized using the parallel islands model of GA. In this version of the GA, the population is distributed into sub-populations (islands) where they evolve quasi-independently towards the solution, and some individuals in each generation migrate from one island to another. A nearly linear speedup was observed using this technique and the resultant computational power enabled them to discover a good solution with only 41 base stations that covered 79.13% of the initial covered surface.

The optimal base station location problem for micro-cells in an urban city environment is solved using Simulated Annealing in Anderson and McGeehan (1994). An empirically determined model for path loss in urban areas given in Erceg et al. (1992), which takes as parameters the heights above ground of the mobile and base station antennas, was used to calculate coverage. While the authors concede that the assumptions made regarding path loss are not necessarily valid in reality, they are sufficient to demonstrate the applicability of SA to this problem.

The service area is modeled as a Manhattan-like grid of streets dividing the area into square city blocks (10 × 10m). A power goal is set such that at each grid point i a certain minimum level of radio reception P_i^* is guaranteed. The cost function is then defined as follows:

$$f = \sum_{i \in A} (P_i^* - P_i)^2 \tag{18.15}$$

where A is the set of all grid points where the actual received power P_i is less than the minimum requirement P_i^*.

Each possible solution is a location x. The neighborhood is defined by allowing the base-stations to make random moves while satisfying any constraints on the location such that the current temperature T at each stage defines the magnitude of these moves. The initial temperature T_o is chosen to be 80 m, the cooling schedule is such that $T' = 0.8T$, and at each temperature 100 moves are allowed. The algorithm is stopped when the difference between the maximum and minimum costs at a given temperature is equal to the maximum change in an accepted move at the same temperature. Since the units of cost and temperature are different, the metropolis criterion is modified as follows:

$$P_{x \to x'} = \min\left(1, \exp\left[-\frac{\left(\frac{f(x') - f(x)}{f(x)}\right)}{\left(\frac{T}{T_o}\right)^2}\right]\right) \tag{18.16}$$

This SA technique was tested on two problems – one involving the location of a single transmitter in a 4 × 4 grid. In this simple problem instance the optimum location could be determined by an exhaustive analysis of all possible grid locations. In all runs (with different initial starting positions) the optimization arrived at a final position within a few meters of the optimum location. Another problem with an 8 × 8 grid and base stations was performed and a figure showing the solution and resultant coverage is presented in the paper. The authors observe that different runs yield different base-station locations with the approximately the same overall costs indicating that near-optimum solutions are not unique.

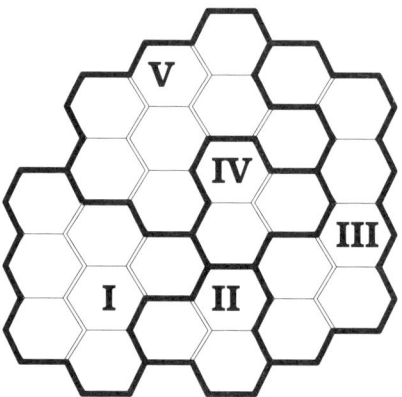

Figure 18.3 Registration areas in a cellular system.

18.3.4   Mobility Management

As mentioned before, one of the goals of PCS is to ensure that users are able to communicate with each other wherever they may be. Users should be able to roam at will, crossing cell boundaries system-wide while they are mobile without losing an ongoing call or being prevented from initiating or receiving calls. This necessitates mobility management, since the system needs to track the location of the user in order to provide service. For mobility management purposes, the system is divided into *Registration Areas* (RA) consisting of several cells, as seen in Figure 18.3. The user's location is maintained in a location database, such as the Visitor Location Register in GSM mentioned before, and this database is updated each time users enter or leave the RA. This update is referred to as a *location update* or LU. When a mobile station is called, its current RA is determined from such a database, and all base stations in the RA broadcast *paging* messages over a reserved forward control channel (FOCC) to inform the mobile unit of this call request.

The LUs result in increased network traffic, and also place a load on the distributed location databases, resulting in greater complexity of the databases' implementation. The paging messages also consume radio resources and their traffic is often limited by the bandwidth of the FOCC (a condition referred to as *paging bound*). It is thus desirable to design the system in such a way as to reduce both LU and paging costs. There is a trade-off between the LU and paging costs which varies with the size of the RA: If the RA is large, there are fewer inter-RA crossings resulting in a lower LU cost, but the number of base stations that need to be paged increases correspondingly. Genetic Algorithms have been used to design optimal RA to reduce LU cost while maintaining paging bound constraints in Wang et al. (1998). Within each RA, only one mobile station responds to the multiple pages transmitted by base stations in the RA. To reduce the paging cost, it is possible to subdivide the RA into paging zones, ordered based on a pre-determined probability of locating the mobile user at different locations within the RA. The optimal planning of such paging zones using GA is treated in Junping and Lee (1997).

The RA planning in Wang et al. (1998) is based on the assumption that the cell planning has already been done and the LU and paging traffic have been estimated for each cell. It is assumed that the paging bound B for each RA is fixed. The RAs are planned so that they each contain a disjoint set of cells. The task is to design the RAs to minimize the following cost function:

$$f = \frac{\alpha_1}{2} \sum_{i=1}^{M} LU_i + \alpha_2 \sum_{i=1}^{M} \max(0, (P_i - B)) \qquad (18.17)$$

where P_i and LU_i are the paging traffic and LU traffic respectively in the ith registration area, and M is the total number of registration areas in the system. The second term in equation 18.17 is non-zero if the paging bounding is exceeded in the ith registration area due to an increased number of cells in it. α_1 and α_2 are constants used to weigh the relative importance of the LU cost and the paging cost.

A binary representation chosen for the chromosomes is based on the cell borders. The borders are numbered sequentially and the corresponding bit is one if that particular cell border is to be a border between two adjoining RAs, and zero otherwise. This representation facilitates the evaluation of the location update cost. The first term in equation 18.17 can be rewritten for this border notation as follows:

$$f = \frac{\alpha_1}{2} \sum_{j=1}^{n} v_j w_j + \alpha_2 \sum_{i}^{M} \max(0, (P_i - B)) \qquad (18.18)$$

where n is the total number of borders, w_j the crossing intensity of the jth border and v_j the value of the jth bit of the chromosome being evaluated (i.e. a determination of whether the jth border is an inter-RA border). It is also shown in the paper how additional terms may be incorporated into the cost function to deal with useless location updating (when no calls arrive between successive LUs) and presets that arbitrarily force certain borders in the chromosome to be 0 or 1. The standard GA operators of mutation and crossover are used. The mutation rate (initially 0.02) is halved after several generations to increase convergence. The GA is terminated after 1000 generations.

The GA is compared with hill-climbing for five systems with varying size (19 to 91 cells) where the paging cost for each cell and crossing intensity for each border were generated with a normal distribution. The result of RA planning over 100 runs shows that GA's consistently outperform hill-climbing providing improvements ranging from 10–30%. The authors conclude that GAs appear to be a valuable approach to the planning of registration areas in PCS networks.

A tracking strategy utilizing the mobility patterns of individual mobile stations is used in Junping and Lee (1997) to plan the partition of RAs into paging zones to minimize paging cost. First a multilayered model is developed based on mobile phone usage for different times of activity during a typical workday for each user – work, home, social. Each mobile user's activity is monitored for each such layer to develop statistical mobility patterns that describe the chance of locating each user in a particular cell during a particular time of the day. Then the cells in the RA are partitioned into paging zones for each user k, and layer j such that each zone consists of cells with similar likelihood of locating that user:

$$P_{k,j} = \{P_{k,j,l}\} \tag{18.19}$$

These zones are arranged in order of descending likelihood of locating the user denoted by prob($P_{k,j,l}$) – the probability of locating the kth user in activity layer j in the lth zone. Now when a call is received for the user in the RA, the zones are paged in this order – which is more efficient than paging the entire area. The cost function derived for each paging zone assignment is based on the mobile location probability for each zone in a partition:

$$f = N(P_{k,j,1}) + \sum_{l} \left(1 - \sum_{i=1}^{l-1} prob(P_{k,j,i})\right) \times N(P_{k,j,l}) \times \alpha \times \beta \tag{18.20}$$

where $N(P_{k,j,l})$ represents the number of cells in the lth zone, α the traffic in the FOCC per page, and β the traffic in the MSC per page. Since paging zones are unique for each user, this optimization must be carried out for each user separately.

The encoding chosen for the GA utilizes an integer representation. The cells in the RA are numbered sequentially and correspond to a locus on the chromosome. The value of the chromosome at that locus is the paging zone to which the cell belongs. Thus the string '11231' indicates that the first, second and fifth cells are in zone 1, the third cell in zone 2 and the fourth cell in zone 3. Zone 1 has the highest mobile location probability. The cells for a given paging zone need not be adjacent as seen in Figure 18.4. The standard genetic operators of crossover and mutation are used. The algorithm is tested for different numbers of paging zones (from 1 to 4) and it is observed that the greater the number of zones, the lower the paging cost. Even a 2-zone partition is observed to have a degree of magnitude lower cost than the traditional area-wide paging scheme.

18.3.5 Call Management

Another issue of mobile communication system planning is call management. One aspect of this is the determination of a call-admission policy – deciding under what conditions a new call to a mobile in a particular cell should be accepted. Allocating channels to every user whenever they are available (the admit-if-possible or AIP strategy) is not necessarily an optimal strategy because this may result in an inability to accept hand-off calls from a neighboring cell.

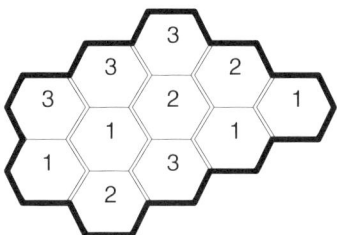

Figure 18.4 Cells belonging to three paging zones within a registration area (for a unique user).

Figure 18.5 Linear cellular system $N = 4$, $k = 1$ (two neighbors considered).

The problem of evolving an admission call policy for each cell based only on local state information (restricted to the cell and its neighbors) is considered in Yener and Rose (1997) using genetic algorithms. In a mobile communication system it is generally considered better to block a new call than drop an ongoing one. The cellular system performance (which depends upon the call admission policy) is the cost function to be minimized:

$$f = P_b + \omega P_h \tag{18.21}$$

where P_b is the new call blocking probability and P_h the hand-off dropping probability. The factor ω is used to indicate the extent to which dropped calls are considered less desirable than blocked calls.

In each cell, the number of states refers to how many of its C channels are occupied. For a linear cellular system with N cells, the total number of global states is $(C+1)^N$ for each decision. The policy is constructed to consist of three binary decisions: (1) admit a new call (2) admit a hand-off call from the left cell; and (3) admit hand-off from right cell. In a global admission policy there is one decision for each global state (thus, $3(C+1)^N$ decisions in all). The states in a global admission policy satisfy the Markov property since each succeeding state depends only upon the previous state of the network. Markov Decision Process (MDP) based techniques such as Dynamic Programming (Sacco, 1987) may be used to solve such a problem but the state space becomes too large for moderate sized networks. If a local policy is considered, and only state information from $2k$ nearest neighbors are used, the state space is $3(C+1)^{2k+1}$. In this case, however, the states are no longer Markov states and MDP techniques are no longer applicable. The state space is still very large: for an example with 9 channels, a policy with $k = 1$ (using information from the current cell and both immediate neighbors in a 1-D linear cellular system)needss 3000 bits ($\sim 1 \times 10^{903}$ possibilities).

A standard GA is used in Yener and Rose (1997), where each chromosome represents a local policy, with admit or reject decisions for the new call and handoff requests for each local state of the system. The local policies are evaluated via Monte Carlo simulation using the cost function described in equation 18.21. The algorithm was tested on a small 4-cell system for both $k = 0$ and $k = 1$ (seen in Figure 18.5) to compare it with the optimal solution for a global policy obtained with MDP and the AIP strategy. As the factor ω increases, performance is observed to become worse since more global information is needed to better determine hand-off possibilities. But in all cases the solutions provided with GA give much better system performance than the AIP strategy. For a larger system with 16 cells and 9 channels the GA (for cases $k = 0$, and $k = 1$) was compared with AIP strategy and the best single-threshold handoff reservation policy (having a threshold t such that if cell occupancy $s > t$, new calls will not be accepted). Again the GA-based solutions (even for $k = 0$) were much better than either of these. Upon examination of the GA-based solutions for $k = 0$, the authors observed that they tended to suggest two-threshold policies with thresholds t_1 and t_2 such that new calls are accepted if the occupancy s is less than t_1 or greater than t_2, but not if it is between the two. The authors suggest an explanation for the result: when the cell is nearly empty ($s < t_1$) new calls may be admitted without fear of not having room for handoff calls. When it is a little full ($t_1 \leq s \leq t_2$) hand-offs are expected and so some channels are reserved for handoff. When the cell is nearly completely occupied ($s > t_2$), it becomes less likely that the neighboring cells are occupied and hence there is a reduced handoff possibility and new calls are accepted again.

The GA was thus able to identify a novel policy structure. Based on a comparison of performances, the authors suggest that since the state space is considerably smaller for policies with $k = 0$ than $k = 1$ (30 bits vs. 3000 bits to represent the policy), and they offer nearly similar results, it is advisable to search for single-cell occupancy based policies.

18.3.7   Optimal Multi-user Detection in CDMA

The applications presented thus far are all geared towards high-level design of mobile communication networks. We now consider the difficult low-level optimization problems in mobile communications that have been tackled using global search techniques. One such problem is that of optimal multi-user detection in Code Division Multiple Access (CDMA) systems. Spread spectrum CDMA is a fairly new wide-band technique used to share the available bandwidth in a wireless communication system. In CDMA the message signal is multiplied by a large-bandwidth *spreading signal* – a pseudo-noise sequence with a chip rate much greater than the rate of the message. Each user is assigned a unique PN codeword that is nearly orthogonal to the other codewords. For the receiver to detect the message signal intended for the user, it must know the corresponding unique codeword used by the transmitter. In the single-user receiver the received waveform is input to a correlating matched filter and the signals of the other users are treated as noise (along with noise due to the channel). However, because the other codewords are still somewhat correlated with that of the receiving user's codeword, this receiver suboptimal in minimizing errors in detection. It has been shown that the optimal receiver would be one that detects the messages for all users simultaneously, i.e. one that performs simultaneous multi-user detection (Verdu, 1998). Formally, considering a synchronous CDMA channel with Additive White Gaussian

Noise (AWGN) characteristics, given that the kth user (of the K users sharing the channel) is assigned a codeword $s_k(t), t \in [0,T]$, the signal $r(t)$ at the receiver is:

$$r(t) = \sum_{k=1}^{K} b_k s_k(t) + \sigma n(t), \quad t \in [0,T] \quad (18.22)$$

where $n(t)$ is a unit spectral density white Gaussian process and $b_k \in \{-1,1\}$ represents the transmitted bits in the message for the kth user. The output of a bank of matched filters that correlate this signal with the codewords for each user is given by:

$$y_k = \int_0^T r(t) \cdot s_k(t) dt \quad (18.23)$$

It has been shown that the output vector $y = [y_1, y_2 ... y_k]$ thus obtained may be used to determine the transmitted message vector $b = [b_1, b_2 ... b_k]$ in an optimal manner (Verdu, 1998). If H is the nonnegative matrix of cross correlation between the assigned waveforms:

$$H_{ij} = \int_0^T s_i(t) \cdot s_j(t) dt \quad (18.24)$$

then

$$y = Hb + n \quad (18.25)$$

where n is a Gaussian K-vector with covariance matrix equal to $\sigma^2 H$. The optimal multiuser detection procedure determining the message that was most likely sent is to determine the vector b that maximizes the following expression:

$$f = 2y^T b - b^T Hb \quad (18.26)$$

This combinatorial optimization problem is known to be NP-hard, as the search space of the possible multi-user message increases exponentially with the number of users.

Wang et al. (1998a) describes the application of GA to optimal multi-user detection. It is assumed that the bits are being decoded sequentially (one bit for each user at any time) using the Viterbi algorithm. To speed up the search and reduce the computational complexity the authors suggest reducing the number of bits by utilizing the information given by the output of the matched filter bank. A confidence measure based on the probability of a bit error in single-user detection is used to determine if some of the bits may be considered as detected. The bits b_k are thus divided into two groups – the group of detected bits and the group of unknown bits. The search space is then restricted to finding values only for the unknown bits. In the genetic algorithm, each chromosome consists of the number of bits that are still unknown. The standard mutation operation, uniform crossover and an elitist selection scheme were used. A total of 20 chromosomes are present in each generation and for each step of the detection process only 6 generations were performed (only 120 evaluations are performed in the optimization to detect the unknown bits).

To examine the proposed detector, Monte Carlo simulations were carried out comparing the GA-based detector with the linear decorrelating detector and the linear MMSE detector for $k = 20$ users. The authors show that the performance of the GA-based detection is very close to that of the MMSE detector and better than the decorrelating detector, and further that the proposed detector is actually computationally faster than these two linear detectors.

A very brief paper (Juntti et al., 1997) also describes the application of genetic algorithms to this problem. The authors reported using Monte Carlo simulations to study the performance of the genetic detector on a problem with 20 users and spreading sequences of length 31. The algorithms were compared for different types of initial guesses (random and those generated by sub-optimal detectors). They concluded that the algorithm was not robust when random initial guesses were used, but was able to improve the detection performance when combined with such sub-optimal detectors.

18.3.7 TDMA Frame Pattern Design

Besides CDMA, another multiple access scheme very commonly used in mobile communication systems is Time Division Multiple Access (TDMA). In TDMA the information for different users is carried in separate time slots in each frame of data. One of the goals of PCS is to be able to carry multimedia multi-rate information on the network. This is easily done in TDMA because users have the flexibility of determining different rates and error control coding for traffic of different kinds of information – such as systems that carry both voice and data.

One way of integrating voice and data in TDMA is to divide the frame into fixed sub-frames for each. However, this *frame pattern* has the significant disadvantage of not being able to adapt to traffic changes. A dynamic scheme that allows a data user to borrow voice slots when needed improves utilization. In schemes where a contention-based transmission technique is used for data, the positions of available data slots within a frame affect the performance of the system.

In Chang and Wu (1993) the problem of designing frame patterns to maximize throughput is considered and solved using a simulated annealing algorithm. It is assumed that each frame consists of N independent slots, N_t of which are assigned for data service (including nominal data slots and unassigned voice slots) while the rest are assigned for voice traffic. The frame pattern of the integrated TDMA system is defined as

$$\vec{x} = (x_1, x_2, \ldots x_{N_t}) \tag{18.27}$$

where $x_i \geq 1$ represent the number of slots between successive data slots. It is assumed that the slotted ALOHA random access protocol (Bertsekas and Gallager, 1992) is used for data transmission, that the queuing behavior of data for a given pattern is able to reach steady state and that data traffic is a Poisson process with mean arrival rate G packets/slot. For a given frame pattern, the mean data throughput is the function to be maximized. To perform minimization, the cost function may be taken as the negative of this throughput measure:

$$f = -\frac{G}{N_t} \sum_{i=1}^{N_t} x_i \cdot \exp(-G \cdot x_i) \qquad (18.28)$$

The number of frame patterns for a given N_t and N can be very large indeed. For a TDMA frame with $N = 60$, and $N_t = 10$, there are as many as 7.5×10^{10} possibilities. Since a brute force technique is not possible in this huge search space, global search techniques are ideal for this NP-hard problem.

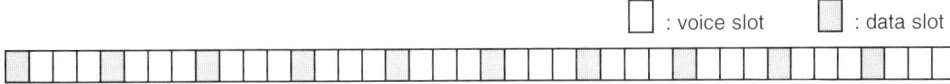

Figure 18.6 Optimal TDMA frame pattern ($N = 40$, $N_t = 10$, $x = [4,4,4,4,4,4,4,4,4,4]$).

For the SA, the initial temperature T_o is chosen to be $10\sigma_\infty$ (the standard deviation of the cost of explored states when T is infinite). The neighborhood function is defined by a simple move (add 1 to one of the x_i, and subtract it from another randomly) and a complex move which is simply a succession of simple moves. A further adaptive scheme is used to determine the selection probabilities of the moves based on data collected during the execution of the algorithm. The standard metropolis criterion is used. The selected cooling schedule is similar to that in equation 18.9, with an additional term ensuring a minimum rate of cooling. The algorithm is stopped when it is found that all the accessed states are of comparable costs at the given temperature.

The algorithm was tested for various loads G for a frame with $N = 40$ and N_t values of 5, 10 and 15. As anticipated, it was shown that there is an optimal frame pattern for a given G and N_t but none for all G and N_t. Due to the restriction of integer solutions for the optimal frame pattern, it was found that the optimal frame pattern is insensitive to small changes in G and thus there are a finite number of total optimal frame patterns. The maximum throughput of 0.368 occurs at $G = N_t/N = 0.25$ when the frame pattern is uniformly distributed – each slot separated from the succeeding one by an inter-distance of 4 when $N_t = 10$ (shown in Figure 18.6). Comparing the results with random frame patterns, throughput gains ranging from 10% to 120% were observed by the authors.

18.3.8 Data Equalization

One major problem with digital communications channels is the distortion of data due to Inter-Symbol Interference (ISI) (Proakis, 1995). The limited bandwidths and frequency selective fading (as observed in mobile radio environments) of these channels cause transmitted symbols to be smeared in time and interfere with adjoining bits. The effect of ISI can be reduced by *equalization*. An equalizing filter compensates for the characteristics of the channel and corrects the distortion, reducing errors in communication.

Most equalizers utilize filters where the output is non-recursively dependent upon present and previous values of the input and exhibit a Finite Impulse Response (FIR). Recursive Infinite Impulse Response (IIR) filters with feedback have outputs that depend not only on previous values of the input but also those of the output. Though susceptible to instability, IIR filters can perform better when the ISI extends over a long duration and typically require fewer weights (filter coefficients) to be adjusted to perform as well.

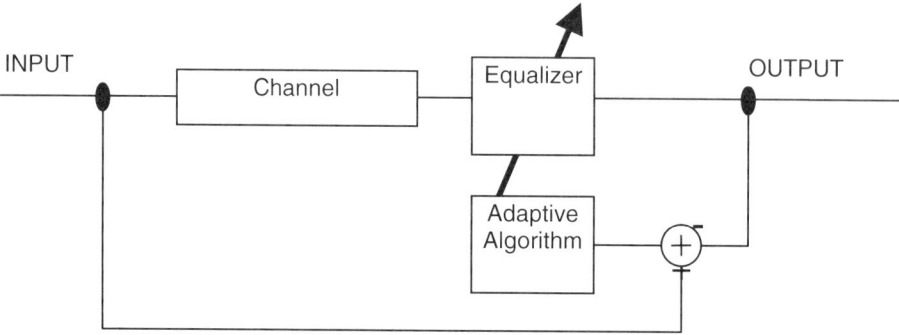

Figure 18.7 Adaptive data equalization.

Figure 18.7 shows the block diagram of an adaptive equalization scheme. The problem of determining the filter coefficients of an adaptive IIR data equalizer is a difficult one. A GA-based algorithm for such equalization is presented in White and Flockton (1994). The chromosomes in the GA represent the filter coefficients (both feedback and feed-forward). The GA used is a simple binary encoded one. The cost function to be minimized is the Mean Square Error (MSE) between the output of the equalizer and the input to the channel, which is known since a training data sequence is used to run the adaptive algorithm. The authors use a 6th order filter and test it on three different channel instances with varying characteristics. In these channels the bit-error rate due to ISI in the data transmitted without equalization ranged from 26% to 32%. After the adaptive equalization was performed, this error dropped to 0% in the first two cases, and 3% in the third case (which had channel noise added in besides the ISI distortion), demonstrating the effectiveness of the genetic adaptive algorithm.

18.3.9 Blind Channel Estimation

Adaptive channel equalizers utilize a training sequence at the start of transmission. But there are cases when the equalizers may have to start or restart without a training sequence – this is referred to as blind equalization. In the fading/multipath channels characteristic of mobile communications, such a restart may be required when there is temporary path loss due to a severe fading. Under such conditions blind channel identification schemes attempt to determine the coefficients of the channel from the statistical properties of the received signals, specifically Higher Order Cumulants (HOC). This knowledge of the channel may then be used to design the equalizer. The application of GA and SA to the problem of blind channel identification using HOC fitting is presented in Alkanhal and Alshebeili (1997) and Chen et al. (1997; 1998).

If the message transmitted is $s(k)$, the received signal $r(k)$ can be modeled as the result of ISI and noise as follows (Chen et al., 1997; 1998):

$$r(k)=\sum_{i=0}^{n_a} a_i s(k-i)+e(k) \tag{18.29}$$

The goal is then to determine the channel coefficients a_i (the channel is assumed to be of order n_a) using only $r(k)$ and some knowledge of the statistical properties of $s(k)$. It has been shown in Chen et al. (1997) that the problem can be expressed as the minimization of the following function:

$$f = J_4(a) = \sum_{\tau=-n_a}^{n_a} \left(C_{4,r}(\hat{\tau},\tau,\tau) - \gamma_{4,s} \sum_{i=\max\{0,-\tau\}}^{\min\{n_a,n_a-\tau\}} a_i a_{i+\tau}^3 \right)^2 \tag{18.30}$$

where $C_{4,r}(\hat{\tau},\tau,\tau)$ is the time estimate of the diagonal slice of the fourth-order cumulant for $r(k)$ and $\gamma_{4,s} = C_{4,s}(\tau_1,\tau_2,\tau_3)|_{\tau_1=\tau_2=\tau_3=0}$ is the kurtosis of $s(k)$.

In Chen et al. (1997), the minimization is approached using a version of genetic algorithms called μGA – where the population size is much smaller than usual and the population is reinitialized randomly while retaining the best individual found up to that point after each run converges. The standard crossover operator is used, but no mutation operation is performed since the reinitialization procedure itself introduces diversity in the process. This process is repeated until no further improvement is seen. A 16-bit number between –1 and 1 represents each channel coefficient, and the encoding is such that these coefficients are arranged linearly in the chromosome.

The algorithm was tested on two channels with known coefficients. To compute the time estimate of the fourth-order cumulant, 50,000 noisy received data samples were used. The final results were the average of 100 different runs where each run utilized a different sequence of received data samples and different initial populations. It was observed that the solutions always converged quickly to a value close to the global optimal channel estimate. The authors remark that the method is accurate and robust as demonstrated by the small standard deviations of the estimated channel coefficients over the different runs. They compare these results with results from a prior SA-based approach in terms of the number of function evaluations required for optimization. It is noted that the SA-based approach would have taken nearly six times as long for these problems. In a related paper (Alkanhal and Alshebeili, 1997) on GA-based blind channel estimation, third-order cumulant statistics are utilized instead and the simple genetic algorithm is used.

The application of an Adaptive Simulated Annealing (ASA) algorithm to blind channel identification with HOC fitting is described briefly in Chen et al. (1998). The function to be minimized is the same as given in equation 18.30. The ASA algorithm, described in this paper, is a technique to speed up the convergence rate of simulated annealing by using estimates of the partial derivatives of the cost function (taken with respect to the solution coefficients) to modify the annealing rate every few moves. The algorithm was tested by applying it to a channel with known coefficients. As expected, ASA was observed to converge to the optimal solution much faster than the standard SA.

18.4 Conclusion

This chapter has surveyed the application of three global search techniques – genetic algorithms, simulated annealing, and tabu search — to difficult optimization problems in the subject of mobile communications. The papers described are all very recent, and they cover a broad spectrum of problems in mobile communications, from network planning and design issues to lower level problems like detection of transmitted data and estimation of channel properties.

We have attempted to provide some detail on the implementations of the various techniques – this becomes a key issue with these algorithms as they are very general in their simplest form and there are a large number of parameters that may be selected to create different implementations. It must be observed, however, that these techniques appear to be quite robust and offer good results for a wide range of implementations as seen for example in the papers on channel assignment. There is, moreover, no empirical evidence of the inherent superiority of one of these techniques over the others for any given global optimization problem. In the few cases where the results of one technique are compared with another, the question of whether each technique was implemented to ensure the fastest convergence possible arises and remains unresolved because of the very nature of these algorithms. Researchers would do well not to ignore the possibilities inherent in each of these techniques and make their decisions on which technique to use based on at least a cursory comparison of their efficiency and ease of implementation.

The speed of convergence becomes important when these techniques are applied to time-critical problems where fast, online solutions are required. So far they have not been applied to such problems in wireless telecommunications because of the large number of function evaluations that are often required to find optimal or near-optimal solutions using these techniques. One of the trends of future research in this area would be the extension of the use of these algorithms to such optimization problems as well. In time, more problems in mobile communications may be identified where such techniques outperform existing schemes. It is already certain that the early success of these approaches will lead to their integration into the design process for future PCS networks.

19

An Effective Genetic Algorithm for the Fixed Channel Assignment Problem

George D. Smith, Jason C.W. Debuse,

Michael D. Ryan and Iain M. Whittley

19.1 Introduction

Channel Assignment Problems (CAPs) occur in the design of cellular mobile telecommunication systems (Jordan, 1996; Katzela and Naghshineh, 1996; MacDonald, 1979); such systems typically divide the geographical region to be serviced into a set of cells, each containing a base station. The available radio frequency spectrum is divided into a set of disjoint channels; these must be assigned to the base stations to meet the expected demand of each cell and to avoid electromagnetic interference during calls.

There are many different approaches to assigning channels to these base stations to achieve these objectives. The most commonly used technique is the Fixed Channel Assignment (FCA) scheme, in which channels are assigned to each base station as part of the design stage and are essentially fixed in that no changes can be made to the set of channels available for a cell. There are also Dynamic Channel Assignment (DCA) schemes, in which all of the available channels are stored in a common pool; each new call that arrives in a cell is assigned a channel from this pool as long as it does not interfere with existing channels that are currently in use; see Katzela and Naghshineh (1996). There are also hybrids of FCA and DCA schemes as well as other variants, such as Channel Borrowing methods. It has been shown that DCA schemes beat FCA schemes, except under conditions of heavy traffic load (Raymond, 1991). Since networks are under increasing pressure to meet ever increasing demand, an effective solution to the FCA problem is therefore extremely desirable. The algorithms presented in this paper have been developed for the FCA problem.

Customers of the network rely on the base station of the cell in which they are situated to provide them with a channel through which they can make a call. In solving the FCA problem, network designers must allocate channels to all the cells in the network as efficiently as possible so that the expected demand of each cell is satisfied and the number of violations of Electro-Magnetic Compatibility constraints (EMCs) in the network is minimised. This paper considers three types of electro-magnetic constraints:

- Co-channel (CCC); some pairs of cells may not be assigned the same channel.
- Adjacent channel (ACC); some pairs of cells may not be assigned channels which are adjacent in the electromagnetic spectrum.
- Co-site (CSC); channels assigned to the same cell must be separated by a minimum frequency distance.

These and further constraints may be described within a compatibility matrix C, each element C_{ij} of which represents the minimum separation between channels assigned to cells i and j, see for example Ngo and Li (1998), Funabiki and Takefuji (1992) and Gamst and Rave (1982). In addition to satisfying these interference constraints, solutions may be required to minimise the total number of channels used or the difference between the highest and lowest channel used, to make more efficient use of the electromagnetic spectrum and therefore to allow for future network expansion.

Solving instances of the FCA problem is not a trivial exercise. If we consider a simple network that exhibits just one of the constraints described above, the CCCs, then the problem is identical to Graph Colouring. The Graph Colouring problem is known to be NP-Complete, (Garey and Johnson, 1979) and consequently it is very unlikely that a polynomial time algorithm exists that can solve all instances of the FCA problem.

There have been many approaches to the solution of the FCA problem, including heuristic techniques such as genetic algorithms (Ngo and Li, 1998; Crisan and Mühlenbein, 1998; Valenzuela et al., 1998; Lai and Coghill, 1996; Cuppini, 1994), simulated annealing (Duque-Antón et al., 1993, Clark and Smith, 1998), local search (Wang and Rushforth, 1996), artificial neural networks (Funabiki and Takefuji, 1992; Kunz, 1991) and various greedy and iterative algorithms (Sivarajan et al., 1989; Box, 1978; Gamst and Rave 1982).

Heuristic search techniques tend to adopt one of two strategies to solve the FCA problem:

- A direct approach that uses solutions that model the network directly, i.e. they contain information about which channels are assigned to which cells.
- An indirect approach whose solutions do not model the network directly. Typically the solutions represents a list of all the channels required to satisfy the demand of the network. Algorithms such as those described by Sivarajan et al. (1989) and Clark and Smith (1998) are used to transform the indirect solutions into real network models that can be used to evaluate the quality of the proposed solutions.

Standard genetic algorithms using a direct representation have been found to perform quite poorly on the FCA problem. Cuppini (1994) and Lai and Coghill (1996) attempt to solve only relatively trivial FCA problems. Ngo and Li (1998) successfully apply their GA to more difficult problems but they report run times of over 24 hours for a single run of some of the simpler problem instances. In addition, they also employ a local search algorithm which fires when the GA gets stuck in a local optimum. In short, the literature does not provide much evidence that an efficient and scalable channel assignment system could be based on a standard GA. This chapter describes a genetic algorithm that adopts a direct approach to solving the FCA problem. The GA is unusual in that it is able to utilise partial (or delta) evaluation of solutions, thereby speeding up the search. In addition, we compare the results of the GA with those of a simulated annealing algorithm that uses the same representation and fitness function as the GA, as well as similar neighbourhood move operators. More details of the algorithms and results presented here can be found in Ryan *et al.* (1999).

Details of the representation, fitness function and operators used by the GA and SA are presented in section 19.2, as are some additional enhancements used by the GA. The benchmark problems on which the algorithms have been tested are shown in section 19.3 and experimental results are presented in section 19.4. Finally, a summary of the work is presented in section 19.5, including details of current work using other heuristic algorithms.

19.2 The Hybrid Genetic Algorithm

19.2.1 A Key Feature of the Hybrid GA

Designing a genetic algorithm for the FCA problem using a direct approach that will execute in a reasonable amount of time is very difficult. The main obstacle to efficient optimisation of assignments using a traditional genetic algorithm is the expense of evaluating a solution. Complete evaluation of a solution to the FCA problem can be extremely time consuming. Algorithms based on neighbourhood search, such as simulated annealing, can typically bypass this obstacle using delta evaluations. Each new solution created by the neighbourhood search algorithm differs only slightly from its predecessor. Typically the contents of only a single cell are altered. By examining the effects these changes have on the assignment, the fitness of the new solution can be computed by modifying the fitness of its predecessor to reflect these changes. A complete evaluation of the assignment is avoided and huge gains in execution times are possible.

Unfortunately, such delta evaluations are difficult to incorporate in the GA paradigm. At each generation a certain proportion of the solutions in a population are subject to crossover. Crossover is a binary operator that combines the genes of two parents in some manner to produce one or more children. The products of a crossover operator can often be quite different from their parents. For example, consider the genetic fix crossover operator employed by Ngo and Li (1998). So long as the two parents are quite different from each other, their children are also likely to be quite dissimilar from both parents. Consequently it is generally impractical to use delta evaluation to compute the fitness of offspring from their parents. After a child has been produced by crossover it must be completely re-evaluated to determine its fitness.

In the light of this shortcoming, a GA using a simple crossover operator, such as the one employed by Ngo and Li (1998), which requires a large amount of time to evaluate a single solution and which does not appear to guide the population towards a speedy convergence, will be comprehensively outperformed by a local search algorithm, such as simulated annealing, that uses delta evaluations. To be competitive with local search techniques, a GA must utilise operators that allow it either to converge very quickly so few evaluations are required or to explore the search space efficiently using delta evaluations. Section 19.2.3 describes a greedy crossover operator that uses delta evaluations to explore a large number of solutions, cheaply, in an attempt to find the best way to combine two given parents.

19.2.2   Representation

The solution representation employed by the hybrid GA is based on the basic representation used by Ngo and Li (1998). Each solution is represented as a bit-string. The bit-string is composed of a number, n, of equal sized segments, where n is the number of cells in the network. Each segment represents the channels that are assigned to a particular cell. Each segment corresponds to a row in Figure 19.1. The size of each segment is equal to the total number of channels available, say m. If a bit is switched on in a cell's segment, then the channel represented by the bit is allocated to the cell. Each segment is required to have a specific number of bits set at any one time which is equal to the cell's *demand*, i.e. the number of channels that must be assigned to this cell. Genetic or other operators must not violate this constraint. The length of a solution is thus equal to the product of the number of cells in the network and the number of channels available, i.e. m times n. Figure 19.1 shows a diagram of the basic representation used.

Channel number

Cell number		1	2	3	...	m-1	m
	1	0	1	0	...	1	0
	2	1	0	0	...	0	1
	3	0	0	1	...	1	0
	...	...	...	...	...	...	...
	n-1	1	0	1	...	0	1
	n	0	1	0	...	1	0

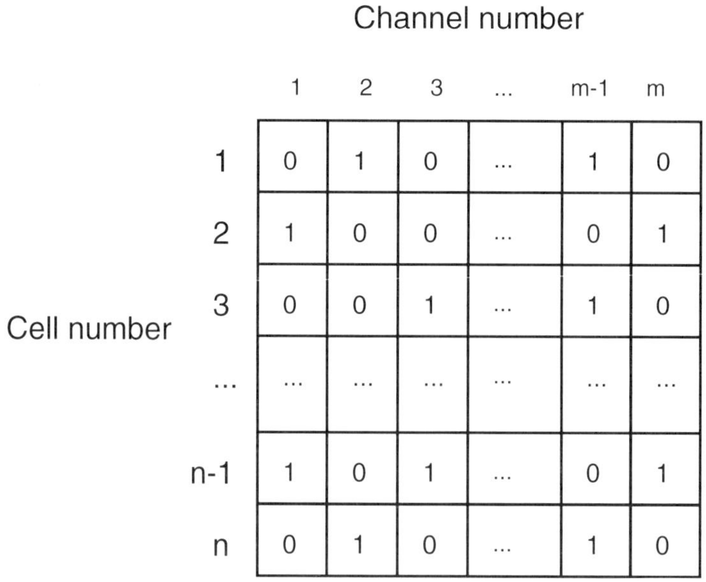

Figure 19.1 Representation of assignment used by the GA and the SA.

In fact, Ngo and Li implement a variation of this representation, in which they only store the offsets based on a minimum frequency separation for CSC. Although we have not implemented this extension here, our initialisation procedures and genetic operators ensure that the CSCs are not violated. However, Ngo and Li achieved a significant reduction in the size of the representation and hence the search space through the use of these offsets. This is worth considering for future work.

19.2.3 Crossover

A good crossover operator for the FCA problem must create good offspring from its parents quickly. Producing good quality solutions will drive the GA towards convergence in a reasonable number of generations, thus minimising the amount of time the GA will spend evaluating and duplicating solutions. Mutation can be relied on to maintain diversity in the population and prevent the GA from converging too quickly. A research group at the University of Limburg (see Smith et al., 1995) devised such a crossover operator for the Radio Link Frequency Assignment Problem (RFLAP), which proved to be very successful. In essence, they used a local search algorithm to search for the best uniform crossover that could be performed on two parents, to produce one good quality child. Once found, the best crossover was performed and the resulting child took its place in the next generation. Whilst the FCA problem and the RFLAP are significantly different to prevent this particular crossover operator being employed in the former, this research does illustrate how a similar sort of operator may be applied to achieve good results for the FCA problem.

The crossover operator employed here uses a greedy algorithm to attempt to find the best combination of genes from two parents to produce one good quality child. The greedy algorithm is seeded with an initial solution consisting of two individual solutions to the FCA problem. The greedy algorithm works by maintaining two solutions. It attempts to optimise only the solution with the best fitness. It achieves this by swapping genetic information between the two solutions. Information can only be swapped between corresponding cells in each of the solutions. When a swap is performed, two channels are selected, one from each solution. The channels are then removed from the solution from which they were originally selected and replaced by the channel chosen from the other solution.

The manner in which these swaps are performed is defined by a neighbourhood. The greedy algorithm explores a neighbourhood until it finds an improving swap, i.e. a swap that leads to an improvement in the fitness of the parent targeted for optimisation. When such a swap is found both solutions are modified and the neighbourhood is updated. The greedy algorithm continues to explore the remainder of the neighbourhood searching for more improving moves. The process continues until the entire neighbourhood is explored. At this juncture the solution being optimised is returned as an only child.

The neighbourhoods are constructed in the following fashion. Each solution is essentially a sequence of sets, one for each cell in the network. Each set contains a certain number of channels that are assigned to the cell corresponding to this set. Two new sequences of sets are created by performing set subtractions on each of the sets in both parents. These new sequences of sets, referred to as the *channel lists*, again contain a set for each cell. Each set in channel list 1 contains channels which have been assigned to the cell represented by this set, in the first parent but not to the corresponding cell in the second parent and vice versa for channel list 2. The neighbourhood is then constructed from these lists in the following manner: (See Figure 19.2)

```
for each cell c
   for each channel i in cell c in channel list 1
      for each channel j in cell c in channel list 2
         Generate move which swaps channels i and j in cell c
```

Since the parents and the channel lists are represented as bit-strings the set subtractions can be efficiently performed as a sequence of ANDs and XORs. The order in which the greedy algorithm explores the moves in the neighbourhood is important. Experimentation has shown that the moves are best explored in a random fashion. Consequently if the crossover operator is applied to the same parents more than once there is no guarantee that the resulting children will be identical.

The process of neighbourhood construction is depicted in Figure 19.2. Figure 19.2(a) illustrates the two parents. These are real solutions to Problem 1 as described in section 19.3. This toy problem has only four cells which have demands of 1, 1, 1 and 3 respectively. The cell segments are denoted by the numbers appearing above the solutions. The solutions are not depicted in bit-string form for reasons of clarity. Performing the set subtraction operations described above yields two channel lists that are displayed in Figure 19.2(b). Finally, Figure 19.2(c) depicts all the moves which are generated from the channel lists. These moves define the neighbourhood of all possible moves.

Computing the channel lists is a very important part of the crossover operation. It guarantees that each move in the neighbourhood will alter the two solutions, maintained by the crossover operator, in some way. Hence, moves that will not effect the solution we are trying to optimise will not be generated and consequently we will waste no time evaluating the solutions they produce. Interestingly, this aspect of the crossover operator does have an advantageous side effect. As the size of the neighbourhood depends upon the size of the channel lists, the number of solutions evaluated by a crossover operator depends on the similarity between the parents upon which it was invoked. As the population of the GA begins to converge, the crossover operators performs less work and the GA improves its speed.

There is one huge advantage of using the neighbourhood described above. Each new solution explored differs only slightly from its predecessor. Consequently, it is entirely practical for the greedy algorithm to employ delta evaluations allowing it to search its neighbourhoods incredibly quickly. So rather than performing one slow evaluation on two solutions as a normal crossover operator would do, it performs quick evaluations on many solutions. The GA can now search the solution space cheaply in the fashion of a local search algorithm.

The effect that delta evaluation has on our hybrid GA is dramatic. Some experiments were performed on the first problem set, described in section 19.4, to assess the impact of delta evaluation on the genetic search. The results of these experiments demonstrated that the GA runs about 90 times faster when using delta evaluations. This result illustrates the most important feature of the hybrid GA presented in this paper. Its ability to explore the search space very efficiently allows the GA to produce effective assignments for large and complicated networks in a reasonable amount of time.

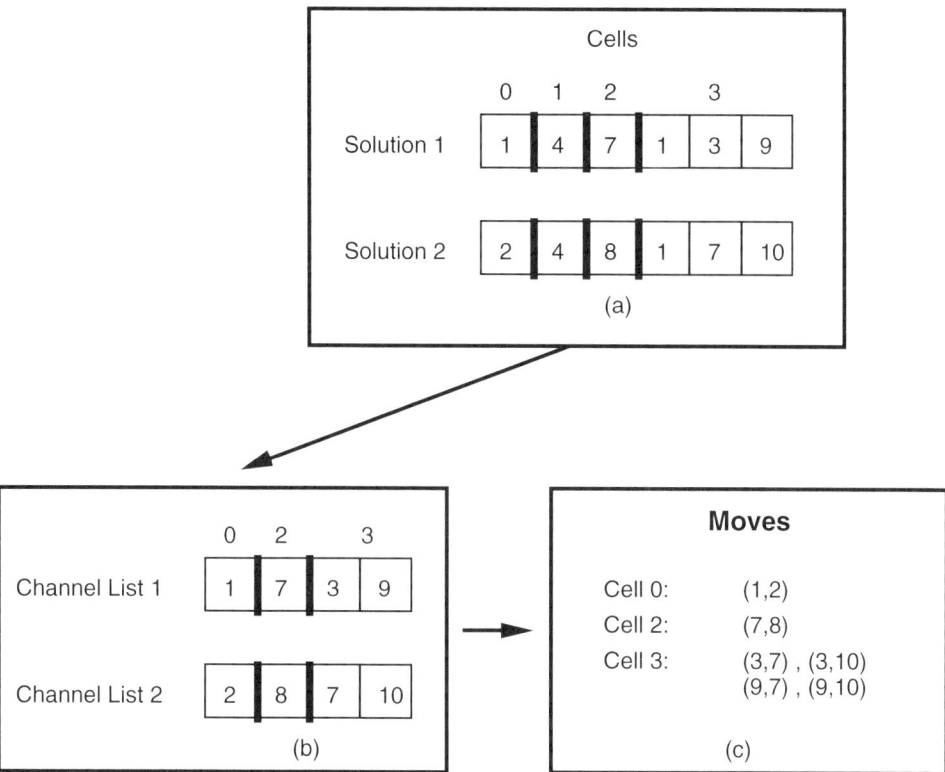

Figure 19.2 Crossover neighbourhood construction.

Finally, relatively low crossover rates of 0.2 and 0.3 have been found to work well with this crossover operator. Due to the greedy nature of the operator, higher crossover rates cause the GA to converge prematurely.

19.2.4 Mutation

The nature of the crossover operator described above tends to cause the GA's population to converge very quickly. Mutation plays an essential role in the hybrid GA by maintaining sufficient diversity in the population, allowing the GA to escape from local optima. Mutation iterates through every bit in a solution and modifies it with a certain probability. If a bit is to be modified, the associated cell is determined. A random bit is then chosen in the same cell that has an opposite value to the original bit. These bits are then swapped and the process continues. The mutation operator cannot simply flip a single bit because this would violate the demand constraints of the cell. A maximum of 100 mutations is permitted per bit-string. Without this limit, mutation would cause the GA to execute very slowly on some of the larger problems.

19.2.5 Other GA Mechanisms

The hybrid GA is loosely based on Goldberg's simple GA (Goldberg, 1989). Each generation the individuals in the population are ranked by their fitness. Solutions are selected for further processing using a roulette wheel selection. A certain proportion of solutions for the next generation will be created by the crossover operator described above. The remaining slots in the next generation are filled by reproduction. Mutation is only performed on solutions produced by reproduction. Allowing mutation an opportunity to mutate the products of crossover was found to have a negative impact on the genetic search. Due to the highly epistatic nature of our representation, a single mutation can have a very detrimental impact on the fitness of the solutions produced, after great effort, by the crossover operator. Were mutation applied to all solutions, an excellent assignment produced by crossover could be corrupted completely before it has a chance to enter the next generation.

19.2.6 Fitness

The fitness of a solution to the FCA problem is determined by the *number* of electromagnetic constraints that it violates. It does not include any information as to whether the network demand is satisfied because this constraint is enforced by the representation and the operators used. More precisely, the fitness, $F(S)$, of a solution, S, is given by

$$F(S) = \sum_{i=0}^{n-1} \sum_{p=0}^{m-1} \left(\sum_{\substack{j=0 \\ j \neq i \\ C_{ij} > 0}}^{n-1} \sum_{\substack{q=p-(C_{ij}-1) \\ 0 \leq q < m}}^{p+(C_{ij}-1)} S_{jq} \right) S_{ip} + \sum_{i=0}^{n-1} \sum_{p=0}^{m-1} \left(\sum_{\substack{q=p=1 \\ 0 \leq q < m}}^{p+(C_{ii}-1)} S_{iq} \right) S_{ip}$$

where S_{ij} is 1 if channel j has been assigned to cell i, otherwise it is 0, C_{ij} is the minimum separation between a channel assigned to cell i and a channel assigned to cell j. The letters n and m represent the number of cells and the number of available channels in the network respectively. The first part of this equation is responsible for computing ACC and CCC violations. The second part calculates the number of CSC violations. $F(S)$ has a minimum value of zero when all the constraints are satisfied.

19.2.7 Heuristic Enhancements

A number of problem specific enhancements can be made to the basic genetic operators, described in the previous section, to improve the performance of the GA on the FCA problem instances. These enhancements are described below.

Ignore good channels
One important enhancement can be made to the crossover operator in an attempt to improve its efficiency. During its execution, the crossover operator constructs a neighbourhood which defines the work that it is to perform. However, this neighbourhood is going to be

used by a greedy algorithm which will only perform an improving move. Since we are just optimising the first solution, we do not need to bother considering channels which are assigned to it without violation. Replacing these channels cannot possibly lead to an improvement in the solution as they were not responsible for any interference in the first place. Thus we can omit these channels from the list of channels that can be swapped out of the first solution. Determining which channels in the first channel list are involved in violations is actually quite expensive and involves a partial evaluation of the first solution. However, if it can prevent the crossover operator performing more than a few swaps, some performance gains might be made. It should be noted that, even though delta evaluation is used to recompute the value of solutions after a swap has occurred, the process is still quite slow.

Eliminating CSCs

Ngo and Li (1998) demonstrate that it is possible to eliminate CSCs completely from the search process for some problems. They achieve this by modifying their representation so that CSCs could not be violated, using a special technique called minimum-separation encoding. While we do not employ this technique, the hybrid GA attempts to employ a similar heuristic without altering the representation. It constructs an initial population that does not contain any CSC violations. This can be achieved by ensuring that for each solution all the channels assigned to a single cell are separated from each other by the CSC frequency separation for that cell. The GA then ensures that neither the crossover nor the mutation operator can perform a swap that can violate a CSC. This heuristic allows us to effectively reduce the size of the search space for certain problems.

19.2.8 Simulated Annealing

Finally, in this section, we present the simulated annealing algorithm that was used to contrast the performance of the GA. Simulated annealing (SA) is a modern heuristic search method that is often applied to combinatorial optimisation problems, including the FCA problem; see Duque-Antón (1993). In SA, the search typically starts at a randomly generated solution. At each iteration, a neighbourhood move is suggested, i.e. a change of value for one or more variables, and that move is accepted if it is better. If the suggested move is worse than the current move, it is accepted with a probability that depends on a temperature parameter. Initially, at high temperatures, worse moves are accepted with a relatively high probability, but as the temperature drops, this probability reduces to zero. This basic mechanism reduces the possibility of becoming trapped in a local optimum. The reader is referred to Reeves (1993) for a fuller description of the basic SA algorithm.

In this application, the SA uses the same representation employed by the GA; see section 19.2.2. SA does not employ a crossover operator, but instead uses a neighbourhood operator which randomly selects a channel that has been allocated to a cell and attempts to replace it with the best unused channel that is not currently assigned to the cell in question.

Delta evaluation is also used in the SA. An initial temperature of 0.1 is used, together with a geometric cooling rate of 0.92 and a temperature step length of 20,000 iterations.

19.3 Benchmark Problems

All the problems described within this document are defined in terms of the following, as previously described in Funabiki and Takefuji (1992); see also Gamst and Rave (1982).

1. A set of n cells.
2. A set of m channels.
3. An $n \times n$ compatibility matrix C as described above.
4. An n element demand vector D, each element d_i of which represents the number of channels required by cell i.

Algorithms designed to solve these problems must generate solutions that completely satisfy the network demand whilst minimising the number of Electro-Magnetic Constraint (EMC) violations. The following objectives must therefore be met by the algorithms applied within this chapter; we will assess their performance in terms of meeting these.

- The traffic demand must be met.
- The resulting interference must be minimised.
- Efficient use should be made of the available spectrum, i.e. we should minimise the number of channels used for an interference-free solution, if possible.

19.3.1 Problem Set 1

Table 19.1 shows some widely used benchmark examples from the literature. Problem 1 is a trivial problem used mostly as illustration, see Sivarajan *et al.* (1989) and Ngo and Li (1998). Problem 2 is a realistic channel-assignment problem from Kunz (1991). Problems 3 to 11 are a related set of problems taken from a variety of sources, see Ngo and Li (1998) and Wang and Rushforth (1996). Although the problems in this last set all have 21 cells in common, they differ in their traffic demand vector D and the compatibility matrix C; see Wang and Rushforth (1996) for full specifications of these problems.

There is one other major difference to the approach that our techniques have. Typically, algorithms will attempt to minimise the number of channels used (or the span) whilst satisfying the traffic demand of each cell and avoiding any interference. In real-world problems, however, one is often given the number of channels that are available as a constraint. The objective, therefore, is to meet the demand and avoid interference by using no more than the number of channels available. If this is not possible, the solution will either not meet the demand or will involve some interference somewhere in the network.

For the problems in Problem set 1, the total number of channels available for each problem has been determined by lower bounds (on the number of channels needed) that have been published for these problems in the literature; see Gamst (1986), Funabiki and Takefuji (1992) and Sivarajan *et al.* (1989). Taking this value, therefore, an interference-free solution to any of the problems in this set will meet all three objectives stated above.

Table 19.1 The first set of benchmark problems – Problem set 1.

Problem	Total cells	Total channels	Total demand
1	4	11	6
2	25	73	167
3	21	221	470
4	21	309	470
5	21	533	481
6	21	533	481
7	21	381	481
8	21	309	470
9	21	414	481
10	21	258	470
11	21	529	470

19.3.2 Problem Set 2

The second set of problems which we use are taken from the first problem set for the NCM2 Workshop on optimisation methods for wireless networks; see Labit (1998) and Ryan *et al.*, (1999). These problems are shown in Table 19.2. For each of these problems, channels 1–291 and 343–424 are available for use within the network; a total of 373 channels are therefore available.

Table 19.2 The second set of benchmark problems – Problem set 2.

Problem	Total cells	Total channels	Total demand
12 P1-100	100	373	677
13 P1-170	170	373	1231
14 P1-214	214	373	1221
15 P1-718	718	373	5101

19.3.3 Problem Set 3

The third set of problems which we use are taken from the second problem set for the ncm2 workshop on optimisation methods for wireless networks, see Labit (1998); these problems are shown in Table 19.3. These problems differ from Problem set 2 in the following ways.

1. Weightings exist for each possible cochannel and adjacent channel interference; these range from 1 to 10. Interference levels that are at level 4 or below are acceptable, but should be kept to a minimum.
2. Each sector contains two antennae. The minimum separation between channels on the same antenna is 17, whilst separations between channels on different antennae in the

same sector are defined by the cochannel and adjacent channel weightings. To allow our approach to take advantage of this situation, the minimum separation between channels at the same sector is reduced from 17 to 9. This means that, for each sector, the first channel was assigned to the first antenna, the second channel to the second, the third channel to the first and so on. The minimum separation between channels at each of the two antennae within the same sector was effectively 9, which meant that cochannel and adjacent channel interference would not occur; the minimum separation between channels at the same transmitter was therefore 18.

3. Within some problems, only 50% of the channels in the range 343–424 may be used. This situation was therefore represented within the problem file by including every other channel in this range. This meant that channels 343, 345, ... ,423 were made available. However, the last channel (423) was increased to 424 so that the spacing between the available channels was maximised.

Table 19.3 The third set of benchmark problems – Problem set 3.

Problem	Total cells	Total channels	Total demand
16 P2-50	50	373	677
17 P2-85	85	332	1231
18 P2-107	107	332	1221
19 P2-359	359	332	5101

19.4 Experimental Results

19.4.1 Problem Set 1

The figures shown in Table 19.4 represent the results of applying the GA and SA over 10 runs to each problem. For each problem, the column headed *Best Fit* shows the fitness of the best solution found over 10 runs, with a 0 indicating an interference-free solution. The column headed *Avg Fit* is the average of the fitness values of the best solutions found in each of the 10 runs, while the column headed *Time* is the average value of the times taken to find the best solution in each of the 10 runs.

The results obtained by the hybrid GA for the first problem set are very encouraging. Referring to Figure 19.4, we can se that the GA can produce interference free solutions to nine out of the eleven problems in seconds. The run times for the GA are admirable, especially when compared to previous GAs that adopted a direct method for solving the FCA problem. For instance, Ngo and Li (1998) report average run times of 20,000 and 90,000 seconds for Problems 3 and 4, respectively. Problems 3 to 8 and Problem 11 are described as being CSC limited, i.e. it is the presence of CSC constraints in the network that make these problems difficult. The GA finds them easy because it does not permit CSC violations in its solutions. The SA does not employ such an approach and consequently it has difficulties with these problems. On the other hand, Problems 9 and 10 are made difficult by the presence of ACC and CCC constraints. The SA outperforms the GA on these problems, but is still not able to find interference-free solutions in general, although one run of the SA did find an interference-free solution for Problem 10.

Table 19.4 GA and SA results for problem set 1. Problems marked * have already been solved to feasibility.

Problem	GA			SA		
	Best	Mean	Time	Best	Mean	Time
1*	0	0	0	0	0	0
2*	0	0	13	0	0	1
3*	0	0	22	1	1.4	176
4*	0	0	84	1	1.2	415
5*	0	0	1	1	1.5	202
6*	0	0	25	4	4.7	324
7*	0	0	4	3	3.6	134
8*	0	0	6	1	1	71
9	16	17.6	401	9	10	553
10	3	5.4	631	0	0.3	442
11*	0	0	19	1	1	77

Solutions with zero fitness have been reported for Problems 1 through to 8 and for Problem 11. For Problems 9 and 10, the best solutions found in the literature were produced by the adaptive local search algorithm of Wang and Rushforth (1996). These solutions required 433 and 263 channels for Problems 9 and 10, respectively. However, the best solution produced by the SA for Problem 10 is superior to the adaptive local search solution, yielding a fitness of 0 with only 258 channels.

19.4.2 Problem Set 2

Turning to Problem set 2, benchmark solutions with zero fitness were reported for Problems 12 (P1-100), 13 (P1-170) and 15 (P1-718) at the NCM Workshop web page Labit (1998), but no solution with zero fitness has been reported for Problem 14 (P1-214). Unfortunately, benchmark times were not available. The results presented in Table 19.5 illustrate that, although the SA has matched this performance, the GA was not able to find a solution with zero fitness for Problem 15.

Table 19.5 GA and SA results for Problem set 2. Problems marked with * have already been solved to feasibility.

Problem	GA			SA		
	Best	Mean	Time	Best	Mean	Time
12*	0	0	474	0	0	14
13*	0	0	2155	0	0	216
14	57	59.6	10564	25	26.9	14610
15*	12	n/a	36011	0	0	1439

19.4.3 Problem Set 3

Recall that, for problems in Problem set 3, weightings exist for each possible cochannel (CCC) and adjacent channel (ACC) interference; these range from 1 to 10. Interference levels that are at level 4 or below are acceptable, but should be kept to a minimum. To deal with the interference weightings, an initial version of each problem was created within which all possible interference constraints were included within the compatibility matrix. If the solution produced violated interference constraints, a second version of the problem was created within which all possible non-CSC interference constraints above level 1 were included within the compatibility matrix, together with all possible CSC interference constraints. Similarly, if the solution produced for this modified version of the problem violated interference constraints, a third version of the problem was created within which all possible non-CSC interference constraints above level 2 were included within the compatibility matrix, along with all possible CSC constraints.

This process continued until a solution was produced for which no interference violations occurred on the modified problem, or the remaining non-CSC constraints were all above level 4. The results produced using this approach are presented in Table 19.6 for SA and in Table 19.7 for the GA; the *Time* column describes the time taken to reach the best solution found within the run performed on each problem. Within experiments performed using this problem set, both the SA-based algorithm and the GA were run just once.

For each problem and for each technique (GA, SA), the number of CCC violations at each level is reported, followed by the number of ACC violations. In addition, figures in bold and in brackets refer to the benchmark results reported on the NCM2 workshop web page, Labit (1998). Referring to Table 19.6, we can see that the SA technique has produced an improvement over the benchmark solution for Problem 16 (P2-50), showing only constraint violations at Level 1.

Table 19.6 Results of SA experiments on Problem set 3.

Problem		Violation levels				# Ch. used	Time (s)
		1	2	3	4		
16	CCC	131 (**29**)	0 (**37**)	0 (**12**)	0 (**2**)	373	1852
	ACC	95 (**29**)	0 (**51**)	0 (**12**)	0 (**10**)		
17	CCC	446 (**394**)	438 (**354**)	507 (**430**)	0 (**0**)	331	203
	ACC	218 (**240**)	0 (**0**)	0 (**0**)	0 (**0**)		
18	CCC	28	177	359	549	332	78
	ACC	161	138	109	85		
19	CCC	1071 (**1035**)	1810 (**1192**)	2366 (**760**)	215 (**10**)	332	1284
	ACC	32 (**30**)	104 (**42**)	176 (**72**)	193 (**42**)		

For Problem 18, the solution produced when all possible non-CSC interference constraints above level 4 were included within the compatibility matrix, together with all possible CSC interference constraints, had 61 constraint violations. None of the constraints violated were CSCs. No benchmark solution for this problem had been reported on the workshop web page by the time of publication. Our algorithm was re-run using a version of Problem 18 within which all possible non-CSC interference constraints above level 5 were included within the compatibility matrix together with all possible CSC interference constraints; the results of this are shown in Table 19.6. Only violations at levels 1 to 4 are included within the table; the solution also violated 842 non-CSC constraints at level 5, of which 727 were CCCs and 115 were ACCs.

Table 19.7 contains the results produced by the GA based approach; the number of CCC violations at each level is reported, followed by the number of ACC violations. No interference free solutions were produced for Problem 19.

Table 19.7 Results of GA experiments on Problem set 3.

Problem		Violation levels				# Ch. used	Time (s)
		1	2	3	4		
16	CCC	72 (**29**)	101 (**37**)	0 (**12**)	0 (**2**)	371	510
	ACC	63 (**29**)	92 (**51**)	0 (**12**)	0 (**10**)		
17	CCC	445 (**394**)	402 (**354**)	531 (**430**)	0 (**0**)	332	5215
	ACC	231 (**240**)	0 (**0**)	0 (**0**)	0 (**0**)		
18	CCC	33	179	363	539	332	356
	ACC	170	132	110	99		

Our approach to these problems could clearly be improved upon in a number of ways. For example, constraints at each level could be removed in a more intelligent manner, such as identifying network areas within which violations are occurring and concentrating on removing the constraints from there. The ideal approach would obviously be to incorporate the weightings within the fitness function, so that no manual intervention is required.

19.5 Conclusions

This paper describes a hybrid GA which we believe has made significant advances in the field of channel assignment through genetic search. The most distinguished quality of the hybrid GA is its ability to explore the search space of FCA problems very efficiently, through the use of delta (partial) evaluations. This technique improved the speed of our GA by a factor of 20. Consequently, the GA can be applied to large and complicated networks, producing good results in a reasonable amount of time.

However, whilst the GA does compare very favourably with previous GA solutions to the FCA problem, it is not as efficient and effective as a well-tuned SA algorithm in solving larger problems. However, the ability of the GA to find interference-free solutions in a robust manner for medium-sized problems shows promise. Further research is required to allow the GA to match or better the performance of SA for larger problems. Possible enhancements to the GA might include a distributed version, as this is where the strength of the GA lies, or the introduction of a pyramid architecture as described in Smith *et al.* (1995).

References

Aarts, E.H.L. and Korst, J.H.M. (1989) *Simulated Annealing and Boltzmann Machines*, Wiley.

Aarts, E.H.L. and Lenstra, J.K. (1997) *Local Search in Combinatorial Optimization*, Wiley.

Abuali, F.N., Schoenefeld, D.A. and Wainwright, R.L. (1994) Terminal assignment in a communications network using genetic algorithms, in *Proceedings of the ACM Computer Science Conference*, 74–81.

Abuali, F.N., Schoenefeld, D.A. and Wainwright, R.L. (1994a) Designing telecommunications networks using genetic algorithms and probabilistic minimum spanning trees, in *Proceedings of the 1994 ACM Symposium on Applied Computing*, 242–246.

Aggarwal, K.K., Chopra, Y.C. and Bajwa, J.S. (1982) Reliability evaluation by network decomposition, *IEEE Transactions on Reliability*, **R-31**, 355–358.

Aiyarak, P., Saket, A.S. and Sinclair, M.C. (1997) Genetic programming approaches for minimum cost topology optimisation of optical telecommunication networks, in *Proceedings of the 2nd IEE/IEEE International Conference on Genetic Algorithms in Engineering Systems: Innovations and Applications (GALESIA'97)*, 415–420.

Alba, E. and Troya, J.M. (1996) Genetic Algorithms for Protocol Validation, in *Proceedings of the 4th conference on Parallel Problem Solving (PPSN IV)*, Springer: Berlin.

Alkanhal, M.A. and Alshebeili, S.A. (1997) Blind estimation of digital communication channels using cumulants and Genetic Algorithms, *International Conference on Consumer Electronics*, 386–387.

Al-Qoahtani, T.A., Abedin, M.J. and Ahson, S.I. (1998) Dynamic Routing in Homogeneous ATM Networks using Genetic Algorithms, in *Proceedings of the 1998 IEEE Conference on Evolutionary Computation*, 114–119.

Amano, M., Munetomo, M., Takai, Y. and Sato, Y. (1993) A Method with Genetic Operators for Adaptive Routing, in *Proceedings of the 1993 Hokkaido Conference of the Electrical Engineering Related Areas*, (in Japanese), 336.

Amin, S., Olafsson, S. and Gell, M. (1994) A Neural Network Approach for Studying the Switching Problem, *BT Technology Journal*, **12**(2), 114–120.

Anderson, H.R. and McGeehan, J.P. (1994) Optimizing Microcell Base Station Locations Using Simulated Annealing Techniques, *IEEE 44th Vehicular Technology Conference*, **2**, 858–862.

Asumu, D. and Mellis, J. (1998) Performance in planning: smart systems for the access network, *BT Technology Journal*, October.

Atiqullah, M.M. and Rao, S.S. (1993) Reliability optimization of communication networks using simulated annealing, *Microelectronics and Reliability*, **33**, 1303–1319.

Axelrod, R. (1984) *The Evolution of Co-operation*, Penguin Books: New York.

Bäck, T. (1993) Optimal Mutation Rates in Genetic Search, in *Proceedings of the 5th Inernational Conference on Genetic Algorithms*, Morgan Kaufmann, 2–9.

Bäck, T. (1994) Selective pressure in evolutionary algorithms: a characterization of selection mechanisms, in *Proceedings of the First IEEE Conference on Evolutionary Computation*, IEEE Press, 57–62.

Bäck, T. (1996) *Evolutionary Algorithms in Theory and Practice: Evolution Strategies, Evolutionary Programming, Genetic Algorithms*, Oxford University Press: Oxford.

Bäck, T., Fogel, D.N. and Michalewicz, Z. (1997) *Handbook of Evolutionary Computation*, IOP Publishing and Oxford University Press.

Bäck, T., Hammel, U. and Schwefel, H.-P. (1997) Evolutionary Computation: Comments on the History and Current State, *IEEE Transactions on Evolutionary Computation*, **1**(3), 1–17.

Balakrishnan, A., Magnanti, T.L., Shulman, A. and Wong, R.T. (1991) Models for Planning Capacity Expansion in Local Access Telcommunication Networks, *Annals of Operations Research*, **33**, 239–284.

Ball, M. and Van Slyke, R.M. (1977) Backtracking algorithms for network reliability analysis, *Annals of Discrete Mathematics*, **1**, 49–64.

Ball, M.O. and Provan, J.S. (1983) Calculating bounds on reachability and connectedness in stochastic networks, *Networks*, **13**, 253–278.

Beizer, B. (1990) *Software Testing Techniques*, second edition, van Nostrand Rheinold: New York.

Bellman, R.E. (1957) *Dynamic Programming*, Princeton University Press.

Beltran, H.F. and Skorin-Kapov, D. (1994) On minimum cost isolated failure immune networks, *Telecommunications Systems*, **3**, 183–200.

Bentall, M., Turton, B.C.H. and Hobbs, C.W.L. (1997) Benchmarking the Restoration of Heavily Loaded Networks using a Two Dimensional Order-Based Genetic Algorithm, in *Second IEE/IEEE International Conference on Genetic Algorithms in Engineering Systems: Innovations and Applications*, IEE: London, 151–156.

Bertsekas, D. and Gallager, R. (1992) *Data Networks*, Prentice Hall, Englewood Cliffs, NJ..

Bhatti, S.N. and Knight, G. (1998) Notes on a QoS information model for making adaptation decisions, in *Proceedings HIPPARCH 98 – 4th International Workshop on High Performance Protocol Architecture*.

Bilchev, G. and Olafsson, S. (1998) Comparing Evolutionary Algorithms and Greedy Heuristics for Adaption Problems, in *Proceedings of the 1998 IEEE International Conference on Evolutionary Computation,* IEEE Press, 458–463.

Bilchev, G. and Olafsson, S. (1998a) Modeling, Simulation and Optimisation of a Distributed Adaptive Filing System, in *International Conference on Optimization Techniques and Applications*, Perth, 595–602.

Bilchev, G. and Parmee, I. (1996) Constraint Handling for the Fault Coverage Code Generation Problem: An Inductive Evolutionary Approach, in *PPSN IV*, Lecture Notes in Computer Science, **1141**, Springer-Verlag: Berlin, 880–889.

Black, U. (1995) *TCP/IP and Related Protocols*, second edition, McGraw-Hill.

Blessing, J.J. (1998) Applying Flow Analysis Methods to the Problem of Network Design, in *Proceedings of the 30th IEEE Southeastern Symposium on Systems Theory*, 424–428.

Blickle, T. (1993) Tournament selection, in Bäck, T., Fogel, D.B. and Michalewicz, Z. (eds.), *Handbook of Evolutionary Computation*, C2.3:1.

Bolla, R., Davoli, F. and Marchese, M. (1997) Bandwidth allocation and admission control in ATM networks with service separation, *IEEE Communication Magazine*, **35**(5), 130–137.

Booker, L.B., Goldberg, D.E. and Holland, J.H. (1989) Classifier systems and genetic algorithms, *Artificial Intelligence*, **40**, 235–282.

Boorstyn, R.R and Frank, H. (1977) Large-Scale Network Topological Optimization, *IEEE Transactions on Communications*, **25**(1), 29–47.

Bousono, C. and Manning, M. (1995) The Hopfield Neural Network Applied to the Quadratic Assignment Problem, *Neural Computing & Applications Journal*, **3**(2), 64–72.

Box, F., (1978) A heuristic technique for assigning frequencies to mobile radio nets. *IEEE Transactions on Vehicular Technology*, **27**(2), 57–64.

Box, D., Schmidt, D.C. and Tatsuya, S. (1992) ADAPTIVE – An object oriented framework for flexible and adaptive communication protocols, in *Proceedings of the 4th IFIP Conference on High Performance Networking*.

Brecht, T.B. and Colbourn, C.J. (1988) Lower bounds on two-terminal network reliability, *Discrete Applied Mathematics*, **21**, 185–198.

Brittain, D. (1999) *Optimisation of the Telecommunications Access Network*. Doctoral Thesis, University of Bristol, UK

Brittain, D., Williams, J.S. and McMahon, C. (1997) A genetic algorithm approach to planning the telecommunications access network, in *Proceedings of the 7th International Conference on Genetic Algorithms (ICGA'97)*, Morgan Kaufmann, 623–628.

Brown, J.I., Colbourn, C.J. and Devitt, J.S. (1993) Network transformations and bounding network reliability, *Networks*, **23**, 1–17.

Budd, T.A. (1981) Mutation analysis: ideas, examples, problems and prospects, in Chandrasekaran, B. and Radichi, S. (eds.), *Computer Program Testing*, North Holland, 129–148.

Calegari, P. and others (1997) Genetic approach to radio network optimization for mobile systems, *IEEE 47th Vehicular Technology Conference*, **2**, 755–759.

Carse, B. (1997) *Artificial evolution of fuzzy and temporal rule based systems*. PhD Thesis, University of the West of England, Bristol, UK.

Carse, B., Fogarty, T.C. and Munro, A. (1995) Adaptive distributed routing using evolutionary fuzzy control, in Eshelman, L.J. (ed.) *Proceedings of the Sixth International Conference on Genetic Algorithms*, Morgan Kaufmann, 389–396.

Carse, B., Fogarty, T.C. and Munro, A. (1996) Evolving fuzzy rule based controllers using genetic algorithms, *Fuzzy Sets and Systems*, **80**, 273–293.

Cedeno, W. and Vemuri, V.R. (1997) Database Design with Genetic Algorithms, in Dasgupta, D. and Michalewicz, Z. (eds.), *Evolutionary Algorithms in Engineering Applications*, Springer-Verlag: Berlin, 189–206.

Celli, G., Costamagna, E. and Fanni, A. (1995) Genetic Algorithms for telecommunication network optimization, in *IEEE International Conference on Systems, Man and Cybernetics, Intelligent Systems for the 21st Century*, 1227–1232.

Chang, C.J. and Cheng, R.G. (1994) Traffic control in an ATM network using fuzzy set theory, in *Proceedings IEEE INFOCOM'94*, 1200–1207.

Chang, C.J. and Wu, C.H. (1993) Optimal frame pattern design for a TDMA mobile communication system using a Simulated Annealing algorithm, *IEEE Transactions on Vehicular Technology*, **42**(2), 205–211.

Chen, W.-K. (1990) *Theory of Nets: Flows in Networks*, Wiley.

Chen, H.M. (1996) *A study on integrated flow control in ATM networks*, Undergraduate Thesis, Dept. of Automation, Tsinghua University, China.

Chen, H.M. (1998) *A study on flow control and pricing in ATM networks*, Graduate Thesis, Department of Automation, Tsinghua University, China.

Chen, S., Wu, Y. and McLaughlin, S. (1997) Genetic algorithm optimization for blind channel identification with higher order cumulant fitting, *IEEE Transactions on Evolutionary Computation*, **1**(4).

Chen, S., Luk, B.L. and Liu, Y. (1998) Application of adaptive Simulated Annealing to blind channel identification with HOC fitting, *Electronics Letters*, **34**(3), 234–235.

Chng, R.S.K., Botham, C.P., Johnson, D., Brown, G.N., Sinclair, M.C., O'Mahony, M.J. and Hawker, I. (1994) A Mult-Layer Restoration Strategy for Reconfigurable Networks, in *GLOBECOM '94*, 1872–1878.

Chong, S., Li, S.Q. and Ghosh, J. (1995) Predictive dynamic bandwidth allocation for efficient transport of real-time VBR video over ATM, *IEEE Journal on Selected Areas in Communications*, **13**(1), 12–23.

Chopra, S. and Rao, M.R. (1994) On the Steiner Tree Problem I & II, *Mathematical Programming*, **64**, 209–246.

Chopra, Y.C., Sohi, B.S., Tiwari, R.K. and Aggarwal, K.K. (1984) Network topology for maximizing the terminal reliability in a computer communication network, *Microelectonics & Reliabiliy*, **24**, 911–913.

Clarke, L. (1976) A system to generate test data and symbolically execute programs, *IEEE Transactions on Software Engineering*, **2**(3), 215–222.

Clark, T. and Smith, G.D. (1998) A practical frequency planning technique for cellular radio, in *Artificial Neural Networks and Genetic Algorithms - ICANNGA*, G.D. Smith, N.C. Steele, R.F. Albrecht (eds.), Springer-Verlag: Wien, 312–316.

Clementi, A. and Di Ianni, M. (1996) On the Hardness of Approximating Optimum Schedule Problems in Store and Forward Networks, *IEEE/ACM Transactions on Networking*, **4**(2), 272–280.

Coit, D.W. and Smith, A.E. (1996) Reliability optimization of series-parallel systems using a genetic algorithm, *IEEE Transactions on Reliability*, **45**, 254–260.

Coit, D.W., Smith, A.E. and Tate, D.M. (1996) Adaptive penalty methods for genetic optimization of constrained combinatorial problems, *INFORMS Journal on Computing*, **8**, 173–182.

Comer, D.E. (1995) *Internetworking With TCP/IP, Vol I: Principles, Protocols, and Architecture*, third edition, Prentice-Hall: Englewood Cliffs, NJ.

Cooper, M.G and Vidal, J.J (1994) Genetic design of fuzzy controllers: the cart and jointed pole problem, in *Proceedings of the Third IEEE International Conference on Fuzzy Systems*, IEEE Piscataway NJ, 1332–1337.

Cormen, T., Leiserson, C. and Rivest, R. (1990) *Introduction to Algorithms*, MIT Press: Cambridge, MA, 348.

Corno, F., Prinetto, P., Rebaudengo, M. and .Sonza Reorda, M. (1996) GATTO: A genetic algorithm for automatic test pattern generation for large synchronous sequential circuits, *IEEE Transactions on Computer-aided Design of Integrated Circuits and Systems*, **15**(8), 991–1000.

Costamagna, E., Fanni, A. and Giacinto, G. (1995) A Simulated Annealing algorithm for the optimization of communication networks, in *URSI International Symposium on Signals, Systems, and Eletctronics, ISSSE '95*, 40–48.

Cox, L.A., Davis, L. and Qiu, Y. (1991) Dynamic Anticipatory Routing in Circuit-Switched Telecommunications Networks, in Davis, L. (ed.), *Handbook of Genetic Algorithms*, Van Nostrand Reinhold: New York, 124–143.

Crisan, C. and Mühlenbein, H. (1998) The breeder genetic algorithm for frequency assignment, in *Parallel Problem Solving from Nature – PPSNV*, A.E. Eiben, T. Bäck, M. Schoenauer, H.P. Schwefel (eds.), Springer-Verlag: Berlin, 897–906.

Cuihong, H. (1997) Route Selection and Capacity Assignment in Computer Communication Networks Based on Genetic Algorithm, in *IEEE Conference on Intelligent Processing Systems*, 548–552.

Cuppini, M. (1994) A genetic algorithm for channel assignment problems. *Communication Network*, **5**(2), 157–166.

Davis, L. (ed.) (1991) *The Handbook of Genetic Algorithms*, Van Nostrand Reinhold: New York.

Davis, L., Orvosh, D., Cox, A. and Qiu, Y. (1993) A genetic algorithm for survivable network design, in *Proceedings of the Fifth International Conference on Genetic Algorithms*, 408–415.

Dawkins, R. (1976) *The Selfish Gene*, Oxford University Press: Oxford.

De Jong, K. (1975) *An analysis of the behaviour of a class of genetic adaptive systems*, PhD Thesis, Department of Computer and Communications Sciences, University of Michigan, USA.

Deb, K. and Agrawal, S. (1998) Understanding Interactions among Genetic Algorithm Parameters. in *Foundations of Genetic Algorithms*, **5**, Morgan Kaufmann.

Deeter, D.L. and Smith, A.E. (1997) Heuristic optimization of network design considering all-terminal reliability, in *Proceedings of the Reliability and Maintainability Symposium*, 194–199.

Deeter, D.L. and Smith, A.E. (1997) Economic design of reliable networks, *IIE Transactions*, to appear.

DeMillo, R.A. and Offutt, A.J. (1991) Constraint-based automatic test data generation, *IEEE Transactions on Software Engineering*, **17**, 900–910.

Dengiz, B., Altiparmak, F. and Smith, A.E. (1997) Efficient optimization of all-terminal reliable networks using an evolutionary approach, *IEEE Transactions on Reliability*, **46**, 18–26.

Dengiz, B., Altiparmak, F. and Smith, A.E. (1997a) Local search genetic algorithm for optimization of highly reliable communications networks, in Bäck, T. (ed.), *Proceedings of the Seventh International Conference on Genetic Algorithms (ICGA 97)*, 650–657.

Dengiz, B., Altiparmak, F. and Smith, A.E. (1997b) Local search genetic algorithm for optimization of highly reliable communications networks, *IEEE Transactions on Evolutionary Computation*, **1**(3), 179–188.

Dijkstra, E.W. (1959) A Note on Two Problems in Connection with Graphs, *Numerische Mathematik* **1**, 269–271.

Dowsland, K.A. (1995) Simulated Annealing, in Reeves, C.R. (ed.), *Modern Heuristic Techniques in Combinatorial Optimization*, McGraw-Hill: New York, 20–69.

Duque-Antón, M., Kunz, D. and Rüber, B. (1993) Channel assignment for cellular radio using simulated annealing. *IEEE Transactions on Vehicular Technology*, **42**(1), 14–21.

Dutta, A. and Kim, Y.K. (1996) A Heuristic approach for capacity expansion of packet networks, *European Journal of Operational Research*, **91**, 395–410.

Dutta, A. and Mitra, S. (1993) Integrating Heuristic Knowledge and Optimization Models for Communication Network Design, *IEEE Transactions on Knowledge and Data Engineering*, **5**(6), 999–1017.

Dziong, Z. and Mason, L.G. (1996) Fair-efficient call admission control policies for broadband networks – A game theoretic framework, *IEEE/ACM Transactions on Networking*, **4**(1), 123–136.

Elbaum, R. and Sidi, M. (1995) Topological Design of Local Area Networks using Genetic Algorithms, in *Proceedings of Infocom 95*, IEEE Press: Los Alamitos, CA, 64–71.

Elshoff, J.L. (1976) An analysis of some commercial PL/1 programs, *IEEE Transactions on Software Engineering*, **2**(2), 113–120.

Erceg, V. and others (1992) Urban/suburban out-of-sight propagation modeling, *IEEE Communications Magazine*, 56–61, June.

Ernst, A.T. and Krishnamoorthy, M. (1996) Efficient Algorithms for the Uncapacitated Single Allocation p-HUB Median Problem, *Location Science*, **4**, 139–154.

Fall, K. and Pasquale, J. (1994) Improving Continuous-Media Playback Performance with In-Kernel Data Paths, in *Proceedings of the IEEE International Conference on Multimedia Computing and Systems (ICMCS)*, 100–109.

Fenton, N. and Pfleeger, S.L. (1997) *Software Metrics: a Rigorous and Practical Approach*, second edition, International Thomson Computer Press: London.

Fish, R.S., Loader, R.J. and Graham, J.M. (1998) DRoPS: Kernel Support for Runtime Adaptable Protocols", in *Proceedings of the 24th Euromicro Conference*, 1029–1036.

Fish, R.S., Ghinea, G. and Thomas, J.P. (1999) Mapping quality of perception to quality of service for a runtime adaptable communication system, in *Proceedings of SPIE ACM SIGMultimedia conference on Multimedia Computing and Networks*.

Fisher, M., Rea, T., Swanton, A., Wilkinson, M. and Wood, D. (1996) Techniques for Automated Planning of Access Networks, *BT Technology Journal*, **14**, 121–127.

Fitzpatrick, J.M. and Greffenstette, J.J. (1988) Genetic Algorithms in Noisy Environments, *Machine Learning*, **3**, 101–120.

Fogel, D.B. (1995) *Evolutionary Computation: Towards a New Philosophy of Machine Intelligence*, IEEE Press: Piscataway, NJ,

Ford, L.R. and Fulkerson, D.R. (1962) *Flows in Networks*, Princeton University Press, Princeton, New Jersey.

Frank, H. and Frisch, I.T. (1971) *Communication, Transmission, and Transportation Networks*, Addison-Wesley: Reading, MA.

Fratta, L., Gerla, M. and Kleinrock, L. (1973) The Flow Deviation Method: An Approach to Store-and-Forward Communication Network Design, *Networks*, **3**(2), 97–133.

Funabiki, N. and Takefuji, Y. (1992) A neural network parallel algorithm for channel assignment problems in cellular radio networks, *IEEE Transactions on Vehicular Technology*, **41**(4), 430–437.

Gamst, A. and Rave, W. (1982) On frequency assignment in mobile automatic telephone systems, in *IEEE Global Telecommunications Conference, Miami*, Paper B3.1.1, 309–315.

Garey, M.R. and Johnson, D.S. (1979) *Computers and Intractability: A Guide to the Theory of NP-Completeness*, W.H. Freeman & Co.: San Francisco, CA.

Garey, M.R., Johnson, D.S. and Tarjan, R.E. (1976) The Planar Hamiltonian Circuit Problem is NP-Complete, *SIAM Journal of Computing*, **5**, 704–714.

Gavish, B. (1991) Topological Design of Telecommunication Networks – Local Access Design Methods, *Annals of Operations Research*, **33**, 17–71.

Gavish, B. (1992) Topological Design of Computer Communication Networks – The Overall Design Problem, *European Journal of Operational Research*, **58**, 149–172.

Gavish, B. and Neuman, I. (1989) A System for Routing and Capacity Assignment in Computer Communication Networks, *IEEE Transactions on Communications*, **37**(4), 360–366.

Geman, S. and Geman, D. (1984) Stochastic relaxation, Gibbs distribution, and the Bayesian restoration of images, *IEEE Transactions on Pattern Analysis and Machine Intelligence*, **6**, 721–741.

Gerla, M. and Kleinrock, L. (1977) On the Topological Design of Distributed Computer Networks, *IEEE Transactions on Communications*, **25**(1), 48–60.

Gerla, M., Frank, H., Chou, W. and Eckl, J. (1974) A Cut Saturation Algorithm for Topological Design of Packet Switched Communication Networks, in *Proceedings of the IEEE National Telecommunications Conference*, 1074–1085.

Gibson, J.D. (ed.) (1996) *The Mobile Communications Handbook*, CRC Press.

Glover, F. (1963) Parametric Combinations of Local Job Shop Rules, Chapter IV of *ONR Research Memorandum*, **117**, GSIA, Carnegie Mellon University, USA.

Glover, F. (1965) A Multiphase Dual Algorithm for the Zero-One Integer Programming Problem, *Operations Research*, **13**(6), 879.

Glover, F. (1989) Tabu Search – Part I, *ORSA Journal on Computing*, **1**, 190–206.

Glover, F. (1989a) Tabu Search – Part II, *ORSA Journal on Computing*, **2**, 4–32.

Glover, F. (1998) A Template for Scatter Search and Path Relinking, in Hao, J.-K., Lutton, E., Ronald, M., Schoenauer, M. and Sayers, D. (eds.), *Artificial Evolution*, Lecture Notes in Computer Science, **1363**, Springer-Verlag: Berlin, 13–54.

Glover, F. (1999) Scatter Search and Path Relinking, in Corne, D., Dorigo, M. and Glover, F. (eds.), *New Ideas in Optimization*, McGraw-Hill: London.

Glover, F. and Laguna, M. (1995) Tabu Search, in Reeves, C.R. (ed.), *Modern Heuristic Techniques in Combinatorial Optimization*, McGraw-Hill: London, 70–150.

Glover, F. and Laguna, M. (1997) *Tabu Search*, Kluwer Academic.

Glover, F., Lee, M. and Ryan, J. (1991) Least-cost network topology design for a new service: an application of a tabu search, *Annals of Operations Research*, **33**, 351–362.

Goldberg, D.E. (1989) *Genetic Algorithms in Search, Optimization and Machine Learning*, Addison-Wesley: Reading, MA.

Goldberg, D.E., Deb, K. and Clark, J.H (1992) Genetic Algorithms, noise, and the sizing of populations, *Complex Systems*, **6**, 333–362.

Gomory, R.E. and Hu, T.C. (1961) Multi-terminal Network Flows, *SIAM Journal of Applied Mathematics*, **9**(4), 551–570.

Graham, D.G. (1991) Software Testing Tools: A New Classification Scheme, *Journal of Software Testing Verification and Reliability*, **1**(3), 17–34.

Greenberg, H.J. and Pierskalla, W.P. (1970) Surrogate Mathematical Programs, *Operations Research*, **18**, 924–939.

Grefenstette, J.J. (1990) *A User's Guide to GENESIS*, Version 5.0.

Griffith, P.S., Proestaki, A. and Sinclair, M.C. (1996) Heuristic topological design of low-cost optical telecommunication networks, in *Proceedings of the 12th UK Performance Engineering Workshop*, 129–140.

Grover, G., Kershenbaum, A. and Kermani, P. (1991) MENTOR: An Algorithm for Mesh Network Topological Optimization and Routing, *IEEE Transactions on Communication*, **39**, 503–513.

Hale, W.K. (1980) Frequency Assignment: Theory and Applications, in *Proceedings of the IEEE*, **68**(12), 1497–1514.

Hamilton, W.D. (1967) Extraordinary sex ratios, *Science*, **156**, 477–488.

Hao, J.-K., Dorne, R. and Galinier, P. (1998) Tabu search for frequency assignment in mobile radio networks, *Journal of Heuristics*, **4**(1), 47–62.

Hardin, G. (1984) The tragedy of the commons, *Science*, **162**, 1243–1248.

Hedrick, C. (1988) *RFC-1058: Routing Information Protocol*, Network Working Group.

Henderson, A.M. (1997) High capacity SDH networks for new operators – Using the National Electricity Distribution Network, Core and ATM Networks, *Networks and Optical Communications III NOC'97 Conference*, IOS Press: Amsterdam.

Hierons, R.M. (1997) Testing from a Finite State Machine: Extending Invertibility to Sequences, *The Computer Journal*, **40**(4), 220–230.

Hiramatsu, A. (1991) Integration of ATM call admission control and link capacity control by distributed neural networks, *IEEE Journal on Selected Areas in Communications*, **9**(7), 1131–1137.

Hoang, D.B. and Ng, M.J.T. (1983) Network Modelling and Optimal Flow Assignment in Computer Networks, in *Proceedings of the 19th IREECON International*, 27–29.

Holland, J.H. (1975) *Adaption in Natural and Artificial Systems*, University of Michigan Press: Ann Arbor, MI.

Holland, J.H. (1976) Adaptation, in Rosen, R. and Snell, F.N. (eds.), *Progress in Theoretical Biology*, volume 4, Academic Press: New York.

Hornung, S., Frost, P., Kerry, J. and Warren, J. (1992) Flexible Architecture and Plant for Optical Access Networks. in *International Wire and Cable Symposium 1992*, 53–58.

Hopcroft, J.E. and Ullman, J.D. (1973) Set merging algorithms, *SIAM Journal of Computers*, **2**, 296–303.

Hsu, I. and Walrand, J. (1996) Dynamic bandwidth allocation for ATM switches, *Journal of Applied Probability*, also available via http://www.path.berkeley.edu/~wlr/.

Hu, T.C. (1982) *Combinatorial Algorithms*, Addison-Wesley: Reading, MA.

Huang, R., Ma, J. and Hsu, D.F. (1997) A Genetic Algorithm for Optimal 3-connected Telecommunication Network Designs, in *Third International Symposium on Parallel Architectures Algorithms and Networks*, 344–350.

Huberman, B. (ed.) (1988) *The Ecology of Computation*, Elsevier: Amsterdam.

Hui, J. (1990) *Switching and Traffic Theory for Integrated Broadband Networks*, Kluwer Academic.

Hunt, J. (1995) Testing control software using genetic algorithms, *Engineering Applications of Artificial Intelligence*, **8**(6), 671–680.

Ida, K., Gen, M. and Yokota, T. (1994) System reliability optimization of series-parallel systems using a genetic algorithm, in *Proceedings of the 16th International Conference on Computers and Industrial Engineering*, 349–352.

IEC 61508 (1997) *Functional Safety of Electrical/Electronic/Programmable Electronic Safety-related Systems*, Version 4.

Ince, D.C. (1987) The automatic generation of test data, *Computer Journal*, **30**(1), 63–69.

Ince, D.C. and Hekmatpour, S. (1986) An empirical evaluation of random testing, *Computer Journal*, **29**(4), 380.

Jack, C., Kai, S.-R. and Shulman, A. (1992) Design and Implementation of an Interactive Optimization System for Telephone Network Planning, *Operations Research*, **40**(1), 14–25.

Jaimes-Romero, F.J., Munoz-Rodriguez, D. and Tekinay, S. (1996) Channel assignment in cellular systems using Genetic Algorithms, in *IEEE 46th Vehicular Technology Conference*, **2**, 741–745.

Jamin, S., Danzig, P., Shenker, S. and Zhang, L. (1996) A measurement-based admission control algorithm for integrated services packet networks, *IEEE/ACM Transactions on Networking*.

Jamin, S., Danzig, P., Shenker, S. and Zhang, L. (1998) Design, implementation, and end-to-end evaluation of a measurement-based admission control algorithm for controlled-load service, in *Proceedings of the 8th International Workshop on Network and Operating System Support for Digital Audio and Video*.

Jan, R.H. (1993) Design of reliable networks, *Computers and Operations Research*, **20**, 25–34.

Jan, R.H., Hwang, F.J. and Cheng, S.T. (1993) Topological optimization of a communication network subject to a reliability constraint, *IEEE Transactions on Reliability*, **42**, 63–70.

Johnson, D.S. and McGeoch, L.A. (1997) The Traveling Salesman Problem: A Case Study in Local Optimization, in Aarts, E.H.L. and Lenstra, J.K.L. (eds.), *Local Search in Combinatorial Optimization*, Wiley: New York, 215–310.

Jones, T. and Forrest, S. (1995) Fitness Distance Correlation as a measure of Problem Difficulty for Genetic Algorithms, in *Proceedings of Sixth International Conference on Genetic Algorithms*. Morgan Kaufmann,

Jones, B.F., Eyres, D.E. and Sthamer, H.-H. (1998) A strategy for using genetic algorithms to automate branch and fault-based testing, *Computer Journal*, **41**(2), 98–107.

Jones, B.F., Sthamer, H.-H. and Eyres, D.E. (1996) Automatic structural testing using genetic algorithms, *Software Engineering Journal*, **11**, 299–306.

Jones, B.F., Sthamer, H.-H., Yang, X. and Eyres, D.E. (1995) The automatic generation of software test data sets using adaptive search techniques, in *3rd International Conference on Software Quality Management*, British Computer Society/Computational Mechanics Publications, 435–444.

Jordan, S. (1996) Resource allocation in wireless networks, *Journal of High Speed Networks*, **5(1)**, 23–34.

Junping, S. and Lee, H.C. (1997) Optimal mobile location tracking by multilayered model strategy, in *Third IEEE International Conference on Engineering of Complex Computer Systems*, 86–95.

Juntti, M.J., Schlosser, T. and Lilleberg, J.O. (1997) Genetic algorithms for multiuser detection in synchronous CDMA, in *IEEE International Symposium on Information Theory*, 492.

Karr, C. (1991) Design of an adaptive fuzzy logic controller using a genetic algorithm, in Belew, R. and Booker, L. (eds.), *Proceedings of the Fourth International Conference on Genetic Algorithms*, Morgan Kaufmann, 450–457.

Kartalopolus, S.V. (1994) Temporal Fuzziness in Communication Systems, in *Proceedings of the World Congress on Computational Intelligence*.

Karunanithi, N. and Carpenter, T. (1997) SONET Ring Sizing with Genetic Algorithms, *Computers in Operations Research*, **24**(6), 581–591.

Katona, G. (1968) A theorem on finite sets, in *Theory of Graphs*, Academia Kiado, Budapest, 187–207.

Katzela, I. and Nahshineh, M. (1996) Channel assignment schemes for cellular mobile telecommunication systems: a comprehensive survey, *IEEE Personal Communications*, **3**(3), 10–31.

Kernighan, B.W. and van Wyk, C.J. (1998) Timing trials or the trials of timing: Experiments with scripting and user-interface languages, *Software – Practice and Experience*, **28**(8), 819–843.

Kershenbaum, A. (1993) *Telecommunications Network Design Algorithms*, McGraw-Hill: New York.

Khanna, A. and Zinky, J.L. (1989) The revised ARPANET routing metric, in *Proceedings of SIGCOMM 89 Symposium*, 45–56.

King-Tim, K., Kit-Sang, T., Cheung-Yan, C., Kim-Fung, M. and Kwong, S. (1997) Using Genetic Algorithms to Design Mesh Networks, *IEEE Computer*, **30**(8), 56–61.

Kinzel, J., Klawonn, F. and Kruse, R. (1994) Modifications of genetic algorithms for designing and optimising fuzzy controllers, in *Proceedings of the First IEEE International Conference on Evolutionary Computation*, IEEE Press: Piscataway, NJ, 28–33.

Kirkpatrick, S., Gelatt jr., C.D. and Vecchi, M.P. (1983) Optimization by Simulated Annealing, *Science*, **220**, 671–680.

Kirkwood, I.M.A., Shami, S.H. and Sinclair, M.C. (1997) Discovering fault tolerant rules using genetic programming, in *Proceedings of the International Conference on Artificial Neural Networks and Genetic Algorithms*, Springer-Verlag: Wien, 285–288.

Kleinrock, L. (1964) *Communication Nets: Stochastic Message Flow and Delay*, McGraw-Hill: New York.

Kleinrock, L. (1970) Analytic and Simulation Methods in Computer Network Design, in *AFIPS Conference Proceedings*, **36**, 568–579.

Knuth, D.E. (1971) An empirical study of FORTRAN programs, *Software – Practice and Experience*, **1**, 105–133.

Ko, K.T., Tang, K.S., Chan, C.Y. and Man, K.F. (1997) Packet Switched Communication Network Design using GA, in *Second IEE/IEEE International Conference on Genetic Algorithms in Engineering Systems: Innovations and Applications*, IEE: London, 151–156.

Ko, K.T., Tang, K.S., Chan, C.Y., Man, K.F. and Kwong, S. (1997a) Using Genetic Algorithms to Design Mesh Networks, *IEEE Computer*, **30**(8), 56–61

Koh, S.J. and Lee, C.Y. (1995) A tabu search for the survivable fiber optic communication network design, *Computers and Industrial Engineering*, **28**, 689–700.

Konak, A. and Smith, A.E. (1998) A general upper bound for all-terminal network reliability and its uses, in *Proceedings of the Industrial Engineering Research Conference*, Banff, Canada, May, CD-ROM.

Konak, A. and Smith, A.E. (1998a) An improved general upper bound for all-terminal network reliability, under revision for *IIE Transactions*.

Koza, J.R. (1992) *Genetic Programming: On the programming of computers by means natural selection*, MIT Press: Cambridge, MA.

Koza, J.R. (1994) *Genetic Programming II: Automatic Discovery of Reusable Programs*, MIT Press: Cambridge, MA.

Kruse, R., Gebhardt, J. and Klawonn, F. (1996) *Foundations of Fuzzy Systems*, Wiley.

Kruskal, J.B. (1956) On the Shortest Spanning Tree of a Graph and the Traveling Salesman Problem, *Proceedings of the American Mathematical Society*, **7**, 48–50.

Kruskal, J.B. (1963) The number of simplices in a complex, in *Mathematical Optimization Techniques*, University of California Press: Berkeley, CA, 251–278.

Kumamoto, T., Tanaka, K. and Inoue, K. (1977) Efficient evaluation of system reliability by Monte Carlo method, *IEEE Transactions on Reliability*, **R-26**(5), 311–315.

Kumar, A., Pathak, R.M., Gupta, M.C. and Gupta, Y.P. (1992) Genetic algorithm based approach for designing computer network topology, in *Proceedings of the 21st ACM Annual Computer Science Conference*, 358–365.

Kumar, A., Pathak, R.M., Gupta, Y.P. and Parsaei, H.R. (1995) A genetic algorithm for distributed system topology design, *Computers and Industrial Engineering*, **28**, 659–670.

Kumar, A., Pathak, R.M. and Gupta, Y.P. (1995a) Genetic algorithm based reliability optimization for computer network expansion, *IEEE Transactions on Reliability*, **44**, 63–72.

Kunz, D. (1991) Channel assignment for cellular radio using neural networks. in *IEEE Transactions on Vehicular Technology*, **40**(1), 188–193.

Kuvayev, L., Giles, C., Philbin, J. and Cejtin, H. (1997) Intelligent Methods for File System Optimization, *4th National Conference on AI and 9th Innovative Applications of AI Conference*, MIT Press: Cambridge, MA, 528–533

Labit, P. (1998) Frequency assignment test problems, University of Montreal, www.crt.umontreal.ca/~brigitt/telecom/test_probs/.

Lai, W.K. and Coghill, G.C. (1996) Channel assignment through evolutionary optimization, *IEEE Transactions on Vehicular Technology*, **45**(1), 91–96.

Lazar, A.A. and Semret, N. (1997) *Auction for network resource sharing*, Technical Report CU/CTR/TR 468-97-02, Columbia University, USA.

Lazar, A.A. and Semret, N. (1998) *Design and analysis of the progressive second price auction for network bandwidth sharing*, Technical report CU/CTR/TR 487-98-21, Columbia University, USA.

Lazar, A.A. and Semret, N. (1998a) *Market pricing of differentiated Internet services*, Technical Report CU/CTR/TR 501-98-37, Columbia University, USA.

Lee, C.C. (1990) Fuzzy logic in control systems: fuzzy logic controller, *IEEE Transactions on Systems, Man and Cybernetics*, **20**, 404–435.

Lee, H.W. and Mark, J.W. (1995) Capacity allocation in statistical multiplexing of ATM sources, *IEEE/ACM Transactions on Networking*, **3**(2), 139–151.

Lee, M. and Takagi, H. (1993) Integrating design stages of fuzzy systems using genetic algorithms, in *Proceedings of the Second IEEE International Conference on Fuzzy Systems*, IEEE: San Francisco, CA, 612–617.

Lee, Y., Lu, L., Qiu, Y. and Glover, F. (1994) Strong Formulations and Cutting Planes for Designing Digital Data Service Networks, *Telecommunication Systems*, **2**, 261–274.

Lee, Y., Qiu, Y. and Ryan, J. (1996) Branch and Cut Algorithms for a Steiner Tree-Star Problem, *INFORMS Journal on Computing*, **8**(3), 1–8.

Leung, Y., Li, G. and Xu, Z.B. (1998) A Genetic Algorithm for Multiple Destination Routing Problems, *IEEE Transactions on Evolutionary Computation*, **2**(4), 150–161.

Lewontin, R.C. (1960) Evolution and the theory of games, *Journal of Theoretical Biology*, **1**, 382–403.

Li, V.O.K. and Qiu, X. (1995) Personal Communication Systems (PCS), *Proceedings of the IEEE*, **83**(9), 1210–1243.

Linkens, D.A. and Nyongesa, H.O. (1995) Genetic algorithms for fuzzy control: Part 1: Offline system development and application, *IEE Proceedings on Control Theory and Applications*, **142**(3), 161–176.

Liska, J. and Melsheimer, S.S. (1994) Complete design of fuzzy logic systems using genetic algorithms, in Schaffer, D. (ed.), *Proceedings of the Third IEEE International Conference on Fuzzy Systems*, IEEE: Piscataway, NJ, 1377–1382.

Luenberger, D.G. (1973) *Introduction to linear and nonlinear programming*, Addison-Wesley: Reading, MA.

Luke, S. and Spector, L. (1996) Evolving graphs and networks with edge encoding: Preliminary report, *Late Breaking Papers at the Genetic Programming 1996 Conference*, 117–124.

Luss, H. (1982) Operations Research and Capacity Expansion Problems: A Survey, *Operations Research*, **30**, 907–947.

MacDonald, V.H. (1979) The cellular concept, *Bell Systems Technical Journal*, **58**(1), 15–41.

Mah, B.A. (1996) *INSANE - An Internet Simulated ATM Networking Environment*, available at http://HTTP.CS.Berkeley.EDU/~bmah/Software/Insane/.

Mamdani, E.H. (1976) Advances in linguistic synthesis of fuzzy controllers, *International Journal of Man-Machine Studies*, **8**, 669–678.

Mann, J. (1995) *Applications of Genetic Algorithms in Telecommunications*, MSc Thesis, University of East Anglia, UK.

Mann, J.W. and Smith, G.D. (1996) A Comparison of Heuristics for Telecommunications Traffic Routing, in Rayward-Smith, V.J, Osman, I.H., Reeves C.R. and Smith, G.D. (eds), *Modern Heuristic Techniques in Combinatorial Optimization*, Wiley: Chichester, 235–253.

Mann, J.W. and Smith, G.D. (1997) The Ring Loading and Ring Sizing Problem, in Smith, G.D., Steele, N.C. and Albrecht, R.F. (eds.), *Proceedings of the International Conference on Artificial Neural Networks and Genetic Algorithms*, Springer-Verlag: Wien, 289–293.

March, S.T. and Rho, S. (1995) Allocating Data and Operations to Nodes in Distributed Database Design, *IEEE Transactions on Knowledge and Data Engineering*, **7**(2), 305–317.

Maruyama, K. (1978) Designing Reliable Packet Switched Communication Networks, *Proceedings of the IEEE ICCC*, 493–498.

Maynard Smith, J. (1989) *Evolution and the Theory of Games*, Cambridge University Press: Cambridge.

Maynard Smith, J. and Price, A. (1973) The logic of animal conflict, *Nature*, **246**, 15–18.

Mazda, F. (ed.) (1996) *Analytical Techniques in Telecommunications*, Reed Educational & Professional: Oxford.

McGregor, P. and Shen, D. (1977) Network Design: An Algorithm for the Access Facility Location Problem, *IEEE Transactions on Communications*, **25**(1), 61–73.

McQuillan, J., Richer, J. and Rosen, E. (1980) The new routing algorithm for the ARPANET, *IEEE Transactions on Communications*, May.

Megson, G.M., Fish, R.S. and Clarke, D.N.J. (1998) Creation of Reconfigurable Hardware Objects in PVM, in Alekandrov, V. and Dongarra, J. (eds.), *Recent Advances In Parallel Virtual and Message Passing Interfaces*, Lecture Notes in Computer Science, **1497**, Springer-Verlag: Berlin, 215–223.

Michalewicz, Z. (1991) A step towards optimal topology of communications networks, in *Proceedings of the Conference on Data Structures and Target Classification*, 112–122.

Milgrom, P. (1981) A theory of auctions and competitive bidding, *Econometrica*, **50**(5), 1089–1122.

Miller, W. and Spooner, D. (1976) Automatic generation of floating point test data, *Transactions on Software Engineering*, **2**(3), 223–226.

Min-Jeong, K. (1996) A comparison of two search algorithms for solving the channel assignment problem, in *5th IEEE International Conference on Universal Personal Communications*, **2**, 681–685.

Minoux, M. (1987) Network Synthesis and Dynamic Network Optimization, *Annals of Discrete Maths*, **31**, 283–324.

Mitchell, M. (1991) *An Introduction to Genetic Algorithms*, MIT Press: Cambridge, MA.

Montgomery, D.C. (1991) *Design and Analysis of Experiments*, third edition, Wiley: Chichester, UK.

Moscato, P. (1999) An Introduction to Memetic Algorithms, in Corne, D., Dorigo, M. and Glover, G. (eds.), *New Ideas in Optimization*, McGraw-Hill: London.

Moscato, P. and Norman, M.G. (1992) A "Memetic" Approach for the Travelling Salesman Problem – Implementation of a computational ecology for combinatorial optimisation on message-passing systems, in *Proceedings of the International Conference on Parallel Computing and Transputer Applications*, IOS Press: Amsterdam.

Moy, J. (1989) *RFC-1131: The OSPF Specification*, Network Working Group.

Muhlenbein, H. (1992), How genetic algorithms really work: I. Mutation and hillclimbing, in R. Manner and B. Manderick (eds.), *Proceedings of the 2nd International Conference on Parallel Problem Solving from Nature*, Elsevier: Amsterdam, 15–25.

Muhlenbein, H. and Schlierkamp-Voosen, D. (1994) The Science of Breeding and its application to the Breeder Genetic Algorithm. *Evolutionary Computation*, **1**, 335–360.

Mukherjee, B. (1997) *Optical Communication Networks*, McGraw-Hill: New York.

Müller, F. and Wegener, J. (1998) A comparison of static analysis and evolutionary testing for the verification of timing constraint, in *4th IEEE Real-timeTechnology and Applications Symposium*.

Muller, C., Magill, E.H., Prosser, P. and Smith, D.G. (1993) Artificial Intelligence in Telecommunications, in *IEEE Global Telecommunications Conference, GLOBECOM '93*, **2**, 883–887.

Munakata, T. and Hashier, D.J. (1993) A genetic algorithm applied to the maximum flow problem, in Forrest, S. (ed.), *Proceedings of the Fifth International Conference on Genetic Algorithms*, Morgan Kaufmann, 488–493.

Munetomo, M., Takai, Y. and Sato, Y. (1997) An Adaptive Network Routing Algorithm Employing Path Genetic Operators, in *Proceedings of the Seventh International Conference on Genetic Algorithms*, Morgan Kaufmann, 643–649.

Munetomo, M., Takai, Y. and Sato, Y. (1998) A Migration Scheme for the Genetic Adaptive Routing Algorithm, in *IEEE International Conference on Man, Machine and Cybernetics*, **3**, 2774–2779.

Munetomo, M., Takai, Y. and Sato, Y. (1998a) An Adaptive Routing Algorithm with Load Balancing by a Genetic Algorithm, *Transactions of the Information Processing Society of Japan*, **38**(2) (in Japanese), 219–227.

Murai, Y., Munetomo, M. and Takai, Y. (1996) A Source Routing Algorithm by Genetic Algorithm, *IPSJ SIG Notes*, **96**(108) (in Japanese), 43–48.

Ndousse, T.D. (1994) Fuzzy neural control of voice cells in ATM networks, *IEEE Journal on Selected Areas in Communications*, **12**(9), 1488–1494.

Nel, D.L. and Colbourn, C.J. (1990) Combining Monte Carlo estimates and bounds for network reliability, *Networks*, **20**, 277–298.

Nemhauser, G.L. and Wolsey, L.A. (1988) *Integer and Combinatorial Optimization*, Wiley: New York.

Ng, M.J.T. and Hoang, D.B. (1983) Optimal Capacity Assignment in Packet-Switching Networks, *Australian Telecommunication Research*, **17**(2), 53–65.

Ng, M.J.T. and Hoang, D.B. (1987) Joint Optimization of Capacity and Flow Assignment in a Packet-Switched Communications Network, *IEEE Transactions on Communications*, **35**(2), 202–209.

Ngo, C.Y. and Li, V.O.K. (1998) Fixed channel assignment in cellular radio networks using a modified Genetic Algorithm, *IEEE Transactions on Vehicular Technology*, **47**(1), 163–172.

Oates, M. (1998a) Autonomous Management of Distributed Information Systems using Evolutionary Computing Techniques, invited paper to *CASYS'98: the 2nd International Conference on Computing Anticipatory Systems*, AIP Conference Procs., **465**, 269–281.

Oates, M. and Corne, D. (1998b) Investigating Evolutionary Approaches to Adaptive Database Management against various Quality of Service Metrics. in Eiben, A.E., Bäck, T., Schoenauer, M. and Schwefel, H.-P. (eds.), *PPSN-V, the Fifth International Conference on Parallel Problem Solving from Nature*, Lecture Notes in Computer Science, **1498**, Springer-Verlag: Berlin, 775–784.

Oates, M. and Corne, D. (1998c) QoS based GA Parameter Selection for Autonomously Managed Distributed Information Systems, in *Proceedings of the 13th European Conference on Artificial Intelligence*, Wiley, 670–674.

Oates, M., Corne, D. and Loader, R. (1998) Investigating Evolutionary Approaches for Self-Adaption in Large Distributed Databases. in *Proceedings of the 1998 IEEE International Conference on Evolutionary Computation*, IEEE Press, 452–457.

Oates, M., Corne, D. and Loader, R. (1999) Skewed Crossover and the Dynamic Distributed Database Problem, in *Proceedings of the 4th International Conference on Artificial Neural Networks and Genetic Algorithms*, Springer-Verlag, Berlin, 280–287.

Oates, M., Corne, D. and Loader, R. (1999a) Investigation of a Characteristic Bimodal Convergence-time/Mutation-rate Feature in Evolutionary Search, in *Proceedings of the 1999 Congress on Evolutionary Computation*, **3**, IEEE Press, 2175–2182.

Oates, M., Corne, D. and Loader, R. (1999b) Variation in Evolutionary Algorithm Performance Characteristics on the Adaptive Distributed Database Management, in *Proceedings of the 1999 Genetic and Evolutionary Computation Conference*, Morgan Kaufmann, 480–487.

Oates, M., Corne, D. and Loader, R. (2000) Visualisation of non-ordinal multi-dimensional landscapes, in John, R. and Birkenhead, R. (eds.), *Soft Computing Techniques and Applications*, Springer-Verlag, 178–185.

O'Dare, M.J. and Arslan, T. (1994) Generating test patterns for VLSI circuits using a genetic algorithm, *Electronics Letters*, **30**(10), 778–779.

Offutt, A.J. (1992) Investigation of the software testing coupling effect, *ACM Transactions on Software Engineering Methodology*, **1**, 5–20.

Olafsson, S. (1995) A General Method for Task Distribution on an Open Heterogeneous Processor System, *IEEE Transactions on Systems, Man and Cybernetics*, **25**(1), 43–58.

Olafsson, S. (1995a) On the stabilities of strategies in competitive systems, *International Journal of Systems Science*, **26**(6), 1289–1312.

Olafsson, S. (1996) Resource Allocation as an Evolving Strategy, *Evolutionary Computation*, **4**(1), 33–55.

Olafsson, S. (1997) Games on Networks, *Proceedings of the IEEE*, **85**(10).

Olafsson, S. (1997a) Self Adaptive Network Utilisation, in *Proceedings of the International Workshop on Applications of Neural Networks to Telecommunications*, **3**, Lawrence Erlbaum, 1–11.

Olsen, A.S. (1997) Adaptive Feedback Compensation for Distributed Load-Based Routing Systems in Datagram Packet-Switched Communication Networks, *Computer Communication Review*, **27**(3), 83–99.

O'Mahony, M., Sinclair, M.C. and Mikac, B. (1993) Ultra-high capacity optical transmission network: European research project COST 239, *Informacija Telekomunikacije Automati*, **12**(1–3), 33–45.

Ortega, A. and others (1995) *Algorithmes de Coloration des Graph et d'Affectation des Fréquences, Approche Géométrique et Analytiques, Heuristiques, Recuit Simulé, Programmation par Contraintes, Algorithme Génétiques*, Technical Report NT/PAB/SRM/RRm/4353, CNET, Paris.

O'Sullivan, M., Vössner, S. and Wegener, J. (1998) Testing temporal correctness of real-time systems: a new approach using genetic algorithms and cluster analysis, *EuroSTAR 98, European Software Testing and reliability Conference*, Munich.

Painton, L. and Campbell, J. (1995) Genetic algorithms in optimization of system reliability, *IEEE Transactions on Reliability*, **44**, 172–178.

Parodi, A. and Bonelli, P. (1993) A new approach to fuzzy classifier systems, in Forrest, S. (ed.), *Proceedings of the Fifth International Conference on Genetic Algorithms*, Morgan Kaufmann, 223–230.

Paul, H. and Tindle, J. (1996) Passive Optical Network Planning in Local Access Networks, *BT Technology Journal*, **14**(2), 110–115.

Paul, H., Tindle, J. and Woeste, C. (1994) A Unified Model for the Local Access Network. *Communications Networks Symposium*, Centre for Communications Network Research, Manchester Metropolitan University.

Paul, H., Tindle, J. and Ryan, H.M. (1996) Experiences with a Genetic Algorithm Based Optimisation System for Passive Optical Network Planning in the Local Access Network, *European Conference of Networks and Optical Communications – NOC'96*, 105–112.

Perros, H.G. and Elsayed, K.M. (1996) Call admission control schemes: a review, *IEEE Communication Magazine*, **34**(11), 82–91.

Peterson, L.L. and Davie, B.S. (1996) *Computer Networks – A Systems Approach*, Morgan Kaufmann.

Pfleeger, S.L. (1991) *Software Engineering: the Production of Quality Software*, Second Edition, Macmillan: New York.

Phadke, M.S. (1989) *Quality Engineering Using Robust Design*, Prentice-Hall: Englewood Cliffs, NJ.

Pham, D.T. and Karaboga D. (1991) Optimum design of fuzzy logic controllers using genetic algorithms, *Journal of Systems Engineering*, **1**, 114–118.

Pierre, S. and Elgibaoui, A. (1997) Improving communication network topologies using Tabu Search, *22nd Annual Conference on Local Computer Networks (LCN'97)*, 44–53.

Pierre, S. and Legault, G. (1996) An Evolutionary Approach for Configuring Economical Packet Switched Computer Networks, *Artificial Intelligence in Engineering*, **10**, 127–134.

Pierre, S. and Legault, G. (1998) A genetic algorithm for designing distributed computer network topologies, *IEEE Transactions on Systems, Man and Cybernetics-Part B: Cybernetics*, **28**(2), 249–258.

Pierre, S., Hyppolite, M.A., Bourjolly, J.M. and Dioume, O. (1995) Topological design of computer communication networks using simulated annealing, *Engineering Applications of Artificial Intelligence*, **8**, 61-69.

Plagemann, T., Gotti, A. and Plattner, B. (1994) CoRA – A Heuristic for protocol Configuration and Resource Allocation, *IFIP Workshop for High-Speed Networks*.

Poon, K.F., Asumu, D., Tindle, J. and Ryan, H.M. (1997) Artificial Intelligence Methods for Telecommunication Network Planning, in *European Conference of Networks and Optical Communications – NOC '97*.

Poon, K.F., Tindle, J., Brewis, S. and Asumu, D. (1998) Optimal Design of Copper Based Telecommunication Networks using AI Methods, in *EuroCable Conference EC'98*.

Poon, K.F., Tindle, J. and Brewis, S. (1998a) An AI method for Optimal Design of Copper Telecommunications Networks, in *Proceedings of the World Multiconference on Systems, Cybernetics and Informatics, SCI '98*, **2**.

Postel, J. (1980) *User Datagram Protocol (UDP)*, RFC 768. Internet Engineering Task Force. Gennaio.

Postel, J. (1981) *Transmission Control Protocol (TCP)*, RFC 793, Internet Engineering Task Force. Gennaio.

Powell, W.B., Jaillet, P. and Odoni, A. (1995) Stochastic and Dynamic Networks and Routing, in Ball, M.O., Magnanti, T.L., Monma, C.L. and Nemhauser, G.L. (eds.), *Network Routing*, Elsevier Science: Amsterdam.

Prim, R.C. (1957) Shortest Connection Networks and Some Generalizations, *Bell Systems Technical Journal*, **36**, 1389–1401.

Proakis, J.G. (1995) *Digital Communications*, McGraw-Hill.

Provan, J.S. and Ball, M.O. (1983) The complexity of counting cuts and of computing the probability that a graph is connected, *SIAM Journal of Computing*, **12**(4), 777–788.

Puschner, P. and Nossal, R. (1998) Testing the results of static worst-case execution time analysis, *19th IEEE Real-time Systems Symposium (RTSS98)*.

Radcliffe, N.J. (1994) The Algebra of Genetic Algorithms, *Annals of Mathematics and Artificial Intelligence*, **10**, 339–384.

Rappaport, T.S. (1996) *Wireless Communications – Principles & Practice*, Prentice Hall: Englewood Cliffs, NJ.

Raymond, P. (1994) Performance analysis of cellular networks, *IEEE Transactions on Communications*, **39**(3), 542-548.

Read, S.H. (1995) *Standard for Software Component Testing*, Issue 3.0, October, British Computer Society Specialist Interest Group in Software Testing.

Rho, S. and March, S.T. (1994) A Nested Genetic Algorithm for Database Design, in *Proceedings of the 27th Hawaii International Conference on System Sciences*, 33–42.

Roberts, L.G. and Wessler, B.D. (1970) Computer network development to achieve resource sharing, in *Proceedings of the Spring Joint Computing Conference*, AFIPS, **36**, 543–599.

Roper, M. (1997) Computer aided software testing using genetic algorithms, *10th International Quality Week*, San Francisco, CA.

Routen, T. (1994) Genetic Algorithms and Neural Network Approaches to Local Access Network Design, in *Proceedings of the 2nd International Workshop on Modelling, Analysis and Simulation of Computer and Telecommunications Systems*.

Rudolph, G. (1994) Convergence Analysis of Canonical Genetic Algorithms, *IEEE Transactions on Neural Networks*, **5**(1).

Ryan, M.D.C., Debuse, J.C.W., Smith, G.D. and Whittley, I.M. (1999) A hybrid genetic algorithm for the fixed channel assignment problem, in *GECCO'99: Genetic and Evolutionary Computation Conference*.

Sacco, W. (1987) *Dynamic Programming: an elegant problem solver*, Janson Publications.

Sairamesh, J., Ferguson, D. and Yemini, Y. (1995) An approach to pricing, optimal allocations, and quality of service, in *Proceedings of the INFOCOM'95*, also available via the URL: http://www.ics.forth.gr/~ramesh/pricing.html.

Savic, D.A. and Walters, G.A. (1995) An evolution program for pressure regulation in water distribution networks, *Engineering Optimization*, **24**, 197–219.

Semret, N. (1996) *TREX: The resource exchange*, http://argo.ctr.columbia.edu:1024/~nemo/Trex/.

Shubik, M. (1982) *Game Theory in the Social Sciences: Concepts and Solutions*, MIT Press: Cambridge, MA. 336–368.

Schultz, A.C., Grefenstette, J.J. and de Jong, K.A. (1993) Test and evaluation by genetic algorithms, *IEEE Expert*, **8**, 9–14.

Schwartz, M. (1987) *Telecommunication Networks: Protocols, Modelling and Analysis*, Addison-Wesley: Reading, MA.

Schwefel, H.-P. (1981) *Numerical optimisation of computer models*, Wiley: New York.

Sedgewick, R. (1988) *Algorithms*, second edition, Addison-Wesley: Reading, MA.

Seo, K.S. and Choi, G.S. (1998) The Genetic Algorithm Based Route Finding Method for Alternative Paths, in *Proceedings of the International Conference on Systems, Man and Cybernetics*, 2448–2453.

Shahbaz, M. (1995) Fixed network design of cellular mobile communication networks using Genetic Algorithms, in *Fourth IEEE International Conference on Universal Personal Communications*, 163–167.

Shami, S.H., Kirkwood, I.M.A. and Sinclair, M.C. (1997) Evolving simple fault-tolerant routing rules using genetic programming, *Electronic Letters*, **33** (17), 1440–1441.

Shimamoto, N., Hiramatsu, A. and Yamasaki, K. (1993) A Dynamic Routing Control Based on a Genetic Algorithm, in *IEEE Conference on Neural Networks*, 1123–1128.

Shulman, A. and Vachini, R. (1993) A Decomposition Algorithm for Capacity Expansion of Local Access Networks, *IEEE Transactions on Communications*, **41**, 1063–1073.

Sinclair, M.C. (1993) The Application of a Genetic Algorithm to Trunk Network Routing Table Optimisation, in *Tenth UK Teletraffic Symposium on Performance Engineering in Telecommunications*, IEE: London, 2/1–2/6.

Sinclair, M.C. (1995) Minimum cost topology optimisation of the COST 239 European optical network, in *Proceedings of the 2nd International Conference on Artificial Neural Networks and Genetic Algorithms (ICANNGA'95)*, 26–29.

Sinclair, M.C. (1997) NOMaD: Applying a Genetic algorithm heuristic hybrid approach to optical network topology design, in *Proceedings of the 3rd International Conference on Artificial Neural Networks and Genetic Algorithms (ICANNGA'97)*, 299–303.

Sinclair, M.C. (1998) Minimum Cost Routing and Wavelength Allocation using a Genetic-Algorithm/Heuristic Hybrid Approach, in *The Sixth International Conference on Telecommunications*, IEE: London, 67–71.

Sinclair, M.C. (1999) Optical mesh topology design using node-pair encoding genetic programming, in *Proceedings of the Genetic and Evolutionary Computation Conference (GECCO'99)*.

Sivarajan, K.N.., McEliece, R.J. and Ketchum, J.W. (1989) Channel assignment in cellular radio, in *Proc. 39th IEEE Vehicular Technology Conference*, IEEE Vehicular Technology Society, 846–850.

Slobodkin, L.B. and Rapoport, A. (1974) An Optimal Strategy of Evolution, *Quarterly Review of Biology*, **49**, 181–200.

Smith, S.F. (1980) *A learning system based on genetic adaptive algorithms*, PhD Thesis, University of Pittsburgh, USA.

Smith, J.E. and Fogarty, T.C. (1996) Adaptively Parameterised Evolutionary Systems: Self Adaptive Recombination and Mutation in a Genetic Algorithm, *Parallel Problem Solving from Nature IV*, Lecture Notes in Computer Science, Springer-Verlag: Berlin.

Smith, A.E. and Tate, D.M. (1993) Genetic optimization using a penalty function, in *Proceedings of the 5th International Conference on Genetic Algorithms*, 499–505.

Smith, G.D., Kapsalis, A., Rayward-Smith, V.J. and Kolen, A. (1995) Implementation and testing of genetic algorithm approaches to the solution of the frequency assignment problem, *EUCLID CALMA RLFAP Technical Report No 2.1*.

Sprent, P. (1992) *Applied Nonparametric Statistical Methods*, second edition, Chapman & Hall: London.

Srivaree-ratana, C. and Smith, A.E. (1998) Estimating all-terminal network reliability using a neural network, *Proceedings of the 1998 IEEE International Conference on Systems, Man, and Cybernetics*, San Diego, CA, 4734–4740.

Srivaree-ratana, C. and Smith, A.E. (1998) An effective neural network estimate of reliability for use during optimal network design, under review for *INFORMS Journal on Computing*.

Stallings, W. (1994) *Data and Computer Communications*, fourth edition, Macmillan.

Steenstrup, M.E (1995) *Routing in Communications Networks*, Prentice Hall: Englewood Cliffs, NJ.

Steiglitz, K., Weiner, P. and Kleitman, D.J. (1969) The Design of Minimum Cost Survivable Networks, *IEEE Transactions on Circuit Theory*, **16**(4), 455–460.

Stevens, W.R. (1994) *TCP/IP Illustrated: The Protocols*, Addison-Wesley: Reading, MA.

Sthamer, H.-H. (1996) *The automatic generation of software test data using genetic algorithms*, PhD Thesis, University of Glamorgan, Wales, UK.

Storey, N. (1996) *Safety-Critical Computer Systems*, Addison-Wesley: Harlow, UK.

Sutton, R.S. (1984) *Temporal credit assignment in reinforcement learning*, PhD Thesis, University of Massachusetts, Department of Computer and Information Science, USA.

Syswerda, G. (1989) Uniform Crossover in Genetic Algorithms, in Schaffer, D.J. (ed.), *Proceedings of the Third International Conference on Genetic Algorithms*, Morgan Kaufmann: Los Altos, CA, 2–9.

Tanaka, Y. and Berlage, O. (1996) Application of Genetic Algorithms to VOD Network Topology Optimisation, *IEICE Trans. Communications*, **E79-B**(8), 1046–1053.

Tanenbaum, A.S. (1988) *Computer Networks*, second Edition, Prentice-Hall, Englewood Cliffs, NJ.

Tanenbaum, A.S. (1996) *Computer Networks*, third edition, Prentice-Hall: Englewood Cliffs, NJ.

Tanterdtid, S., Steanputtanagul, W. and Benjapolakul, W. (1997) Optimum Virtual Paths system based in ATM Network using Genetic Algorithms, in *International Conference on Information, Communications and Signal Processing*, 596–601.

Tate, D.M. and Smith, A.E. (1995) A genetic approach to the quadratic assignment problem, *Computers and Operations Research*, **22**(1), 73–83.

Thaker, G. and Cain, J. (1986) Interactions between routing and flow control algorithms, *IEEE Transactions on Communications*, **34**(3), 269–277.

Thrift, P. (1991) Fuzzy logic synthesis with genetic algorithms, in Belew, R. and Booker, L. (eds.), *Proceedings of the Fourth International Conference on Genetic Algorithms*, Morgan Kaufmann, 509–513.

Tracey, N., Clark, J. and Mander, K. (1998) The way forward for unifying dynamic test case generation: the optimisation based approach, *Dependable Computing and its Applications*, IFIP.

Turk, M. and Pentland, A. (1991) Eigenfaces for recognition, *Journal of Cognitive Neuroscience*, **3**(1).

Turton, B.C.H. and Arslan, T. (1995) An Architecture for Enhancing Image Processing via Parallel Genetic Algorithms & Data Compression, in *First IEE/IEEE International Conference on Genetic Algorithms in Engineering Systems: Innovations and Applications*, IEE: London, 337–342.

Turton, B.C.H. and Arslan, T. (1995a) A Parallel Genetic VLSI Architecture for Combinatorial Real-Time Applications – Disc Scheduling, in *First IEE/IEEE International Conference on Genetic Algorithms in Engineering Systems: Innovations and Applications*, IEE: London, 493–498.

Turton, B.C.H. and Bentall, M. (1998) Benchmark networks for both Physical and Virtual Networks In *Fifteenth UK Teletraffic Symposium on Performance Engineering in Information Systems*, IEE Press: London.

Valenzuela, C., Hurley, S. and Smith, D. (1998) A permutation based genetic algorithm for minimum span frequency assignment, in *Parallel Problem Solving from Nature – PPSNV*, A.E. Eiben, T. Bäck, M. Schoenauer, H.P. Schwefel (eds.), Lecture Notes in Computer Science, Springer-Verlag: Berlin, 907–916.

van Nimwegen, E. and Crutchfield, J (1998) Optimizing Epochal Evolutionary Search: Population-Size Independent Theory, *Santa Fe Institute Working Paper* 98-06-046 (also submitted to *Computer Methods in Applied Mechanics and Engineering*, special issue on Evolutionary and Genetic Algorithms in Computational Mechanics and Engineering, D. Goldberg and K. Deb, editors).

van Nimwegen, E. and Crutchfield, J. (1998a) Optmizing Epochal Evolutionary Search: Population-Size Dependent Theory, *Santa Fe Institute Working Paper* 98-10-090 (also submitted to *Machine Learning*, 1998).

Ventetsanopoulos, A.N. and Singh, I. (1986) Topological optimization of communication networks subject to reliability constraints, *Problems of Control and Information Theory*, **15**, 63–78.

Verdu, S. (1998) *Multiuser Detection,* Cambridge University Press: Cambridge.

Vickrey, W. (1961) Counterspeculation, auctions and competitive sealed tenders, *Journal of Finance*, **16**.

Vickrey, W. (1962) Auction and bidding games, in *Recent Advances in Game Theory*, Princeton University Press, 15–27.

Wall, L., Christiansen, T. and Schwartz, R.L. (1996) *Programming Perl*, second edition, O'Reilly Associates.

Walters, G.A. and Smith, D.K. (1995) Evolutionary design algorithm for optimal layout of tree networks, *Engineering Optimization*, **24**, 261–281.

Wang, W. and Rushforth, C.K. (1996) An adaptive local-search algorithm for the channel-assignment problem (CAP), in *IEEE Transactions on Vehicular Technology*, **45**(3), 459–466.

Wang, T.P., Hwang, S.Y. and Tseng, C.C. (1998) Registration Area Planning for PCS Networks Using Genetic Algorithms, *IEEE Transactions on Vehicular Technology*, **47**(3), 987–994.

Wang, X.F., Lu, W.S. and Antoniou, A. (1998a) A Genetic Algorithm-based multiuser detector for multiple-access communications, in *IEEE International Symposium on Circuits and Systems – ISCAS'98*, **4**, 534–537.

Watkins, C.J.C.H. (1989) *Learning from delayed rewards*, PhD Thesis, King's College, Cambridge, UK.

Watkins, A.L. (1995) The automatic generation of test data using genetic algorithms, in *Proceedings of the 4th Software Quality Conference*, **2**, 300–309.

Wegener, J., Sthamer, H.-H., Jones, B.F. and Eyres, D.E. (1997) Testing real-time systems using genetic algorithms, *Software Quality Journal*, **6**(2), 127–135.

Weibull, J.W. (1995) *Evolutionary Game Theory*, MIT Press: Cambridge, MA.

Welch, B.B. (1995) *Practical Programming in TCL and TK*, Prentice Hall: Englewood Cliffs, NJ..

White, L.J. and Cohen, E.I. (1980) A domain strategy for computer program testing, *IEEE Transactions on Software Engineering*, **6**(3), 247–257.

White, M.S. and Flockton, S.J. (1994) A genetic adaptive algorithm for data equalization, *Proceedings of the First IEEE Conference on Evolutionary Computation*, **2**, 665–669.

Woeste, C., Tindle, J. and Ryan, H.M. (1996) Cost Effective Equipment Planning for Passive Optical Networks Based on Tabu Search, in *European Conference of Networks and Optical Communications, NOC '96*, IOS Press: Amsterdam.

Wright, S. (1932) The roles of mutation, inbreeding, crossbreeding and selection in evolution, in *Proceedings of 6th International Conference on Genetics* (Ithaca, NY, 1932), **1**, D. F. Jones (ed.) (Menasha, WI: Brooklyn Botanical Gardens), 356–366.

Wurman, R. (1997) *Multidimensional auction design for computational economies*, Unpublished dissertation proposal, University of Michigan, USA.

Xanthakis, S., Ellis, C., Skourlas, C., Le Gall, A. and Katsikas, S. (1992) Application des algorithmes génétic au testes des logiciel, in *Proceedings 5th International Conference on Software Engineering*, Toulouse, France.

Xu, J., Chiu, S.Y. and Glover, F. (1996a) Using Tabu Search to Solve the Steiner Tree-Star Problem in Telecommunications Network Design, *Telecommunication Systems*, **6**, 117–125.

Xu, J., Chiu, S.Y. and Glover, F. (1996b) Probabilistic Tabu Search for Telecommunications Network Design, *Combinatorial Optimization: Theory and Practice*, **1**(1), 69–94.

Xu, J., Chiu, S.Y. and Glover, F. (1998) Fine-Tuning a Tabu Search Algorithm with Statistical Tests, *International Transactions in Operational Research*, **5**, 233–244.

Yeh, M.S., Lin, J.S. and Yeh, W.C. (1994) New Monte Carlo method for estimating network reliability, in *Proceedings of 16th International Conference on Computers & Industrial Engineering*, 723–726.

Yener, A. and Rose, C. (1997) Genetic algorithms applied to cellular call admission: local policies, *IEEE Transactions on Vehicular Technology*, **46**(1), 72–79.

Zadeh, L. (1965) Fuzzy Sets, *Information and Control*, **8**(3), 338–353.

Zeeman, E.C. (1981) Dynamics of the evolution of animal conflicts, *Journal of Theoretical Biology*, **89**, 249–270.

Zhang, L.J. and Li, Y.D. (1995) *Dynamic bandwidth allocation based on fuzzy neural network and genetic algorithm in ATM networks*, Technical Report of network and information research group, Dept. of Automation, Tsinghua University.

Zhang, L.J., Mao, Z.H. and Li, Y.D. (1994) An improved genetic algorithm based on combinative theory and fuzzy reasoning and its applications, in *International Conference on Neural Information Processing*, 180–185.

Zhang, L.J., Mao, Z.H. and Li, Y.D. (1995) Mathematical analysis of crossover operator in genetic algorithms and its improved strategy, in Proceedings of the 1st IEEE International Conference on Evolutionary Computation, IEEE Press, 412–417.

Zhu, H., Hall, P.A.V. and May, J.H.R. (1997) Software unit test coverage and adequacy, *ACM Computing Surveys*, **29**(4), 366–427.

Zhu, L., Wainwright, R.L. and Schoenefeld, D.A. (1998) A genetic Algorithm for the Point to Multipoint Routing Problem with Varying Number of Requests, in *ICEC'98*, 171–176

Zitterbart, M. (1993) A model for flexible high performance communication subsystems, *Journal on Selected Areas in Communications*, **11**, 507–518.

Zongker, D. and Punch, B. (1995) *lil-gp 1.0 User's Manual*.

Zuse, H. (1991) *Software Complexity: Measures and methods*, de Gruyter: Berlin.

Index

2-connectivity, 23, 24, 27–29
2-opt, 45
3-opt, 45

Aarts, 2, 332, 334, 335, 342
Abuali, 17, 21
access network planning, 7, 81, 115, 117, 120, 122, 124, 132, 133
adaptable protocol, 201, 202, 216, 219
adaptive heuristic critic, 141
adaptive routing, 151, 153, 166
ADDMP, 245, 249, 250, 252, 253
ADSL, 80
Advance Mobile Phone Service, 331
Aggarwal, 18, 20
Agrawal, 253
Aiyarak, 100, 104, 110
Al-Qoahtani, 170
Alba, 186
Alkanhal, 355
all-terminal reliability, 18, 20, 21
ALOHA, 352
Alshebeili, 355
Amano, 156
AMPS, 331
APSL, 202, 208, 210
ARPANET, 139, 148
Arslan, 183, 281
ASA, 355
AT&T, 331
Atiqullah, 18, 20
ATM, 138, 168, 170, 176, 187, 308, 317, 326
auction, 308, 316–324, 326, 328, 329
AWGN, 350
Axelrod, 284

B-ISDN, 138, 308
Bäck, 2, 9, 131, 238, 240, 253, 274, 333
backtracking, 24, 35, 171
Balakrishnan, 119
Baldi, 185
Ball, 21, 35

bandwidth allocation, 309, 311
bang-bang control, 142
Base Transceiver Stations, 336
base-station, 346
BCET, 273, 279, 280
BCS, 266–269
BDR, 240, 243, 262
Bellman-Ford algorithm, 139, 151
Beltran, 20
Bentall, 169, 171, 177, 182
Berlage, 236
Bernoulli crossover, 76
Bertsekas, 352
Bhatti, 205
Bilchev, 223, 229, 234, 284
bin-packing, 238
Black, 166
Blessing, 35, 41
Blickle, 285
blind channel identification, 335, 354
Bolla, 311
Bonelli, 143
Booker, 143
Boorstyn, 41, 119
Bousono, 284
branch and bound, 20, 24, 29, 333
branch exchange, 36, 41
Brecht, 21
Brewis, 97
Brittain, 100, 115, 124, 127

cabling, 6, 17, 85, 117
caching, 236, 279
Calegari, 344
call admission control, 308–311, 313, 315–317, 326
call admission policy, 349
call blocking probabilities, 311
Campbell, 20
CAP, 339, 342
capacity assignment problem, 48–52
Carpenter, 173
Carse, 137, 142–144, 153

CAST tool, 281
CBR, 169, 308
CDMA, 335, 350, 352
Cedano, 236
cellular networks, 339
Chang, 309, 352
channel allocation problem, 341, 344
channel-assignment problem, 367
Chen, 101, 307, 310, 311, 314, 324, 355
Cheng, 309
Chengdu, 44
Chinese Cities, 39, 46, 47, 49, 50
Chiu, 57
Chng, 170
Choi, 170
Chopra, 17, 71
Christian, 97
Christos, 114
circuit-switching networks, 152
Clark, 253, 259
Clarke, 277
classifier system, 138, 143
Clementi, 35
co-channel interference, 340, 341
COBOL, 273
Coghill, 339, 340, 359
Cohen, 273
Coit, 20, 23
Colbourn, 21
COM, 119
COMET, 307
common knowledge, 308, 321, 323, 326
community of agents, 284
computer networking, 17
concentrator, 21, 118–120
concentrator-location problem, 122
congestion, 2, 139, 154, 183, 196, 308, 317
connected-nodes encoding, 112
cooling schedule, 240, 243, 334, 342, 345, 353
cooperative game model, 308, 310–313, 316, 326
copper architectures, 81
Corne, 1, 233, 284
COST, 101

Costamagna, 337
Cox, 152, 171
cruise control, 280
Crutchfield, 253
CTR, 237
Cuihong, 176
Cut Saturation, 51

Daimler-Benz, 277, 279, 281
data equalization, 354
data flow diagram, 181
datagram, 138, 144
Davis, 21, 152, 171, 175
Dawkins, 122
De Jong, 8, 254, 286
deadlocks, 186
Deb, 253
Deeter, 30
defuzzification, 140, 141
delta evaluation, 360, 363, 366
DeMillo, 278
Dengiz, 17, 22, 25, 100
depth-first search, 27
Digital AMPS, 331
Dijkstra, 37, 137, 139, 148, 152–155, 157–159, 174, 175
Dimitrakakis, 114
distributed database, 2, 224, 236, 237
distributed routing, 137–139, 144, 147, 150, 153
distributed system, 224, 284, 285, 308
diversification generator, 60, 65, 72, 76
DRoPS, 201–203, 208, 216, 219
DRT, 241
DSPF, 174
DSS, 57, 58
dynamic access configuration, 236
dynamic bandwidth allocation, 309, 311, 313, 316
dynamic channel allocation, 339
dynamic programming, 48,–50, 52, 333

eigenvalues, 299
Elbaum, 36
Electro-Magnetic Compatibility, 358, 367

Elgibaoui, 337
Elsayed, 309
Elshoff, 273
EON, 101, 104, 105, 107–113
epistasis, 176
Ericsson, 331
error correction, 62, 71, 72
Ethernet, 193
ETR, 241
evolution strategies, 2, 9, 274
evolutionarily stable strategy, 285, 291
evolutionary algorithm, 9, 36, 104, 137, 141, 170, 183, 236, 272, 274
evolutionary dynamics, 292, 295
evolutionary game dynamics, 304
evolutionary game theory, 285
evolutionary programming, 2, 9
expectation dynamic, 285, 299, 304
Eyres, 281

fair shares, 308, 310, 313, 315, 326
fault-tolerant, 172
FCA, 339, 358–362, 365, 366, 369, 372
FCC, 58
FCDACS, 144, 146
FCS, 153
feedback mechanisms, 174
FIR filter, 353
fitness distance correlation, 245
fitness landscape projection, 252, 253
fitness landscapes, 124, 242
Fitzpatrick, 131
flow control, 13, 183, 193, 298, 215, 307, 308, 313, 317, 326
flow-equivalence property, 44, 45
flow-equivalent graph, 41, 42, 44, 45
FOCC, 346, 348
forma, 121–123
Forrest, 245
FORTRAN, 273
fractional load distribution, 297, 298
frame relay, 57
Fratta, 47
Fulkerson, 41, 139, 151
Funabiki, 339, 341, 359, 367
fuzzification, 140, 141
fuzzy bifurcated routing, 148, 150

fuzzy classifier system, 138, 143, 144, 146
fuzzy controller, 141–150
fuzzy logic, 2, 10, 11, 12, 138
Fuzzy Membership Functions, 142
fuzzy neural network, 309

Gallager, 352
game theory, 2, 10, 13, 284, 290, 296, 332
Gamst, 359, 367
GARA, 151–166
Garey, 19, 45, 332, 344, 359
Gavish, 35, 119, 169
GENESIS, 109
Genetic Adaptive Routing Algorithm, 151, 153
genetic algorithm, 2, 9, 11, 12, 17, 19, 36, 40, 47, 50,–53, 68, 90, 96, 109, 112, 116, 119–121, 124–133, 138, 141–143, 150, 173, 183, 189, 195, 224, 229, 230, 233, 238, 240, 243, 266, 273, 274, 277–286, 304, 308–316, 326, 332, 333, 340, 349–352, 355, 356, 359, 360
genetic programming, 2, 112, 285
GenOSys, 81, 92, 95, 96
Gerhard, 281
Gerla, 35, 36, 41, 47
Gibson, 331, 338
Global System for Mobile Communications, 331
Glover, 2, 5, 20, 57, 59, 64, 68, 332, 334
Goldberg, 2, 9, 17, 27, 88, 228, 238, 286, 288, 333, 365
Gomory-Hu, 41, 54
GoS, 311, 312
GOTND, 337
gradient descent, 143
Graham, 269
graph theory, 85, 332, 337, 339, 344
greedy algorithm, 20, 34, 171, 228, 230, 231–233, 362, 363, 366
Greenberg, 68
greenfield network, 81, 96
Grefenstette, 109
Griffith, 107

Grover, 36, 41
GSM, 331, 336, 337, 346
GTE, 118, 120

Halstead, 266
Hamilton, 291
Hamming distance, 245, 250–252, 273, 277
hand-off, 348–350
Harbin, 44
Harmen, 281
hillclimbing, 4, 5, 248
HLR, 336
HOC, 354, 355
Holland, 2, 9, 88, 143, 186, 188, 228, 238, 285, 286, 332
homogeneous condition, 334
hop-count, 157, 163
Hopcroft, 24, 27
Hu, 41, 42, 48, 49, 54
Huang, 174
Huberman, 284
Hui, 186, 190
Huimin, 307
Humboldt, 277
Hunt, 280

Ida, 20
IIR filter, 354
impact analysis, 80
information theory, 266
inhomogeneous condition, 334
insane, 187, 193–195
integer programming, 40, 48, 60, 74, 119, 171
Internet Protocol, 166, 187, 193, 198
IPVR, 212
ISDN, 336
ISI, 353, 354, 355
ISO9000, 268

Jacobian, 298
Jaimes-Romero, 339, 340
Jamin, 311
Jan, 17, 20, 22, 24, 25, 27, 28, 29, 32
Java, 323

Jeffrey, 35
Jiefeng, 57
jitter, 167, 201, 202, 209, 210, 212, 216, 218
Johnson, 19, 45, 332, 344, 359
Jones, 245, 265, 273, 274, 277, 278
Jordan, 358
Juntti, 352
Karaboga, 142
Karr, 142
Kartalopolus, 199
Karunanithi, 173
Katona, 21
Katzela, 338, 358
Kernighan, 45, 279
Kershenbaum, 36, 41, 168
Khanna, 137, 139
King-Tim, 35, 36, 41, 43, 50
Kinzel, 142
Kirkpatrick, 332
Kirkwood, 172
Kleinrock, 35, 38, 47
knapsack, 48, 49
Knight, 205
knowledge elicitation techniques, 141
Knuth, 273
Koh, 17, 20
Konak, 22
Korst, 2, 334, 342
Koza, 2, 110, 111, 113, 285
Krishnamachari, 331
Krishnamoorthy, 125
Kruse, 141
Kruskal, 7, 21, 45
Kumar, 18, 20, 100
Kunz, 359, 367
kurtosis, 355

Labit, 368, 370, 371
Laguna, 5, 334
Lai, 339, 340, 359
LAN, 57, 238
landscape visualisation, 253
landscapes, 245, 264
Lazar, 317, 323
LCSAJ, 267, 271

Index 397

leaky bucket, 309, 313
leased lines, 52, 76, 183, 336
Lee, 17, 20, 59, 71, 141, 142, 309, 311, 347, 348
Lenstra, 332, 335
Leung, 174
Lewontin, 290
Li, 307, 331, 339, 359–361, 366–69
Lin-Kernighan, 45
linear programming, 3, 40
linguistic metrics, 266
link-state, 168
Liska, 143
load balancing, 2, 8, 163, 166, 236, 237
load traffic multiplier, 148
local search, 2–88, 25, 29, 71, 73, 119, 122, 124, 125, 127, 131, 250, 332, 333, 341, 343, 359–363, 370
local tariffs, 49
logic circuits, 270
LOTOS, 186
Luss, 119
Lyapunov function, 294

M/M/1 queue, 38, 176
MacDonald, 358
Mann, 169, 170, 173
Maruyama, 36, 41
MARX, 308
Masaharu, 151
Mason, 311
maximum flow problems, 41
maximum spanning tree, 42–44
McCabe, 266
McGeehan, 345
McGeoch, 45
McGregor, 119
McQuillan, 139
MDP, 349, 350
MDS, 344
Mellis, 96, 97
MENTOR algorithm, 36, 41
message queues, 186, 187
Michalewicz, 100
microprotocol, 201, 202, 208, 209
Milgrom, 317, 318, 319, 321
Miller, 277

Min-Jeong, 339, 342–344
minimum path connectedness, 21
Mitchell, 333
mix-source crossover, 192
mixed integer linear programming, 173
MMSE, 352
mobile communications, 331–335, 350, 354
MOGA, 90, 94, 96
Monte Carlo simulation, 18, 21, 22, 28, 36, 350, 352
Montgomery, 176
MOTHRA, 278
MTBF, 267
Muhlenbein, 253
Muller, 332
multi-point crossover, 173
Munakata, 152
Munetomo, 151, 158, 172
Munro, 137
Murai, 156
MWFEG, 42

Nash equilibrium, 285, 320, 321
natural selection, 188, 286
Ndousse, 309
nearest neighbours, 245, 250, 251, 253
neighborhood, 61, 63, 71, 333–335, 341–345, 353
neighbourhood analysis, 245
Nel, 22
Nemhauser, 333
Nepal, 187–198
network adaptation algorithms, 223
network control, 147, 321, 326
network design, 7, 12, 18–25, 30, 32, 35–41, 47–53, 57, 58, 76, 80, 85, 100–107, 176, 358
network diameter, 20, 100, 103, 174
network management, 9, 10, 11, 13, 307
network planning, 6, 80, 84, 86, 89, 94, 96, 97, 115, 116, 119, 120, 133, 356
network simulation, 146, 150, 187
Neumann, 291
neural network, 10, 12, 36, 59, 112, 199, 206–219, 284, 308, 309, 324, 325, 326, 341, 359

Ng, 35, 47
Ngo, 339, 341, 359, 360, 361, 366, 367, 369
Nimwegen, 253
NMT, 331
Nordic Mobile Telephone, 331
NP-complete, 35, 45, 339
NP-hard, 19, 21, 59, 332, 334, 335, 344, 351, 353

Oates, 1, 235, 238, 242, 250, 254–257, 260–263, 284
object-oriented, 82, 83, 89, 132
OCX, 176
Offutt, 278
Olafsson, 223, 236, 283, 284, 291–294, 297, 299, 301, 303
Olsen, 174, 183
OMT, 84, 96
Open Shortest Path First, 152
operating system, 201, 203, 219, 279
optical fibre, 80, 99, 100, 115, 116, 118
optical mesh, 104, 107, 112
optical network, 80, 100, 103, 112, 115, 116, 119
optimal mutation rate, 256, 264
order-based crossover, 171, 175, 181
order-based mutation, 180
Ordinance Survey, 85
orthogonal crossover, 176
OSI, 199

P-FCS1, 143, 144, 146, 147
packet switch networks, 9, 20, 137, 138, 139, 152
paging systems, 331
paging zone, 347, 348, 349
Painton, 20
parallel genetic algorithms, 183
Parmee, 229, 234
Parodi, 143
partial evaluation, 366
Parzen window, 324, 325, 326
PASCAL, 158, 159
passive optical network, 119, 127, 133
path crossover, 154, 155

path mutation, 154
path relinking, 59, 64
PCS, 331, 332, 346, 347, 352, 356
permutation based crossover, 171
Perros, 309
Pfleeger, 266
Phadke, 120
Pham, 142
Pierre, 17, 20, 36, 41, 47, 51, 52, 100, 337
Pierskalla, 68
pipe network, 21
Pirelli, 133
PMX, 174, 340
POI, 337
point-to-point network, 2, 115–117, 127, 131, 133, 167, 238, 241
Poisson distribution, 38, 190, 194, 267
Poisson process, 189, 190, 195, 352
polytope, 59
PON, 89, 97, 100, 116, 117, 123, 127
Poon, 6, 79, 89, 110
POTS, 80
Powell, 120
Proakis, 353
probe connection, 187–195
problem complexity, 258, 261, 264, 332
protocol implementation, 200, 219
protocol performance characteristics, 210
protocol stack, 187, 201
protocols, 9, 12, 151, 185–187, 198, 200, 201, 202, 205, 208, 219
Provan, 21
PSE, 138
PSP, 317, 321, 323
PSTN, 336
PTN, 332
PTP, 116, 127

Qiu, 331
QoS, 199–211, 216, 218, 307–318, 326

Radcliffe, 121, 123
radio communication, 336, 338, 344, 358
radio propagation characteristics, 344
radio wave propagation, 345

Raiffa point;, 312, 315
random meshes, 170
random search, 182, 254, 256, 261, 286
Rao, 18, 20, 71
RAP, 207, 208, 209, 212, 219
Rappaport, 336, 338
Reading Adaptable Protocol, 207, 208
real-time restoration algorithm, 182
real-time video, 138
Rebaudengo, 185
reference set, 60, 66–69, 71, 72, 76
regularity assumption, 317, 321–323, 328
reliability, 18–36, 53, 84, 100, 101, 104, 118, 265, 267
replicator dynamics, 285, 292, 294, 297, 301–305
resource management problems, 152, 284
restoration algorithm, 170, 181
Rho, 236
RIP, 151, 152, 160–163, 166
Roberts, 23, 24, 27
Roper, 278
Routen, 122
router, 148, 149
routing algorithm, 146, 147, 149, 151, 152, 158, 160, 161, 162, 166, 168, 173
routing control centre, 139
routing tables, 2, 3, 9, 139, 151, 152, 167, 168, 171, 187
RSVP, 201
RTR, 123
runtime adaptable protocols, 201
Ryan, 358, 359, 368

Sacco, 333, 349
Sairamesh, 311
Savic, 17
scatter search, 61, 64, 68
scheduling, 59, 202, 317
schema theory, 285, 293, 304
Schlierkamp-Voosen, 240
Schultz, 280
Schwartz, 138
Schwartz, 225, 234
Schwefel, 2, 285
SDH, 80

security, 57
Sedgewick, 55
self-adapting mutation, 124
service provision, 11, 12, 283, 291, 294, 295, 304
Shahbaz, 337
Shimamoto, 170
shortest path routing, 37, 47, 139, 146, 149, 163, 167
Shubik, 312
Shulman, 118, 119
signal processing, 199
simulated annealing, 2, 3, 5, 8, 20, 36, 53, 116, 119, 125–127, 170, 173, 244, 278, 281, 286, 332, 342, 352, 355–360, 366
Sinclair, 99–111, 170, 171
Singh, 20
single point crossover, 23, 174
Sivarajan, 359, 367
skewed crossover, 262
Skorin-Kapov, 20
SKT, 240, 243, 263
Slobodkin, 290
Smith, 1, 17, 18, 20, 22, 23, 30, 33, 36, 124, 143, 169, 170, 173, 284, 285, 290, 291, 292, 359, 361, 372, 99, 114
smoothing, 309
SNA, 57
SNNS, 212, 214, 215
software testing, 267, 270, 273, 274, 279, 281
source routing, 153, 166
source-sink reliability, 18, 21
spanning tree algorithm, 62
SPF, 152, 160, 161, 162, 164, 166
Squillero, 185
Srivaree-ratana, 36
SSE, 213
Stallings, 138
statistical multiplexing, 308
steepest descent, 124, 332, 333, 341, 343
Steiglitz, 36, 41
Steiner node, 59–62, 65, 69, 71–73
Steiner Tree, 58, 71, 76
Stevens, 187
store-and-forward, 138

Storey, 266, 268, 269
STS, 59, 60, 61, 64, 65, 73, 74, 76
subset generation method, 60, 68, 74, 76
Syswerda, 8, 240, 253

tabu list, 334, 335, 342, 343
tabu search, 2, 5, 6, 8, 20, 59, 61, 69, 71, 74–77, 119, 278, 281, 332–334, 342, 343, 344, 356
tabu tenure, 63, 72
TAGGER, 278
Takagi, 142
Takefuji, 339, 341, 359, 367
Tanaka, 236
Tanenbaum, 138, 153, 168
TAR, 237
Tate, 23, 33
TCL, 187
TCP, 186–201
TDMA, 335, 352, 353
tele-working, 265
teleoperation, 199
TELNET, 200
temperature parameter, 5, 366
Thaker, 147, 148
Thomson, 311, 312, 315
Thrift, 142
Tindle, 6, 79, 89, 119
TNT, 240, 243
tournament selection, 105, 240
Tracey, 281
tragedy of the commons, 284, 317
tree-and-branch topologies, 116
tree-star, 59
TREX, 323
Troya, 186
Turton, 114, 167, 169, 183, 233
two-point crossover, 170, 263

Ullman, 24, 27
uniform crossover, 8, 29, 33, 122, 123, 175, 192, 193, 240, 263, 341, 351, 362
union of rings, 46, 47
unit costs, 18, 30, 34
utility function, 284, 296, 297, 300, 301, 304, 305, 310–319, 326, 327

Vachini, 118, 119
VBR, 169, 308
VDM, 271
vector-distance, 152
Vemuri, 236
Ventetsanopoulos, 20
Verdu, 350, 351
video conferencing, 173, 185, 199, 201
video on demand, 2, 236
virtual network, 182
virtual path, 37, 177
virtual path, 37
Viterbi algorithm, 351
VLSI, 281, 332

Walters, 17, 21
WAN, 238
Wang, 347, 351, 359, 367, 370
water systems, 17
Watkins, 141, 278
WCET, 273, 279, 280
Wennink, 233
Wessler,, 23, 24, 27
Whittley, 358
Wicker, 331
wireless communication, 331, 332, 350
Wolsey, 333
Wuhan, 44
Wurman, 321

Xanthakis, 278

Yanda, 307
Yang, 281
Yener, 349

Zadeh, 140, 143
Zeeman, 291
zero-one integer programming, 59
Zhang, 309, 314
Zhu, 173
Zinky, 137, 139
Zitterbart, 200, 205
Zongker, 110
Zuse, 266